MAPPING THE END TIMES

Mapping the End Times

List price $49.95
Discount price $39.96

Purchased by_____

DATE DUE

Critical Geopolitics

Series Editors:
Klaus Dodds, Royal Holloway, University of London, UK
Alan Ingram, University College London, UK
Merje Kuus, University of British Columbia, Canada

Over the last two decades, critical geopolitics has become a prominent field in human geography. It has developed to encompass topics associated with popular culture, everyday life, architecture and urban form as well as the more familiar issues of security, inter-national relations and global power projection. Critical geopolitics takes inspiration from studies of governmentality and biopolitics, gender and sexuality, political economy and development, postcolonialism and the study of emotion and affect. Methodologically, it continues to employ discourse analysis and is engaging with ethnography and participatory research methods. This rich field continues to develop new ways of analysing geopolitics.

This series provides an opportunity for early career researchers as well as established scholars to publish theoretically informed monographs and edited volumes that engage with critical geopolitics and related areas such as international relations theory and security studies. With an emphasis on accessible writing, the books in the series will appeal to wider audiences including journalists, policy communities and civil society organizations.

Other books in this series

Spaces of Security and Insecurity
Geographies of the War on Terror
Edited by Alan Ingram and Klaus Dodds
ISBN 978 0 7546 7349 1

Mapping the End Times
American Evangelical Geopolitics and
Apocalyptic Visions

Edited by

JASON DITTMER
University College London, UK
and
TRISTAN STURM
University of California Los Angeles, USA

ASHGATE

© Jason Dittmer, and Tristan Sturm 2010

All rights reserved. No part of this publication may be reproduced, stored in a retrieval system or transmitted in any form or by any means, electronic, mechanical, photocopying, recording or otherwise without the prior permission of the publisher.

Jason Dittmer, and Tristan Sturm have asserted their right under the Copyright, Designs and Patents Act, 1988, to be identified as the editors of this work.

Published by
Ashgate Publishing Limited
Wey Court East
Union Road
Farnham
Surrey, GU9 7PT
England

Ashgate Publishing Company
Suite 420
101 Cherry Street
Burlington
VT 05401-4405
USA

www.ashgate.com

British Library Cataloguing in Publication Data
Mapping the end times : American evangelical geopolitics
and apocalyptic visions. -- (Critical geopolitics)
1. Evangelicalism--Political aspects--United States.
2. Christianity and politics--United States. 3. End of the
world--Political aspects--United States.
I. Series II. Dittmer, Jason. III. Sturm, Tristan.
320.5'5'0973-dc22

Library of Congress Cataloging-in-Publication Data
Mapping the end times : American evangelical geopolitics and apocalyptic visions / edited by by Jason Dittmer and Tristan Sturm.
    p. cm. -- (Critical geopolitics)
Includes bibliographical references and index.
ISBN 978-0-7546-7601-0 (hardback) -- ISBN 978-1-4094-0083-7 (pbk) --
ISBN 978-0-7546-9983-5 (ebook) 1. Evangelicalism--United States. 2. Christianity and international affairs. 3. Christianity and politics--United States. 4. Eschatology--Political aspects--United States. 5. End of the world. I. Dittmer, Jason. II. Sturm, Tristan.
  BR1642.U5M35 2009
  277.3'082--dc22
                                                                                2009032352

ISBN 978-0-7546-7601-0 (hbk)
ISBN 978-1-4094-0083-7 (pbk)
ISBN 978-0-7546-9983-5 (ebk)

Printed and bound in Great Britain by
MPG Books Group, UK

Contents

List of Figures and Table	*vii*
Notes on Contributors	*ix*
Foreword: Evangelism, Secularism and Pluralist Possibility	
William E. Connolly	*xi*
Acknowledgements	*xvii*

Introduction: Mapping the End Times   1
Tristan Sturm and Jason Dittmer

PART I   CONTESTING THE AMERICAN HOLY LAND

1. "What would Lee do?" Religion and the Moral Landscapes of Southern Nationalism in the United States   27
David Jansson

2. Contests over Latter-day Space: Mormonism's Role within Evangelical Geopolitics as seen through Last-days Novels   49
Ethan Yorgason

3. Obama, Son of Perdition?: Narrative Rationality and the Role of the 44th President of the United States in the End-of-Days   73
Jason Dittmer

PART II   AMERICAN EVANGELICAL EXCEPTIONALISM

4. Apocalyptic Exceptionalism: Rosenberg, Clancy and the Prophecy of Americanism   99
Simon Dalby

5. The 'New World Order' and American Exceptionalism   119
Michael Barkun

6. Imagining Apocalyptic Geopolitics: American Evangelical Citationality of Evil Others   133
Tristan Sturm

PART III MISSIONARY GEOPOLITICS

7	The Problematic Synergy Between Evangelicals and the US State in Sub-Saharan Africa *Hannes Gerhardt*	157
8	Reaching the Unreached in the 10/40 Window: The Missionary Geoscience of Race, Difference and Distance *Ju Hui Judy Han*	183
9	Between Armageddon and Hope: Dispensational Premillennialism and Evangelical Missions in the Middle East *Carolyn Gallaher*	209

Afterword: The Geopolitics of End-Time Belief
in the Era of George W. Bush   233
Paul Boyer

Index   *249*

List of Figures and Table

Map 7.1	Evangelicals' self-projected growth in Africa	168
Figure 8.1	Cultural Distance Diagram	198
Table 6.1	Changing dispensational toponymns, 1970–2008	139

Notes on Contributors

Michael Barkun is a Professor of Political Science at the Maxwell School of Syracuse University. His books include *Culture of Conspiracy: Apocalyptic Visions in Contemporary America* (University of California Press, 2003) and *Religion and the Racist Right* (University of North Carolina Press, 1997).

Paul S. Boyer is Merle Curti Professor of History Emeritus at the University of Wisconsin-Madison and editor-in-chief of the *Oxford Companion to United States History*. His books include *When Time Shall Be No More: Prophecy Belief in Modern American Culture* (Harvard University Press, 1992).

William E. Connolly is Krieger-Eisenhower Professor at Johns Hopkins University where he teaches political theory. His recent publications include *Why I Am Not A Secularist* (University of Minnesota Press, 1999), *Neuropolitics: Thinking, Culture, Speed* (University of Minnesota Press, 2002); *Pluralism* (Duke University Press, 2005) and *Capitalism and Christianity, American Style* (Duke University Press, 2008). His new book, *A World of Becoming* is currently in production.

Simon Dalby is Professor of Geography, Environmental Studies and Political Economy at Carleton University in Ottawa. He is author of *Creating the Second Cold War* (Pinter and Guilford, 1990), *Environmental Security* (University of Minnesota Press, 2002) and *Security and Environmental Change* (Polity, 2009) and coeditor of *Rethinking Geopolitics* (Routledge, 1998), *The Geopolitics Reader* (Routledge, 1998, 2006) and the journal *Geopolitics*.

Jason Dittmer is a Lecturer in Human Geography at University College London. His research interests are in popular geopolitics and the construction of meaning from cultural resources. He is the author of *Popular Culture, Geopolitics, and Identity* (Rowman and Littlefield, 2010) and co-editor of *Aether: The Journal of Media Geography*.

Carolyn Gallaher is in the School of International Service at American University. Her work on right-wing militias in the United States and Ireland has resulted in two books, *On the Fault Line: Race, Class, and the American Patriot Movement* (Rowman and Littlefield, 2003) and *After the Peace: Loyalist Paramilitaries in Post-Accord Northern Ireland* (Cornell University Press, 2007).

Hannes Gerhardt is an Assistant Professor in the Department of Geosciences at the University of West Georgia. His research interests include cultural imaginaries, transnational territoriality, and the intersection between geography and ethics/religion.

Ju Hui Judy Han is a Korea Foundation Postdoctoral Fellow at the University of British Columbia and received her PhD in geography from the University of California at Berkeley. She is currently preparing a book manuscript on contemporary Korean/American Christian missions engaged in world evangelization efforts.

David Jansson is a visiting research fellow at the Department of Social and Economic Geography at Uppsala University in Sweden. His research interests include nationalism, geographic identity, the cultural politics of race and ethnicity, migration, and place branding.

Tristan Sturm is a PhD Candidate in the Department of Geography at UCLA and a Lady Davis Fellow at the Hebrew University of Jerusalem where he currently is doing ethnographic work with American evangelical Christians.

Ethan Yorgason is a recently appointed assistant professor of geography at Kyungpook National University in Daegu, South Korea. Before that appointment he taught for seven and a half years at Brigham Young University Hawaii. His interests lie in the intersections of political, cultural, historical, and regional geography, with much work focusing on religion (particularly Mormonism), the American West, and Eastern Taiwan.

Foreword:
Evangelism, Secularism and Pluralist Possibility

William E. Connolly

To the extent geopolitical discourse has participated in secularism, it has joined that larger cohort, at first in ignoring the risks, dangers and possibilities that evangelism poses to democratic, pluralist, and egalitarian politics and then in addressing those issues in blunt and self-reinforcing ways. This new book on the geo-politics of evangelism transcends those limitations. It reviews the inner complexities of evangelism in America; it addresses dangers posed by the movement; and it searches out positive actualities and possibilities lurking within it. It is a timely and compelling set of chapters.

What was it about secularism as a doctrine of the social sciences that placed a filter between it and the re-emergence of evangelism? First, its devotees have tended to place themselves above the fray as they underplay the extent to which they, too, bring elements of theistic or nontheistic faith with them when they enter the public arena. Second, and as a corollary, they have overplayed the extent to which it is possible to articulate encompassing criteria of democratic participation that are free of any particular creed. This has led them to allow one brand of religion—the Christianity of high Protestantism—to set the unconscious frame in which evangelical and Islamic movements are examined. Third, in setting high Protestantism as the unconscious background, they have also underplayed the role that ritual, church assemblies and the media play in coding the visceral register of intersubjectivity. That is, they have ignored or relegated only to a special set of minorities those thought-imbued intensities that people bring with them into public life. Pretending democracy could simply be 'deliberative' or 'rational', they have acted as if the coding of the visceral register of cultural life occurred only in the 'private realm' where churches and families play a major role. In fact, secularists, as well as every new political movement must either draw upon the intercoded resources of image, rhythm, music, and concepts to engage the visceral register of intersubjectivity or consistently find themselves lagging behind cultural forces that do draw extensively upon such media.

In this study all three of the above assumptions are jostled, as the inner workings, complexities and impact of American evangelism are examined. Indeed, several of the chapters are geo-political in another way. They help us to see how elements of faith travel from one place to another. This has always been true. The faith of

Jesus traveled in progressive waves from small towns in the eastern Mediterranean to northern Africa, to Europe, to the Americas, and to middle Africa. The faith of Mohammed has traveled to most corners of the world. And every other major faith has been geopolitical in its way too.

Evangelism, as it travels through time and space, has morphed in several ways. To take the United States as merely one marker, Judaism, Catholicism, the theory of evolution, smoking, drinking, lipstick, and gambling provided central things to oppose in the 1920s. Today, lipstick and gambling are fine; the movement actively promotes alliances with one wing of Judaism and Catholicism; and Islam, secularism, homosexuality, and abortion have emerged as new cultural objects of attack. Does this mean that the need to have a definitive set of enemies is more important to the right edge of the movement than the need to maintain the same creedal identity over time? Or does it mean that the world has shifted as creedal continuities have remained stable? This is a complicated question, which some of the essays in this collection engage.

My own sense is that every faith community is defined by two dimensions that are interinvolved but not entirely reducible to one another. The first is the *creedal* element, by which I mean the formal doctrinal beliefs enunciated. Say, Jesus as the Son of God at birth, the resurrection, eternal salvation for believers, and the last judgment. The second is the *spiritual* element that infuses the creedal code. Thus evangelists may share a set of formal beliefs but diverge significantly on the extent to which they take a punitive approach to others or support priorities such as the readiness to go to war, the willingness to torture enemies, the readiness to impose their views about abortion and homosexuality on others, and so on. Some partisans infuse evangelical belief with a spirituality of care for others and the earth while others infuse it with a spirit of revenge against those outside the fold. It is true that when George W. Bush and his mother got into an argument—as was widely reported in the media—over whether unbaptized babies can go to heaven or not, a creedal element was involved. But peeking through the creedal difference were divergent spiritual orientations to being, with W. eager to find and elevate doctrinal elements that would condemn those outside the creed he embraces.

This connection without identity between creed and spirituality is profoundly important to politics. It is both a subtlety we must bring to bear when we examine evangelical politics and an issue we must address with respect to ourselves as we build a pluralist assemblage appropriate to the new historical conjuncture. For given the global expansion of capital, the acceleration of pace in many zones of life, and the rapid rates of legal and illegal migration, we today face a veritable minoritization of territorial states occurring at a more rapid pace than heretofore. By the ugly word 'minoritization' I mean a situation in which minorities of many types increasingly populate territorial states, and in which many states find it increasingly impossible to locate an empirical majority that corresponds to the majoritarian standard it has historically advanced. The issue, then, is decreasingly whether each state will in fact be marked by a majority creed and increasingly what kind of spiritual affinities across creedal difference can be invoked to respond to this new condition. Will pluralism grow and blossom? Or will defensive constituencies harden and insist upon treating one minority as the official stance to which all others must adjust?

In the United States what I have called "the evangelical-capitalist resonance machine" has pushed in the second direction. The right edge of capital has formed spiritual alliances with the right edge of evangelism, even though they often diverge from each other significantly in belief. These two forces have bonded with a large section of white, blue collar and lower middle class males to work against pluralism and egalitarianism together. Today, a counter movement will be one that is even more pluralistic than heretofore in the creeds advanced by its participants—I use the word 'creed' to encompass theology or fundamental philosophy of either a theistic or nontheistic type. It will seek to draw a section of evangelicals into a larger movement made up of monotheists and nontheists of multiple types, whose affinities of spirituality draw them toward pluralism, egalitarianism and care for the earth. It will not be a movement in which the participants leave their creeds entirely behind them when they participate in public life. It will be one in which a plurality of creedal orientations participate in a larger assemblage bound together by affinities of spirituality. It will draw strength from the visceral register of intersubjectivity and refined powers of deliberation alike, with each modality infusing the other. Same as it always was.

****

In *Capitalism and Christianity, American Style* (2008), I first sought to identify the spiritual ties between one wing of capital and one wing of evangelicals, addressing the dangers the combination poses. Then I began to think about constituencies within each official grouping with spiritual propensities that may open them to move in a different direction. John Sanders book, *The God Who Risks* (1998), provides one starting point. He is an evangelical who seeks to revise some elements in the evangelical creed while staying within the larger fold. His most novel creedal claim is that scripture shows us how God learns as history proceeds. God maintains the same caring spirituality over time, but learns and adjusts as history proceeds. This, of course, is a shift in evangelical doctrine based upon Sanders's reading of scripture, even as he continues to confess the sanctified birth of Jesus, the resurrection and promise of eternal salvation. But, to me, the key thing is the mode of spirituality that infuses this book. Sanders does not demonize opponents; he seeks expansive engagements within his church; and he opens the door to learn from others just as his God periodically learns from history.

There are other voices emerging within evangelism today that approximate the presumptive generosity apparent in the work of Sanders. Such voices receive some attention in this volume, and they must receive even more in the future.

Another step, perhaps, is for those of us who have come from secular, nontheistic backgrounds to engage sophisticated theologies that make a claim on our creedal attention. In so doing, we not only open up possibilities of connection across various creeds, we also begin to test *ourselves*. In that respect I note the work of the theologian Catherine Keller (2003), whose thinking both enlivens me in some ways and poses challenges to my nontheism that cannot easily be brushed aside. She is, you might say, a Nietzschean with an active theology defining a stance that invites communication with both traditional theologies and atheological stances that had purported to leave such questions behind.

****

In *The Face of the Deep: A Theology of Becoming*, Catherine Keller (2003) argues that hints of a world of becoming in Judeo-Christian scripture have been displaced by later redactions, translations and interpretations. What is (now) the second verse of Genesis captivates her (2003, xv). "When in the beginning Elohim created heaven and earth, the earth was *tohu va bohu*, darkness was upon the face of the *tehom*, and the *ruach elohim* vibrating upon the face of the waters." This verse depicts an unruly world already there when the God acted upon it, giving it shape and definition. The earth was without form. It was infused with wildness, with the *tehom* as salt water, chaos and depth, quivering in itself. Pluripotentiality simmered in it, making it susceptible to creative shaping but not to total control. Even the early idea of spirit, *mrfet*, suggests the idea of vibration, creating strange resonances between Keller's reading of scripture and recent work in complexity theory in the natural sciences. Moreover, those resonances are explored in her book.

Dominant Christian theology, from at least Augustine to today, has sacrificed the early themes of depth, vibration and becoming to the theme of an omnipotent, creator God who stands above the world (what Paul Tillich calls the "supranaturalism" that has handicapped Christianity). As a result of that fateful decision, institutionally affirmed for centuries, imperfection has been sucked out of divine creation, humans are defined as fateful carriers of original sin, a large portion of human agency has been absorbed into divine grace, and positive worldly activism has been blunted in some instances and transmuted into the politics of bellicosity in others. If a perfect God created the world from nothing, all worldly imperfection has either been introduced by us through free will or it is a veil of appearances to be torn away in the future. The doctrine of human free will, in which all of us inherited a will divided against itself after the first agent freely acted against the command of God, protects God from participation in the human predicament, while it persistently sets up one constituency or other to be particularly deserving of divine punishment.

On the other hand, if God acted creatively upon a simmering, unformed, steamy mess already there, a portion of *primordial* guilt is lifted from humanity. We also face a future that is not entirely determined by the past, free will, or grace. We inhabit a world of becoming. Both Genesis and species evolution are now said to occur "at the edge of chaos," with an element of creativity and uncertainty circulating through both. The powers of creativity now become *distributed* between God, humanity and the larger world.

In a world of becoming God and humanity are co-present. "The action of God is its *relation—by feeling and so being felt*, the divine invites the *becoming* of the other, by feeling the becoming of the other the divine itself becomes" (Keller 2003, 198). The feeling of the divine, correspondingly, uplifts us as we pursue the human adventure in a world that is not closed. "*An oscillation between divine attraction and divine reception, invitation and sabbath*" now inhabits desire (ibid., 198). The past works upon the future through uncertain vibrations during the protraction of the present. The present always vibrates with a degree of uncertainty. But some 'moments' of oscillation and perturbation open upon a future that is yet more uncertain, for good and bad. As Keller puts it (2003, 227), "becoming remains incalculable in advance and evanescent in the present. It gets causally interconnected only after the fact. It recapitulates, it decides, it flows beyond itself."

Having translated the unilateral, uncertain grace of an omnipotent God into co-dependency between God, world and humanity, Keller calls upon us to act in favor of human dignity, equality, and presumptive respect for the protean character of diversity. She also calls upon us to accept without existential resentment a world of becoming that exceeds both the powers of God and humanity. She thus challenges one inflation temptation inside secular thought as she does another within Christian theology. She knows that this creed will not be accepted by most. But she thinks it is important to return it to the register of creedal discussion.

It is not too surprising that Keller has written another book (2004) in which she resists apocalyptic demands flowing from evangelical re-enactments of the Book of Revelation, as she challenges the American pursuit of "messianic imperialism". Divinity, creation out of the deep, becoming, risk, worldly participation, new surprises—each component in this theology folds and melds into the others, fomenting a distinctive account of the human predicament and modes of response to it.

I do not concur with Keller's faith in a limited God acting upon a world of becoming, though I respect her reading of scripture and very much appreciate the spirituality she pours into it. I confess a world without divinity, immanent to itself, in which multiple force-fields of different types and speeds periodically interact, and in which the universe itself is open to an uncertain degree. Such a world is worthy of our spiritual embrace, first, because we emerged from it, and, second, because it allows us to participate modestly in modes of real creativity that exceed us.

Such a vision contests the modes of transcendence of each of the three monotheisms, as it also finds recent secular conceptions of science, time, and world mastery to be wanting. It is very probably impossible to prove my creed definitively, even if reasoned arguments can be given on its behalf. That shows to what extent that it, like those of classical evangelism, secular conceptions of human mastery, a learning, providential God, a theology of becoming, is a contestable faith brought with me into the public square. It is how I do so that is important.

Will it be possible to engage positively a large minority of evangelists as more of us acknowledge the extent to which we bring a living faith with us into the public square? And as we acknowledge the elements of comparative contestability in our own faiths? And as we confess the need to engage others on fundamental existential questions? One cannot say with confidence in advance. But it may be a worthy wager to test in everyday life, as we note the modest contribution a smug doctrine of secularism has made to the rise of pugnacious elements within evangelism.

****

The chapters in this book have stimulated such thoughts in me. They engage their topics closely and thoughtfully. They identify key dangers and possibilities lurking within evangelism, secularism and the interstices between them. And they often put themselves on the line as they proceed past the smug approaches to religiosity that have too often damaged academic accounts. It is an excellent and timely set of contributions.

References

Connolly, W. (2008), *Capitalism and Christianity, American Style* (Durham: Duke University Press).
Keller, C. (2003), *The Face of the Deep: A Theology of Becoming* (New York: Routledge).
—. (2004), *God and Power: Counter-Apocalyptic Journeys* (Minneapolis: Fortress Press).
Sanders, J. (1998), *The God Who Risks, A Theology of Providence* (Downers Grove, IL: Intervarsity).

Acknowledgements

The editors would like to thank Val Rose at Ashgate Publishing for her incredible support from the very beginning of this project, and to all at Ashgate who anonymously helped this book get to press. Further thanks are due to the authors of the chapters, who have produced an incredible array of thought-provoking essays. We would also like to thank *Estaño* (Philip Stein) for graciously letting us use his painting for the cover of this volume.

Jason Dittmer would like to thank his wife Stephanie for her endless support and interest, and also Klaus Dodds for commenting on the initial book proposal as well as his chapter in the book. As with all projects, numerous colleagues at UCL and beyond have influenced this project, and so much thanks is due to them (but, unfortunately, no royalties).

Tristan Sturm would like to thank John Agnew for his guidance in academia and life in general, Simon Dalby for his enthusiasm and support, Ben Moffat for confidence, Michael Curry, J. Nicholas Entrikin, and Iain Wallace for suggestions on early drafts of his chapter, and, of course, the brilliant graduate students in the Department of Geography at UCLA for insight and much needed distraction. Of course this would not be possible without his family, Doug, Karen, and Clinton, who have supported and encouraged his long-distance endeavors. His contribution was supported in part by SSHRC of Canada.

Dedications

*Jason   For my mother Arleen,
who always encouraged me to think critically*

*Tristan   To Arunima, for love, patience and my sanity,
and in loving memory of Ethel Sturm and Louis Louchard*

Introduction
Mapping the End Times

Tristan Sturm and Jason Dittmer

Introduction

The leading historian of Israel, Gershom Gorenberg (2002), called evangelical prophecy of the apocalypse "dangerous" for American and Israeli interests. In *American Fascists* (2007), Pulitzer Prize-winning journalist and Harvard Divinity School graduate Christopher Hedges argued that evangelicals resemble the early fascist movements in Italy and Germany at the beginning of the twentieth century, and therefore constitute a gathering threat to American democracy. Edward Said (2002, n.p.) wrote that Christian Zionists were "a menace to the world and furnish[ed] Bush's government with its rationale for punishing evil while righteously condemning whole populations to submission and poverty." Alternatively, Christian principles of forgiveness and humility offer political 'ways forward' that reject the processes of division that bring out the polemical energy in the above quotes.

It has been argued in geopolitical literature recently that "religion is the emerging political language of the time" (Agnew 2006, 183) and "international politics is being increasingly scripted in the spatial grammar of a millennial struggle between Good and Evil" (Bialasiewicz 2006, 720). With four of the last seven U.S. presidents claiming to be 'born-again' (Jimmy Carter, Ronald Reagan, Bill Clinton, and George W. Bush)[1] political geographers should be attentive to the strategies and reasoning practices used in the name of religion. Geographic inquiry into apocalyptic or millennial Christianity is often hinted at but has, with few exceptions, received no focused analysis (but see special issues of the journals *Hérodote* 2005 and *Geopolitics* 2006; Gerhardt 2008).

It is a fundamental tenet of critical geopolitics that culture and politics are co-constitutive. As such, rather than critique theology in this volume we are interested in examining the geographical imaginations of American evangelicals' religious worlds: their fiction, maps, public performances, discourses, prophecy books, and descriptions. Thinking, wanting, waiting, speaking, listening, and looking all (re)construct geopolitical world views associated with Evangelicalism. American

---

1 We say claimed because Gerald Ford declared himself 'born again' in 1976 although he was an Episcopalian because Jimmy Carter's born-again bent was appealing to the electorate (see Lieven 2004). Barack Obama has specifically avoided referencing himself as 'born again' (see the Pulliam and Olsen [2008] interview with candidate Obama).

evangelical geopolitics is a complex topic that is manifested in many different ways and generates varied reactions. Scholars from religious studies, history, sociology, anthropology, and political science who have paid close notice to evangelicals have generally ignored the spatializing processes nested within evangelical discourses. This book attempts to highlight these geographies and spark further discussion of the evangelical geopolitical imagination in both its utopian and dystopian manifestations. We outline the intersection of religious studies with the sub-field of geography called geopolitics in order to faciliate further this engagement.

Geopolitics and Religion

The relationship between geopolitics and religion is a longstanding one, but has recently undergone major changes, especially in academic circles. Of course, religion is at the root of international relations, with the question of religious identity at the core of the Treaties of Westphalia that are often cited as the beginning of modern understandings and practices of sovereignty and the inter-state system (e.g., Morganthau 1960; see also Campbell 1992; Agnew 1998). Similarly, the 'threat' of Islam played a great role in the forging of Europe's geopolitical identity as 'Christendom' during that same era. 'Geopolitics' as a concept is, of course, of much more recent vintage, dating only from the very end of the nineteenth century (Agnew 1998). Concerned primarily with environmental factors, so-called classical geopolitics neglected the role of culture such that religion would have seemed out of place in serious geopolitical analysis. Later incarnations of geopolitics emphasized the role of political economy and *Realpolitik*, and so at least officially were also uninterested in religion, even as religio-cultural assumptions continued to undergird formal theorizing.

Critical geopolitics can be understood fundamentally as the study of spatial discourses associated with global visions of power (Dalby 1990; Ó Tuathail 1996). As critical geopolitics has flourished over two decades, it has been pushed and pulled in many directions that call into question such a simple definition of the project, including nonrepresentational and more humanist approaches. This volume remains fairly closely connected to the more conventional original focus on discourse and the global scale; however this is not because we reject the newer approaches to the study of geopolitics, but rather because we are coming to the study of geopolitics and religion relatively late, and the salience of the topic to conventional questions of American foreign policy and cultural power during the presidency of George W. Bush beg for scholarly attention. That said, many of the chapters presented in this book have deviated from discourse and textual analysis to use interviews, participant observation, and even ethnography, broadly understood, as methods for analysis. Critical geopolitics, in all its forms, has been unified by its political and scholarly commitment to the reversal of academic geography's historic role as handmaiden to empire. Instead of that dubious legacy, critical geopolitics has advanced understandings of the processes that produce hegemonic knowledges of militarism and perpetuated inequality. It was with this goal in mind that this book was envisioned.

The rise of critical geopolitics within political geography offered new possibilities for engagement with the religion, in contrast to formulations in the 'new' geography of religion that are less concerned with critical social theory than with questions concerning subjectivity (Livingstone, Keane, and Boal 1998; Kong 2001; Holloway 2006). Most of these possibilities remained latent and unrecognized until recently. As has been lately noted, scholars of critical geopolitics remained largely dazzled by elite discourses and representations of place for two decades. Like previous scholars of geopolitics, we remained products of our time – fixated on the end of the Cold War and politicians' subsequent search for geopolitical order through discourse. Further, the modernization thesis has long been influential in the social sciences, arguing that religion itself would fall victim to the twin forces of modern life and globalization. It became easy to think of religion as something that mattered in other parts of the world, but was irrelevant in the context of American geopolitics. It took the presidency of George W. Bush to raise the issue of evangelical Christianity (and in particular the unique and powerful strands of evangelicalism indigenous, but not endemic, to the U.S.) to the attention of scholars within critical geopolitics, although the subject had long been of interest to those in other disciplines (some of whom are in this volume). A recent flurry of work in critical geopolitics has just begun to open up this "most exciting frontier" (Sidorov 2006, 340).

An interesting element of this emerging literature is that it has been remarkably focused on evangelicalism, often to the exclusion of other religious perspectives, and this book is no exception (but see de Busser 2006; Dijkink 2006; Megoran 2006; Sidorov 2006; West 2006). That this should occur during a period marked by President Bush's 'Global War on Terror', when much of the developed world's state-power was directed towards analysis and containment of the purported threat of Islamist violence, is remarkable evidence both of the anti-hegemonic bent of critical geopolitics and also of the lingering effects of the crisis of representation. Nevertheless, it would be a benefit to the literature on religion and geopolitics were subsequent studies to broaden out and consider subaltern geopolitical visions that are religiously informed. This is important not because it would serve the needs of the state (that is a path critical geopolitics judiciously avoids) but because these perspectives are just as key as some strands of American evangelicalism in perpetuating injustice and bodily violence (and equally, like some strands of American evangelicalism, may hold the keys to reducing injustice and violence).

Geopolitical Analysis and American Evangelicalism

Critical geopolitics has, since its inception, been systematized as the study of three different kinds of discourse: formal, practical, and popular (Ó Tuathail and Agnew 1992; Sharp 2000). Drawing on Lefebvrean theory, this trio is understood as being only separate in an analytic sense, as they each inform and substantiate the others. Formal geopolitical discourse has been described as the imagined geographies of geopolitical theorists, both from the academy and from think tanks. In regards to American evangelicalism, there are numerous nodes and loci that serve as centers of formal geopolitical thought. These institutes were formed from the late nineteenth century onward as an investment in intellectual infrastructure following the national embarrassment of the Scopes Monkey Trial, and include the Moody Bible Institute, Dallas Theological Seminary, Regent University, and Liberty University. In particular, Tulsa, Dallas, Chicago, and Colorado Springs have emerged as centers of evangelical geopolitical theorizing.

It would be a mistake, however, to think of evangelical geopolitical theory as somehow walled off from more 'mainstream' geopolitical theory. For instance, Samuel Huntington's "clash of civilizations" thesis resonated strongly with many evangelicals because of the centrality Huntington afforded religion and culture in his argument. In addition to the written word, the sphere of formal theorizing also includes the spoken word of individual evangelical pastors across the United States (and around the world). These men (indeed, the job remains strongly gendered in most evangelical parishes) study the Bible individually and produce weekly sermons that often touch on the structures of geopolitics or current events in a biblical context. The de-centered nature of evangelical thought (one of the central tenets is of personal relationships with the Bible and God) means that evangelical geopolitics resists monolithic interpretations. Nevertheless, much attention from the critical geopolitics community has focused on the Christian Zionist lobby in Washington and Jerusalem/Tel Aviv, which have had a tremendous and direct influence on both countries' foreign policies. American evangelicals provide perhaps the largest base of support for pro-Israel interest in the United States, providing the most unqualified, unwavering, and uncritical support for Israel in what is by most accounts an already Israel-friendly populace. Evangelicals often support highly violent foreign policies, such as the 2008–2009 War on Gaza and the possibility of a nuclear first strike against Iran. The reasons for this are highly controversial, as seen in the recent dust-up over Mearsheimer and Walt's book *The Israel Lobby* (2007).

Practical geopolitics continues to be focused on the practitioners of statecraft, although increasing recognition of the importance of other global discourses emanating from prominent NGOs and international organizations is apparent in most critical geopolitical writing. This wider definition applies in the realm of evangelical geopolitics, but the recent George W. Bush presidency has certainly dominated attention from scholars of critical geopolitics for its perceived role in implementing evangelical-influenced policy, both at home and abroad. The 'crusade' against Islamism, Operation: Infinite Justice, Axis of Evil, and other ill-judged rhetorics associated with the 'War on Terror' have dominated the representations of the Bush Administration, but other policies have been just as crucial in identifying the U.S. government with the evangelical agenda, from de-funding international development aid that included birth control, to leading the charge to create a cease-fire in the civil war in Sudan between the Muslim-dominated government and the Christian and Animist south, to ignoring the Muslim-on-Muslim conflict that emerged in Darfur at the same time (Gerhardt 2008). Nevertheless, practical evangelical geopolitics did not begin with George W. Bush. A common argument has been that the U.S. government itself has been at times run by born-again Christians pining for the apocalypse or interpreting world events and acting on these events from this theological perspective (Sturm 2006).

While there is little direct evidence of this claim, and the role of religion in American elections makes clear why some politicians might voice beliefs they do not hold sincerely, there is enough circumstantial evidence to raise questions. For instance, President Reagan and members of his administration have been particularly associated with making apocalyptic and religiously motivated statements, although exactly what Reagan intended and what strategies he employed to garner support among the Christian Right is unclear. However, his official biographer, Edmund Morris (1999, 632–33), wrote that Ezekiel was Reagan's "favorite book of prophecy". Morris's 'fictional' biography is obviously a questionable source; however, Reagan was quoted by the *Washington Post* when he was Governor of California as asking Senator James Mills if he was "familiar with the prophet Ezekiel." Upon replying that he was not, Reagan explained:

> In the thirty-eighth chapter of Ezekiel it says God will take the children of Israel from among the heathen [where] they'd been scattered and will gather them again in the promised land. Ezekiel says that... the nation that will lead all the other powers into darkness against Israel will come out of the north. What other powerful nation is to the north of Israel [besides Russia]? None. But it didn't seem to make sense before the Russian revolution, when Russia was a Christian country. Now it does, now that Russia has become communistic and atheistic, now that Russia has set itself against God. Now it fits the description perfectly. (Mills quoted in Vidal 1993, 1001–2).

Reagan also appointed Christian Zionists to influential political positions including figures like Attorney General Ed Meese, Secretary of Defense Caspar Weinberger, and Secretary of the Interior James Watt. These political outcomes make sense when considered alongside the recent coalition-building politics of the Republican Party, which William Connolly (2005) has described as a cowboy capitalist-Evangelical resonance machine, in which the power of the U.S. state and that of the evangelical movement each feed off of each other without being caused or determined by the other. Michael Lindsay argues in his book, *Faith in the Halls of Power* (2007), that over the past twenty or so years evangelicals have taken to high ranking positions of power, made a vast network of relationships in government, and influenced the White House through political lobby groups. These include the Christian Israel Public Affairs Committee (CIPAC), which was launched as the Christian lobby wing of the Likud party, and the Christian counterpart of the Washington based pro-Israel lobby, the American Israel Public Affairs Committee (AIPAC). Others include John Hagee's Christians United for Israel (CUFI) and the United Coalition for Israel (UCFI) whose biblical imperative is the territorial expansion of Israel. Just as practical evangelical geopolitics did not begin in 2001, it has not come to an end with the George W. Bush regime; indeed evangelical influence is increasingly felt in governmental discourse from the 'global south' to the European Union, as some of the chapters in this volume attest.

Popular geopolitics refers to the discourses found in, and resonating through, everyday life practices. This has usually been conceptualized in critical geopolitics as elite discourses that are disseminated among the masses, thus keeping the discourses analyzed within critical geopolitics strictly an elite affair. However, recent shifts in the sub-discipline have emphasized the role of everyday audiences in re-interpreting these discourses and recognized the continued agency of (in this case) evangelical consumers of popular culture and news media. Among evangelical cultural forms, popular and academic attention has focused on the *Left Behind* book series, which fictionalizes the final seven years of the current age roughly as outlined within American apocalyptic evangelicalism (adhering to an eschatology called dispensational premillennialism, see below). These books have been a runaway best-seller in the United States, selling over 60 million copies and spawning a variety of spin-offs, including movies, video games, audio books, children's versions, and a graphic novel (Dittmer and Spears 2009). The series single-handedly turned its publisher, Tyndale House, into a publishing giant (see Afterword, this volume).[2] As such, this series has received the lion's share of popular and academic attention. However, the series is only the most prominent symbol of the consolidation of the evangelical market using new media such as the Internet. Literary sensations such as William Young's *The Shack* (2007) illustrate the power and possibilities of evangelical audiences' word-of-mouth in the selection and perpetuation of some discourses over others. Hollywood cinematic examples include, although not designed for evangelicals alone, *The Passion of the Christ* (2004), *The Lion, the Witch, and the Wardrobe* (2005), the evangelical big budget film, *2012: Doomsday* (2008), and Roland Emmerich's secular apocalyptic version simply titled, *2012* (2009). Christian popular music and worship songs contain a wealth of doctrinal messages that are inculcated weekly before and after services, at concerts, and in iPods (see McAlister 2008). The focus on evangelical eschatology in much of the work thus far published within critical geopolitics has left open analyses that focus on other forms of geopolitically-informed evangelical popular culture.

---

2 It can be argued that *Left Behind* is the best-selling fiction series by American authors. Certainly other fiction series have sold many millions more, but on a per-book basis (16), selling 62 million in 14 years qualifies it as a contender if we discount the posterity granted to its predecessors and account for books included in the series.

Dispensational Premillennialist Eschatology and its Rise in America

Christian millennialism is not a recent phenomenon, although historically it has experienced varying levels of popularity (Cohn 1957). To put the current movement in context, it is necessary to trace its pedigree, at least at a cursory level. The first Christians believed the kingdom of God was imminent from the beginning (Collins 2000). Since then, some branches of prophetic belief about the end of time have become distinctly American. As Wallace (2006) has suggested, to many American evangelicals God is not an abstract character from which believers glean moral grounding. Rather, for American evangelicals God has a firm grip on history through an earthly engagement with geopolitical events and His central territorial concerns are America and Israel. This leads Wallace to argue that in some ways the movement is more American in its reading of Scripture than it is Christian. That said, McAlister (2003, 782) points out that American evangelicalism of the last two decades is "not your father's fundamentalism," insolated from worldly events. The movement is, rather, transnational in orientation as evidenced by followers' renewed interest in politics and the influence exerted on them by non-American voices (albeit with particular American nationalist ideology [Lieven 2004]).

The Americas were conceived in the fervor of apocalypticism. Christopher Columbus thought American gold and silver was placed in his hands to finance a new crusade against the Muslims in Jerusalem to rebuild the third temple, "the apocalyptic city par excellence" (Kirsch 2006, 170). Columbus thought he was personally written into the Bible to rebuild it. To Columbus, America (or what he thought was Asia) was the New Heaven and New Earth according to divine revelation. As is well known, some of America's earliest colonists saw America itself as the new Israel and Puritans were known to ruminate on the end of the world, which led to what historian Stephen Stein calls "the Americanization of the apocalyptic tradition" (quoted in Boyer 1992, 68; see also Bercovitch 1978; Cherry 1998).[3] During the American Civil War both sides, North and South, considered themselves to have divine favor (see Chapter 1, this volume). Of course both World Wars would be envisioned in the same way, but this time with the United States geographically unified in battle against evil (Marsden 1980). It should not be surprising then that sign watching would explode with fervor during the present "wars and rumors of wars" in the Middle East because the times we are in can always be thought of as horrible enough to be envisaged as the end (Matthew 24:7).

The evangelical movement in America can be traced to revivalist George Whitefield and reformist Jonathan Edwards, and Methodism more generally (Marsden 1980). Mark Noll (2002), one of the leading scholars today on American and Canadian evangelicals, argues that the modern American evangelical

---

3 Although Thomas Jefferson denigrated the Book of Revelation as 'merely the ravings of a maniac' and made his own Bible, cleansed of what he thought was pseudoepigraphia, leaving only the gospels.

movement began with a Protestant fissure between mainline Protestantism and fundamentalism in the early twentieth century. Evangelicals, although often used as a term to encompass all Protestant groups in the United States, were a 'third way' group between mainline Protestantism and fundamentalism. Rejecting the sympathies with modernism of the mainline Protestants and the societal alienation of the fundamentalists, evangelicals choose a middle ground focused on a literal reading of Scripture and a limited reformism in society, perhaps best exemplified in the modern preaching and activism of Billy Graham. Evangelicalism can be defined as stressing the literal authority of the Bible, personal conversion, and being born again. This said, the term evangelical has lost any precision (if it ever had any) with reference to American Protestantism.

Over 60 million Americans are evangelical Christians. According to the 2008 Pew U.S. Religious Landscape Survey, 26.3 percent of American adults claim to be affiliated to an 'Evangelical Protestant Church,' which is up from 14 percent in 1900 (Barrett 1998).[4] Most believe in a biblical 'end to history' and approximately two-thirds of these evangelicals are dispensational premillennialists—they believe Christ will return 'pre'-vious to Christ's 'millennial' rule (Weber 2004, 9). Dispensational premillennialist eschatology posits that the world, since Christ's resurrection, has been on a steady moral decline.[5] In this theology the world will only see accelerated environmental degradation, war, famine, pestilence, and strife until Armageddon and Christ's return. Seven years before the millennium is to begin—a period called the Tribulation—all true believing Christians will be Raptured (Latin for 'caught up') into the sky to witness the coming war of Armageddon from their seats in Heaven (John 14:1–3; 1 Corinthians 15:50–58; 1 Thessalonians 4:13–18).[6] The Rapture, in short, is an escape mechanism. But the most important part of this eschatology for our interests in geopolitics is the linking of biblical allegory with the shape of the world, for example when Palestine became spiritual Israel. It is what Yorgason (this volume) calls, "geo-eschatology."

Many dispensational premillennialists believe that after the Rapture, the Antichrist (Satan's representative on Earth) becomes the leader of an international government. The Tribulation begins with this world government's signing of a covenant (or peace-pact) with Israel. For the first half of the Tribulation (three and

---

4 18.1 percent of adults align themselves with 'Mainstream Protestant Churches', calling into question what counts as 'mainstream', certainly not the majority of Protestantism in this case.

5 Eschatology is that part of theology that deals with last things: birth and death, judgment, Heaven and Hell, and the end of the world.

6 The Rapture is found in a letter Paul writes to Thessalonians, "and the dead in Christ will rise first; then we who are alive, who are left, shall be caught up together with them in the clouds to meet the Lord in the air; and so we shall always be with the Lord" (4:16–17). Heaven is still thought of as skyward consistent with medieval Christian cosmology. In this conception outside the known universe is the "Empyrean of God" (Wertheim 1999, 34).

one-half years) the Antichrist abides by this treaty.[7] Because of the Antichrist's peaceful and charismatic world leadership, he brings détente between Israel and other nations and groups (Daniel 9:27, 11:40–41; Ezekiel 38–39; Matthew 24:15–20). At this time he is concentrating "his efforts on consolidating his power" and letting the Jews re-build Solomon's Temple (Lindsey 1998, 164). Taking advantage of Israel's relaxed borders, Gog, the King of the North with his army of nations known as Magog, Meshech, Tubal, Rosh, Gomer, and Beth-togarmah, descend upon Israel. The Kings of the South—Persia, Ethiopia (Cush), Put—flank Israel from the south, but God intervenes and destroys these armies before they can desecrate the Temple (Daniel 11:40–45; Ezekiel 38–30).[8]

Marking the beginning of the second half of the Tribulation, the Antichrist breaks his covenant with Israel by leading his army into Israel (Daniel 9:27; 11:40–41). He then claims his seat in Solomon's Temple and declares himself God. The ten horned beast (or ten kings of the Roman Empire) of Revelation (13:1) are consolidated under the Antichrist's sovereignty and he sets up his capital in Babylon (Revelation 17:16–17). Christ then returns to the Mount of Olives and slays the armies that gathered against Him in the Jezreel Valley at the battle of Armageddon ('Har Megiddo' or Mount Megiddo). The Antichrist is killed in battle and Satan is cast into an abyss for Christ's one thousand-year rule on Earth (Revelation 20:1–3). At the end of the millennium, Satan frees himself in one last ditch effort to takeover the world but is destroyed and Heaven and Earth are made new again (Revelation 20:7–10; Isaiah 65:17; 66:22; 2 Peter 3:13).

'Post'-millennialism, an eschatology that has periodically appeared since the third century, was systematized by the Unitarian teacher Daniel Whitby (1638–1726). Postmillennialism posits that humans will usher in the Kingdom of God by Christianizing the world's population. Premillennialists, in contrast, believe that only God can bring about the millennium; the events of the future will be too traumatic to be solved by human hands. In the postmillennial schematic, the millennium will be between Christ's first and second coming and not *after* His second coming (Boettner 1977). From around the American Revolution to the 1930s, postmillennialism had almost supplanted premillennialism in America, mirroring the project of optimism and an embrace of modernity as a moral path to earthly perfection embodied in the Second Great Awakening and social reforms which lead to the Social Gospels.[9] Therefore while postmillennialism found expression in the Enlightenment, premillennialism focused on the trauma of modernity. Hunt

---

7 Premillennialists disagree on whether the resurrection-rapture will take place before, during, or after the Great Tribulation; they are pre-tribulationists, mid-tribulationists, and post-tribulationists respectively (Robbins and Palmer 1997).

8 Other dispensational theologians argue that the so called Ezekiel War happens before the Rapture (see LaHaye and Jenkins 1996).

9 Kant was a kind of secular postmillennialist. In his essay, 'The End of All Things' (1794) (Das Ende aller Dinge), Kant advised that Christianity shed 'the end' and convert the approach to it as unending progress that makes the good world better (Shaffer 1995).

(2001, 59) points out that postmillennialism, like premillennialism, constructs a binary, "perceiv[ing] practically everything outside of their boundaries as satanic". However, a key difference is that postmillennialists are not fatalists, they believe that there will be an "exorcism of high-ranking demonic forces from particular geographic areas" in the final establishment of the Millennial Kingdom. But postmillennialism for this reason was seen as too sympathetic to the liberal side in the early twentieth century fundamentalist debates and was subsequently relegated to ever smaller constituencies, taking with it postmillennialism's fiery cultural and political progressivism.[10]

Postmillennialism has undergone a limited (and divisive) revival in the last three decades, however. Fundamentalist postmillennialism (or reconstructionism), a reinterpretation of Whitby's eschatology stemming from Rousas Rushdoony's 850-page *Institutes of Biblical Law* (1973), states that in order for the millennium to begin the world must first be ruled by the laws of the Five Books of Moses (Pentateuch). This group, predominantly conservative neo-Pentecostalists, have recently attempted to implement their biblical blueprint for government through Judge Roy Moore's 2005 bill, the Constitution Restoration Act (co-sponsored by 60 congressmen), which in his words would force "the constitution of every state of the Union [to] acknowledge God and His sovereignty... [as well as the] three branches of the federal government" (Moore 2004). Despite this deeply conservative revival, premillennialism is still the hegemonic eschatology among socially and politically conservative American evangelicals in part because of theo-historical argument that the early church was premillennialist, the complete Christianization of the world is unlikely, and because this eschatology is connected to a much larger publishing and media industry.

Amillennialism is a symbolic or spiritual concept in which events written in the Bible are thought to take place in the spiritual realm rather than on earth. Other ideas falling into the category of amillennialism include those of St. Augustine of Hippo, who interpreted the millennium as beginning from the date Christ was resurrected. In the Middle Ages the relationship between history and eternity, good and evil was complex. Some thought Christ's victory had already taken place with his sacrifice on the cross and the establishment of the Christian church at the Pentecost. From that time the church has been rejoicing the Incarnation and the Old Testament fulfillment of the Heavenly City of Jerusalem as having descended from Heaven.

Such eschatological imaginations, it has been argued, are central to understanding modern American evangelical Christianity and its geopolitics, from the optimism or pessimism they promote to the literalism of God's hand intervening

---

10 Hal Lindsey (1989), a leading American premillennial interpreter of eschatology, in a polemical treatise on why the Holocaust happened, argues that postmillennial ideas of both positing that the Covenant with the Jews was fulfilled in the first century CE and that Christians could make a better world without God's assistance, led Hitler to deny the Jews any role in the Bible and take it upon himself to annihilate European Jews.

in human affairs (see Sturm 2008). Curry-Roper (1990) argues that eschatological doctrine, because it deals with birth and death, beginning and end of time, is a major social decision-making variable. If we are to take ancient origin and heritage myths seriously in geopolitical analysis, then future mythologies, which are just as corrupted and fully formed as the past, should have equal resonance with the present (Rosenberg and Harding 2005). The distinction between what is said and what is done, however, complicates such assertions. Doctrine is no longer a central characteristic of representation in the social sciences for this reason and much recent research takes 'doing,' as opposed to 'believing,' as the primary category of analysis (Asad 1993). Furthermore, because Protestant doctrine is open to individual interpretation, eschatology can have the effect of essentializing belief. That said, religion can be studied both through careful discourse analysis and also through performances of doctrine, spirituality and enchantment of everyday life.

It is on premillennialism that this volume focuses, given its predominance in the American context. In particular, this volume focuses on a particular type of premillennialism called dispensationalism (also known as pre-tribulation premillennialism). Dispensational premillennialists believe that history has been split up into seven historical periods or dispensations, in each of which God deals with humanity differently; the current penultimate epoch being known as the Church Age or 'Great Parenthesis' will be followed by Christ's millennium.[11] The present age is parenthetical of any prophetic fulfillment because the Jews rejected Jesus, and therefore suspended prophetic history. With the reestablishment of Israel in 1948, dispensationalists believe that the prophetic wheels have been greased and God's prophetic hand has once again entered world history, dividing it geographically into regions of friends and of enemies and guiding believers to proselytize busily before the Rapture. To prevent confusion, Mark Hitchcock, a prominent prophecy theologian and popular author, wrote in an email interview (March 13, 2006), "Dispensationalism is a subset of premillennialism. All dispensationalists are premillennialists, but not all premillennialists are dispensationalists. The *sine qua non* or essence of dispensationalism is a consistent, literal interpretation of the Bible and a distinction between Israel and the church. Dispensationalists believe that God has a future program for national Israel and that the OT [Old Testament] covenants will be literally fulfilled with Israel in the future." This runs contrary to the common Christian (largely Catholic) replacement theology which posits that Christians have replaced the Jews as the chosen people of Abraham's Covenant (Provan 1987).

---

11 We take premillennialism as the base theological belief amended by dispensationalism, hence the order dispensational premillennialism. The idea of breaking up biblical time into different epochs is often traced to the Calabrian monk, Joachim of Fiore (1135–1202 CE), who in keeping with the trinity, divided God's treatment of human kind into three epochs: the Age of Law, represented by God; the Age of Grace, marked by Christ's birth and spanning into the present; and the Age of the Spirit, where evil is finally cast from time and space (McGinn 1985).

Although premillennialism has existed in Christianity since the first century of the Common Era, dispensational premillennialism was developed by John Nelson Darby, a member of the Plymouth Brethren who, from 1859 to his death in 1882, toured the United States six times preaching dispensationalism as the true intention of God's word (Boyer 1992; Sandeen 1970). Darby found a receptive audience for his ideas in the United States most likely because of its paranoid style of politics that linked up well with the American frontier and colonial experience (Hofstadter 1967, 3–40). Other pastors and prophecy expounders like Dwight L. Moody and Cyrus Scofield, however, were much more effective in disseminating Darby's thought. Moody's Bible Institute would become one of the most important training grounds for American pastors as well as one of the largest publishing engines concerning evangelical thought. Cyrus Scofield's King James *Reference Bible* published in 1909 by Oxford University Press, which sold over ten million copies, included Scofield's interpretation of Darby's thoughts on eschatology as notes inserted into the center of the page—which had the effect of confusing American readers in distinguishing between what was Scofield's notes and what was Scripture (Boyer 1992).[12] Dispensationalists countered the Social Gospel and modernism by arguing that the millennium was at hand and that God was going to usher it in, not humans via good works. These expounders did not come without their detractors. Shailer Mathews's *Will Christ Come Again* (1917), among others from the University of Chicago Divinity School, pegged dispensational premillennialist thought as an embellishment of scripture that no longer belongs in the modern church (Marsden 1980). Mathews wrote that just because early Christians believed in premillennial prophecy this did not mean that modern Christians should, if they did they "logically ought to include belief in a flat earth…" (cited in Weber 2004, 79).

Dispensationalists believe that Israel (the Jews) was given a covenant for an earthly kingdom while the Church (the Christians) were given a spiritual covenant for a kingdom in Heaven. There are then two historical programs God makes, one for the people of the earth—the Jews—and the other for the people of Heaven—the Christians. This is in part the function of Darby's Rapture: to solve the divisive issue of only having one chosen people on earth at a time. The other function is to spare true believers from the atrocities of the Tribulation. The Darbyites not only read the Bible as literal meaning they also read it in its entirety, giving the Old Testament equal authority to the New Testament, which was more voice than most other Protestants gave it at this time. Because of this Darby had to accommodate the chosen people of the Old Testament, the Jews, as well as the chosen people of the New Testament. Therefore the Rapture divides the two eras. Prior to the

---

12 More recently the Ryrie Study Bible has had a significant impact on disseminating dispensationalism.

Rapture, Christians are God's chosen people and after it the Jews are given a final chance to redeem themselves and accept Christ.[13]

With the growing influence modern evolutionary science had on Protestant churches, dispensationalism and its literal reading of the Bible were increasingly accepted among the so called 'fundamentalist' circles (see Armstrong 2001, 175–78; Stump 2000). Weber (2004, 13) observes that "By World War I, dispensationalism had become synonymous with fundamentalism and Pentecostalism." Indeed, with World War I, the Great Depression, and World War II, the world appeared to many to be falling further into an earthly Hell. Cyrus Scofield's popularization of Russia as the evil 'Rosh' in the Bible via his *Scofield Study Bible* seemed prescient with the Russian Revolution, the creation of the Soviet Union and the later inception of the Cold War. Dispensationalism seemed to be getting its politico-geographic signs right. However, the most influential sign of the End Times was the reestablishment of Israel in 1948. These two geo-predictions of the End Times, the rise of Israel and the Soviet Union, would provide the veracity to convert millions of people to dispensationalist views of history and belief (Boyer 1992).

Many Protestant churches preach dispensationalism, although it should not be confused with denominational belief. While some Baptists, Pentecostalists, Presbyterians, Methodists, and even Anglican Protestants preach dispensationalism, not all do, and Southern Baptists refer to it in their official creedal statements. For the most part, dispensationalism is growing most rapidly in non-denominational (or independent) churches and Pentecostal movements whose missionaries have had lasting theological effects on evangelicals around the world, specifically in the global south. However, there is no consensus within these denominations as to what view of eschatological history is the biblically accurate one. Prophecy is most pervasive in the Greater American South or 'Bible Belt,' which spills over the borders of the Confederacy and the Mason-Dixon to include parts of the Mid-west and the southern halves of Ohio, Indiana, and Illinois. To the west the Bible Belt conceivably includes Oklahoma, parts of Colorado, Arizona, and Southern California, where millions of Southerners migrated north and west in the mid-twentieth century, bringing with them their ideas of individualism and reduced government involvement in both public and private affairs (Williams 2002; Dochuk 2007). However, it can be found in the United States as a whole and

---

13 It is under major debate, and in some cases revision, if the Tribulation begins immediately upon the Rapture or if there is a space of time between the Rapture and the Tribulation. Pre-trib advocates like Mark Rosenberg believe there is a significant period of time. In a sermon he gave in Jerusalem at King of Kings church in 2008 he said, "my father is an architect and watching him I know it takes a long time to build something. We know that the Third Temple is completed upon the Tribulation". Messianic Jews, however, tend to be post-trib advocates so that both Jews and Christians are forced to endure God's wrath, therefore not privileging one over the other.

has enduring movements in Canada, particularly Southern Ontario and Alberta (Marsden 1980).[14]

Within the so-called New Christian Right, the late Jerry Falwell, James Dobson, Tim LaHaye, and Pat Robertson all follow (or followed) Darby's basic dispensational premillennialist schema. R. Scott Appleby (2003, 381) writes, "the selection and reinterpretations of politically charged doctrines or precepts, around which a political movement is built" is classic "fundamentalism." Consistent across denominations and churches is the belief in the Rapture to distinguish between the Church and Israel, a premillennial return of Christ, a violent Tribulation period, a suspension of prophetic time in the present (or Church) age, the conversion of the Jews during the Tribulation period, and a feeling of an imminent millennium (see Crutchfield 1991). Dispensationalism is contested and always changing. Among the new Christian right and even the more liberal, younger movement of American evangelicals, many still adhere to Darby's major tenets.

However, what exists today is the pessimism of premillennialism combined with the missionary enthusiasm of postmillennialism. On the one hand, there is still a binary vision of good evangelicals and the evil armies of Revelation, Ezekiel, and Daniel that will amass at the battles of Gog and Magog and Armageddon in modern day Israel. On the other hand there is a presumption that although time is running out, there is still enough time to save souls whether these souls are the biblical enemy or not (see Chapters 7, 8, and 9, this volume). During the Cold War, the USSR was the enemy of America and of God to most American evangelical Christians. That said, missionary work in the USSR was thought of as the most righteous and spiritually prosperous calling for American evangelicals. Similarly, since the end of the Cold War, Arabs, particularly Muslims, in the Middle East have become one of the most sought after (and most difficult) targets of conversion. This is exemplified by the modern era's most famous evangelical missionary, Brother Andrew (Andrew van der Bijl), who worked in the USSR until the end of the Cold War and since then has worked in the Middle Eastern states spreading the Word of God. His autobiographical books, *God's Smuggler* (1967) and *Light Force* (2004) have sold over ten million copies. What should be clear is that inside the binary premillennial discourse of Arabs as radically Other is another that sees Arabs as fertile ground for souls that most need saving.

This internal conflict between revivalist fatalism and reformer optimism is summed up in Billy Graham's apocalyptic text, *Approaching Hoofbeats: Four Horseman of the Apocalypse*: "every person who is a follower of Christ is responsible to do something for the hungry and sick in the world. We must do what we can, even though we know that God's ultimate plan is the making of a new earth and a new heaven" (quoted in Kirsch 2006, 233). Within this statement we can find the roots of a more moderate dispensationalism that is

---

14 For more on this, Marc Silk with various co-editors has recently published a series of volumes between 2004–2006 on the regional geography of Christian beliefs in the United States.

both less ethnically and religiously dualistic, but more importantly is also less fatalistic concerning human agency (see Burns 1999). Many other evangelicals, particularly youth fed up with negative media portrayals of evangelicalism and world-wide developmental inaction in places like Africa, have moved toward the more liberal 'postmodern' theologies of Jim Wallis, Tony Campolo and Ron Sider and into suburban, stripped plain, and gospel-focused churches like Rick Warren's Saddleback Church in Southern California.

Although it might seem at the start of the Obama Administration that the use of religion to construct Bush-era dualisms is a thing of the past, it is safe to say that religion is a more constant variable in American politics. Obama's religious outreach campaign, led by Joshua DuBois, was the most comprehensive religious initiative for any Democratic campaign. Furthermore, the choice of Rick Warren for the Presidential invocation would suggest a different kind of evangelicalism being formally promoted for America, one keen on recognizing both the Arabic and Hebrew names for Jesus, 'Esa' and 'Yeshua', reflecting a more pacific view of the Israel/Palestine conflict and also wider recognition of an Abrahamic, and not just a Judeo-Christian, root to America's civil religion (see Herberg 1960; Bellah 1967). Warren's use of 'Jesus' rather than 'God' can be interpreted differently however. Jesus was never named in the Inaugural Invocation until the George W. Bush presidency (when Franklin Graham made a point to use Jesus instead of God), Billy Graham used only 'God' for the Bush Sr. and Clinton presidencies in a conscious attempt to include the millions of Americans who were not Christian. Therefore Warren reaffirmed America as an Abrahamic state.

Chapters

For a book entitled *Mapping the End Times*, we admittedly present few maps. Our concern with mapping is not in a literal cartographic sense; rather we are interested in geopolitical imaginations that are not always rendered cartographically but are described textually or through other visualities. Furthermore, this volume's titular reference to the 'apocalypse' is not always a central question of these chapters, as the term is colloquially defined. We instead understand the term apocalypse from its Greek origin *apo-kalyptein*, meaning 'to un-cover' or 'un-veil' God's plan for earth's history, and so the chapters in this volume focus on the role of theology and eschatology in geopolitical imaginations. There are many related issues remaining outside the scope of this volume. We do not claim to be comprehensive or encyclopedic – instead we see this volume as a first step toward the more comprehensive study of other religiously informed geopolitical imaginations. The most obvious are non-evangelical Christian geopolitics, and there are many possibilities for future research in this fruitful field (see Megoran 2006). Geopolitical analysis has been largely centered on Western discourses, and this volume is little different. The reason for this focus is America's hegemony in cultural, economic, political, and specifically religious, world affairs for the last hundred or so years. Denominational (and non-denominational) complexity indicates a myriad of identities, all fluid, and some deviating quite far from the theology presented, above, and it is for this reason that roughly half of the chapters in this book have attempted some type of intersubjective ethnographic analysis rather than simply an intertextual discourse analysis. Ethnography is broadly understood here, from formal interviews, to having lived in the places and with the people written about, to the common occupation of digital or virtual spaces.

Dave Jansson argues in his chapter that evangelicalism plays a role in the internal geopolitics of the United States as well as the more traditional outward-facing version. Through interviews with neo-Confederates and his own concept of "internal orientalism", Jansson interrogates the special role of evangelical religion in Southern identity *vis-à-vis* the rest of the country. For these individuals evangelicalism carries connections to the moral, racial, and political landscapes of the South. His interviews with everyday people render problematic any homogeneity that might otherwise be ascribed to individuals through geopolitical discourse.

Ethan Yorgason's chapter takes the reader into the world of Mormonism, comparing and contrasting it with the evangelical dispensational premillennialism that is more prominently addressed in this volume. Yorgason does this through a comparison of Mormon last days novels to LaHaye and Jenkins's evangelical *Left Behind* series. Yorgason points out the many eschatological and even geopolitical similarities but also points out that the two theologies' common advocacy of conservative American foreign policy should not be surprising. These geopolitical views span the conservative population in the United States, secular or Christian. Although very similar in structure and geopolitical specifics, he concludes that while evangelicals and Mormons share many values like 'the family', they do not see eye-to-eye on many other specifics, such as eschatology, and therefore share a tenuous relationship of convenience.

Jason Dittmer draws on internet ethnography to explore the interconnections between American identity, race, and religion as demonstrated through the 2008 presidential election, in which eventual winner Barack Obama was dogged by rumors that he was both Muslim and the Antichrist. This conspiracy theory was linked to new techno-cultural assemblages such as evangelical Internet forums, in which new knowledges are produced and disseminated. Using the concept of narrativity, Dittmer shows how new identities are continually (re)produced through the intertwining and intermixing of quite contradictory narratives of race, religion, and Americanism. A further implication is that the production of religio-geopolitical knowledges often occurs far distant from, and often in contradiction of, political and religious elites.

In his chapter, Simon Dalby analyzes the techno-thriller genre of literature, which originated in the American secular tradition but has since branched out to include evangelical themes. Using the tools of popular geopolitics and Debrix's concept of tabloid geopolitics, Dalby compares Tom Clancy's *Teeth of the Tiger* with Joel Rosenberg's *The Ezekiel Option*. He argues that they share a notion of American exceptionalism and a sense of geopolitical insecurity. Therefore, he rejects strands of popular culture that celebrate vigilantism and easy morality, instead arguing for the celebration of new virtues that will lead to a more nonviolent world.

Michael Barkun's chapter argues that America has been reinserted into a once internationally-focused dispensational discourse about the End Times. He argues however, that the role America always played in the background has become more prominent and obviously fleshed out with recent efforts by Pat Robertson and John Hagee, among others, to include New World Order conspiracy theories in dispensationalist geopolitical imaginations. This inclusion of an American destiny and meaning, drawing from roots in American exceptionalism and Puritan narratives, has made once-fatalist dispensationalists feel that they can actively change the world, but in limited ways. The tension between human action and divine will is still very present.

Also reaching into the cultural history of dispensationalism's ethnic geographical imaginations is Tristan Sturm's chapter, which brings the reader into the nonfiction world of American evangelical apocalyptic best-sellers. By tracing the designations of evil spaces from the 1970 Cold War publication of Hal Lindsey's *The Late Great Planet Earth*, to the more recent geographic inclusions of Iran into the prophetic lexicon, Sturm fleshes out the many strategies dispensationalists use to legitimate the dissonance between the static ethno-geographies of the Bible and the messy geopolitics of a protean world of new enemies. It is argued that through strategies of othering, privileging space over time, and inter-textual citationality this geo-prophetic back-peddling is an attempt to recast geopolitics as God's making.

Drawing on the political theory of Carl Schmitt and Chantal Mouffe, Hannes Gerhardt traces the continuities and discontinuities between the synergy of the U.S. state and the evangelical mission movement in Africa. Specifically, the identification of the enemy ('the essence of the political') is seen as a key point on which the two actors cannot agree. The United States is rooted in a liberal-universalist governmentality while conservative evangelicals see Islam as a competitor for souls. Gerhardt identifies this disjunction as a potential problem for the U.S. state if it does not disentangle itself from the evangelical movement.

Ju Hui Judy Han's chapter focuses on the geopolitical construct of the 10/40 window, an abstraction created by evangelicals to focus missionary attention on particular parts of the world. Her chapter, based on ethnographic research conducted with a Korean American church, highlights the plural nature of evangelization and the ways in which it is visualized. Evangelization is not merely an extension of American imperialism, but also draws from a variety of other influences, including Third World liberation and generosity for the Other. The ethnic and geographic categories used in practice must thus also be conceptualized as real affective and emotional connections despite their many flaws.

Carolyn Gallaher's chapter, informed by interviews with prominent missionary leaders, discusses the "Middle East Mission paradox". That is, she discusses the incongruity between the official voices of dispensationalism that present fatalism and an anti-Arab and anti-Muslim discourse with the voices of missionary leaders who argue for tempered cultural understating and a self-reflexive missionary ethic about what it means to be American and dispensationalist. Using the concept of performativity, she argues that the "evangelical geopolitical vision" should not be homogenized but should instead be explicated in its full diversity.

In his afterword to this volume, Paul Boyer looks back at past decade, reviewing the role of evangelicalism and prophecy in popular, civic, and governmental culture before looking forward and making some tentative claims about the future relationship between evangelicalism and American geopolitics.

Conclusions

The chapters in this volume attest to the diversity within American evangelical geopolitics. It should be clear that talking about 'evangelicals' or even 'dispensationalists' as a catholic whole is problematic. One tension that is shot through these chapters is that of the will to missionary work and the fatalism of the apocalypse. This tension roughly grafts onto the balancing stories America tells itself about its history between the American universalist (manifest destiny) and the particularist (a response to the experience of their colonial past). These stories act as a guide for the ethics of intervention (missionary work and defense of human rights) versus isolationism (local nationalism) (Hervieu-Leger 2002; see Sturm 2008). This tension, according to Hervieu-Leger (2002, 103), explains all "paradoxes between religion and space." This dualism can be seen as the central question of the definition of Americanism itself. As Michael Kazen and Joseph McCartin (2006, 2) have recently pointed out, there is a tension in Americanism with conflict and cohesion "producing both internal strife and solidarity against foreign enemies." For American evangelicals this tension is both reproduced through a biblical prism that undergirds what events mean and shores up institutional dogma, missionary work, support for Israel, and a performative covenantal American nationalism. Americanism and evangelicalism resonate not just at the scale of cowboy capitalism, but also at the dialectical intersection of intervention and introversion.

Mappings of the apocalypse or the missionary world are important because they reaffirm what devotees want to take for granted about the world. This is an application of an idea commonplace within the critical geopolitics literature. However, as stated in the opening sentences of this introduction, geopolitics (and related fields such as International Relations) has been slow to investigate religious belief systems. It should be clear that an analysis of dispensational geopolitics is important in large part because it has political implications at most analytical scales, from the individual to the international. Further, the deep authority granted within evangelical geopolitics to the male social order is deserving of feminist criticism. Prophecy is almost exclusively a male providence and one could develop a feminist critique of prophecy and authority as the escapist strategy of the Rapture seeks to free Holy men from their bodily existence. We hope that religion gets sustained inquiry with geopolitical and political geography literatures, not simply as related to historical moments and journalistic responses to contemporary conundrums, but rather to take religion as an embedded and lasting element in much geopolitical theorizing.

The rise of the 'new culture wars' in the 1990s and the historical electoral repudiation of the George W. Bush Administration in 2008 may give the impression of an era in which evangelical power in the U.S. rose and then fell in a classic Shakespearean story of overreach (or in a Connolly-esque shift from resonance to dissonance). However, it must first be admitted that it is too soon to tell if the era is over, given the proximity of recent events, particularly with Israel's shift to the political right (seen most recently in the 2009 elections). It must also be clear that this narrative is embedded in a statist view of evangelical (geo)politics, one that takes political power to be congruent with electoral power. This particular narrative then misses the longue durée of evangelical power; evangelicals have been a major force within the United States for much of its history; it is only when evangelicals turn their attention to electoral politics that it enters conventional history books (Noll 2007).

A more nuanced approach to evangelical history reveals a somewhat cyclical pattern of evangelical engagement in the public sphere, as we have shown in this volume, often not reflective of electoral politics. The meaning of the public sphere itself is contested among evangelicals, with some viewing it as beyond the pale, others viewing it as God's battleground for the soul of his chosen country, and yet others falling somewhere in between. As evangelical willingness to engage in public affairs in a highly organized way tends to rise and fall over time as issues become more and less salient and as socioeconomic crises wax and wane, we can surely expect that American evangelicals will continue to play a role in the interactions between the U.S. and the rest of the world, through missionary work, face-to-face NGO diplomacy, lobbying for particular foreign policies, and through everyday performances in the world.

References

Agnew, J. (1998), *Geopolitics: Re-visioning World Politics* (London: Routledge).
—. (2006), 'Religion and geopolitics,' *Geopolitics* 11:2, 183–91.
Appleby, R. S. (2002), 'Fundamentalist and Nationalist Religious Movements,' in Agnew, J., Mitchell, K., and Toal, G. (eds) *A Companion to Political Geography* (Oxford: Blackwell), 378–92.
Andrew, B., Sherril, J. and Sherril, E. (1967), *God's Smuggler* (Washington Depot, CT: Chosen Books).
Andrew, B. and Janssen, A. (2004), *Light Force* (Grand Rapids: Revell).
Armstrong, K. (2001), *The Battle for God* (New York: Ballantine).

Asad, T. (1993), *Genealogies of Religion: Discipline and Reason of Power in Christianity and Islam* (Baltimore: Johns Hopkins University Press).
Barrett, D.B. (1998), 'A Century of Growth,' *Christianity Today*, November, 50–1.
Bellah, R.N. (1967), 'Civil Religion in America', *Journal of the American Academy of Arts and Sciences* 96:1, 1–21.
Bercovitch, S. (1978), *The American Jeremiad* (Madison: University of Wisconsin Press).
Bialasiewicz, L. (2006), "The death of the West': Samuel Huntington, Oriana Fallaci and a new 'moral' geopolitics of births and bodies,' *Geopolitics* 11:4, 701–24.
Boettner, L. (1977), 'Postmillennialism.' In *The Meaning of the Millennium: Four Views* Clouse, R. (ed.) (Downers Grove, IL: InterVarsity Press), 117–42.
Boyer, P. (1992), *When Time Shall Be No More: Prophecy Belief in Modern American Culture* (Cambridge: Belknap Press).
Campbell, D. (1992), *Writing Security: United States Foreign Policy and the Politics of Identity* (University of Minnesota Press, Minneapolis).
Cherry, C.C. (1998), *God's New Israel: Religious Interpretations of American Destiny* (Chapel Hill, NC: University of North Carolina Press).
Cohn, N. (1957), *The Pursuit of the Millennium: Revolutionary Millenarians and Mystical Anarchists of the Middle Ages* (Oxford: Oxford University Press).
Collins, J.J. (2000), *The Encyclopedia of Apocalyptism* Vol. 1. (London: Continuum).
Connolly, W.E. (2005), 'The Evangelical-Capitalist Resonance Machine,' *Political Theory* 33:6, 869–86.
Crutchfield, L. (1991), *The Origins of Dispensationalism: The Darby Factor* (Lanham: University Press of America).
Curry-Roper, J.M. (1990), 'Contemporary Christian Eschatologies and their Relationship to Environmental Stewardship', *Professional Geographer* 42:2, 157–69.
Dalby, S. (1990), *Creating the Second Cold War: The Discourse of Politics* (New York: Guilford).
De Busser, C. (2006), 'From Exclusiveness to Inclusiveness: The Changing Politco-Territorial Situation of Spain and its Reflection on the National Offerings to the Apostle Saint James from the Second Half of the Twentieth Century,' *Geopolitics* 11:2, 300–316.
Dijkink, G. (2006), 'When Geopolitics and Religion Fuse: A Historical Perspective,' *Geopolitics* 11:2, 192–208.
Dittmer, J. (2008), 'The Geographical Pivot of (the End of) History: Evangelical geopolitical imaginations and audience interpretation of *Left Behind*,' *Political Geography* 27:3, 280–300.
Dittmer, J. and Spears, Z. (2009), 'Apocalypse, Now? The Geopolitics of *Left Behind*,' *Geojournal* 74:3, 183–89.

Dochuk, D. (2007), 'Evangelicalism Becomes Southern, Politics Becomes Evangelical: From FDR to Ronald Reagan,' in Noll, M. and Harlow, L. (eds) *Religion and American Politics: From Colonial Period to the Present* 2nd edn (Oxford: Oxford University Press).

Gerhardt, H. (2008), 'Geopolitics, Ethics, and the Evangelicals' Commitment to Sudan,' *Environment and Planning D: Society and Space* 26:5, 911–28.

Gorenberg, G. (2000), *The End of Days: Fundamentalism and the Struggle for the Temple Mount* (Oxford: Oxford University Press).

Hedges, C. (2007), *American Fascists: The Christian Right and the War on America* (New York: Free Press).

Herberg, W. (1960), *Protestant-Catholic-Jew: An Essay in American Religious Sociology* (New York: Anchor Books).

Hérodote (2005), 'Les évangéliques à l'assaut du monde,' 119:4. <http://www.herodote.org/rubrique.php3?id_rubrique=30>, accessed 1/26/07.

Hervieu-Leger, D. (2002), 'Space and religion: new approaches to religious spatiality in modernity,' *International Journal of Urban and Regional Research* 26:1, 99–105.

Hofstadter, R. (1967), 'The Paranoid Style in American Politics', in Hofstadter, R. (ed.) *The Paranoid Style in American Politics and Other Essays* (New York: Vintage), 3–40.

Holloway J., (2006), 'Enchanted Spaces: The Séance, Affect and Geographies of Religion,' *Annals of the Association of American Geographers* 96:1, 182–187.

Hunt, S. (2001), 'The Rise, Fall and Return of Post-Milenarianism', in Hunt, S. (ed.) *Christian Millenarianism: From the Early Church to Waco* (Bloomington: Indiana University Press), 50–61.

Kazen, M. and McCartin, J. (2006), 'Introduction,' in Kazen, M. and McCartin, J. (eds) *Americanism: New Perspectives on the History of an Ideal* (Chapel Hill, NC: University of North Carolina Press), 1–24.

Kirsch, J. (2006), *A History of the End of the World: How the Most Controversial Book in the Bible Changed the Course of Western Civilization* (New York: HarperCollins).

Kong, L. (2001), 'Mapping "New" Geographies of Religion: Politics and Poetics in Modernity,' *Progress in Human Geography* 25:2, 211–33.

LaHaye, T.F. and Jenkins, J.B. (1996), *Left Behind: A Novel of Earth's Last Days.* (Wheaton, IL: Tyndale House).

Lieven, A. (2004), *America Right or Wrong: An Anatomy of American Nationalism.* (New York: Oxford University Press).

Lindsey, H. (1989), *The Road to the Holocaust* (New York: Bantam).

—. (1998), *Planet Earth: The Final Chapter* (Beverley Hills, CA: Western Front).

Lindsay, D.M. (2007), *Faith in the Halls of Power: How Evangelicals Joined the American Elite* (Oxford: Oxford University Press).

Livingstone, D.N., Keane, M.C. and Boal, F.W. (1998), 'Space for Religion: A Belfast Case Study,' *Political Geography* 17:2, 145–70.

Marsden, G. (1980), *Fundamentalism and American Culture: The Shaping of Twentieth Century Evangelicalism, 1870–1925* (Oxford: Oxford University Press).

Mathews, S. (1917), *Will Christ Come Again* (Chicago: American Institute of Sacred Literature).

McAlister, M. (2003), 'Prophecy, Politics, and the Popular: The *Left Behind* Series and Christian Funademantalism's New World Order,' *South Atlantic Quarterly* 102:4, 773–98.

—. (2008), 'What is Your Heart For? Affect and Internationalism in the Evangelical Public Sphere,' *American Literary History* 20:4, 870–95.

McGinn, B. (1985), *The Calabrian Abbot: Joachim of Fiore in the History of Western Thought* (New York: Macmillan).

Megoran, N. (2006), 'God On Our Side? The Church of England and the Geopolitics of Mourning 9/11,' *Geopolitics* 11:3, 561–79.

Mearsheimer, J. J. and Walt, S. M. (2007), *The Israel Lobby and U.S. Foreign Policy* (New York: Farrar, Straus and Giroux).

Moore, R. (2004), 'Judge Roy Moore Introduces Constitution Restoration Act 2004,' *WAFF News*. <http://www.waff.com/Global/story.asp?S=1644862> accessed 1/26/07.

Morganthau, H. (1960), *Politics Among Nations: The Struggle for Power and Peace* (New York: Knopf).

Morris, E. (1999), *Dutch: A Memoir of Ronald Reagan* (New York: Random House).

Noll, M. (2002), *The Work We Have to Do: A History of Protestants in America* (Oxford: Oxford University Press).

—. (2007), 'Introduction.' in Noll, M. and Harlow, L. (eds) *Religion and American Politics: From the Colonial Period to the Present,* 2nd Edition (Oxford: Oxford University Press), 3–19.

Ó Tuathail, G. and Agnew, J. (1992), 'Geopolitics and Discourse: Practical Geopolitical Reasoning and American Foreign Policy,' *Political Geography* 11:2, 190–204.

Ó Tuathail, G. (1996), *Critical Geopolitics: The Politics of Writing Global Space* (Minneapolis: University of Minnesota Press).

Provan, C.D. (1987), *The Church is Israel Now* (Vallecito, CA: Ross House Books).

Pulliam, S. and Olsen, T. (2008), 'Q&A: Barack Obama,' *Christianity Today*. <http://www.christianitytoday.com/ct/2008/januaryweb-only/104–32.0.html?start=2>, accessed 3/5/09.

Robbins, T. and Palmer, S.J. (1997), *Millennium, Messiahs, and Mayhem: Contemporary Apocalyptic Movements* (New York: Routledge).

Rosenberg, D. and Harding, S. (2005), *Histories of the Future* (Durham, NC: Duke University Press).

Rushdoony, R.J. (1973), *The Institutes of Biblical Law* (Nutley, NJ: P and R Publishing).

Said, E. (2002), 'Waiting on a Countervailing Force: Europe Versus America,' *Counter Punch*. <http://www.counterpunch.org/said1116.html>, accessed 3/5/09.

Sandeen, E.R. (1970), *The Roots of Fundamentalism: British and American Millenarianism, 1800–1930* (Chicago: University of Chicago Press).

Shaffer, E. (1995), 'Secular Apocalypse: Prophets and Apocalyptics at the End of the Eighteenth Century,' in Bull, M. (ed.) *Apocalypse Theory and the End of the World* (Oxford: Blackwell).

Sharp, J. (2000), *Condensing The Cold War: Reader's Digest and American Identity* (Minneapolis: University of Minnesota Press).

Sidorov, D. (2006), 'Post-Imperial Third Romes: Resurrections of a Russian Orthodox Geopolitical Metaphor,' *Geopolitics* 11:2, 317–47.

Stump, R. (2000), *Boundaries of Faith: Geographical Perspectives on Religious Fundamentalism* (Lanham, MD: Rowman and Littlefield Publishers).

Sturm, T. (2006), 'Prophetic Eyes: The Theatricality of Mark Hitchcock's Premillennial Geopolitics,' *Geopolitics* 11:2, 231–55.

—. (2008), 'The Christian Right, Eschatology, and Americanism: A Commentary on Gerhardt,' *Environment and Planning D: Society and Space* 26:5, 929–34.

Vidal, G. (1993), *United States: Essays, 1952–1992* (New York: Broadway Books).

Wallace, I. (2006), 'Territory, Typology, Theology: Geopolitics and the Christian Scriptures,' *Geopolitics* 11:2, 192–230.

Williams, P.W. (2002), *America's Religions: From Their Origins to the Twenty-first Century* (Champaign: University of Illinois Press).

Weber, T.P. (2004), *On the Road to Armageddon: How Evangelicals Became Israel's Best Friend* (Grand Rapids: Baker Academic).

Wertheim, M. (1999), *The Pearly Gates of Cyberspace: A History of Space From Dante to the Internet* (New York: W.W. Norton).

West, W.J. (2006), 'Religion as Dissident Geopolitics? Geopolitical Discussions Within the Recent Publications of Fethullah Gulen,' *Geopolitics* 11:2, 280–99.

Young, W. (2007), *The Shack* (Newbury Park, CA: Windblown Media).

PART I
Contesting the American Holy Land

Chapter 1
"What would Lee do?" Religion and the Moral Landscapes of Southern Nationalism in the United States

David Jansson

The idea of 'the South' in U.S. discourse is inextricably linked with religion. 'Southern religion' is understood as different from an often unstated national norm, and religion is often seen as central to the lives of both white and black Southerners. Non-Southerners see white Southerners in particular as more religious than other U.S. residents, as well as more conservative in their religious beliefs (Mathews 2006). Many of those inside the South would agree with these views; as John Shelton Reed (1993, 141) famously quipped, "Even those Southerners who don't go to church at least know which one they're not going to."

In this chapter I explore the connections between religion and Southern nationalism through an analysis of interviews with the nationalist organization the League of the South. I begin with a brief review of the historical role of religion in the southeastern U.S., focusing in particular on the extent to which it has informed white Southern identity as well as Southern nationalism. There is some degree of overlap between the religious experiences of white and black Southerners; Boles (1985, 31), for example, notes that even during the antebellum period, there was, to a surprising degree, "a biracial religious culture" in the South, and an adherence to evangelical Protestantism "has applied at least as much to the black population as to the white" (Prunty 1977, 4). In fact, in most Southern states, the percentage of black residents who belong to Baptist or Methodist churches is significantly higher than the corresponding percentage of the white population (Frank 1999). But as Southern nationalists are almost exclusively white, a result of Confederate history and subsequent battles over civil rights, this chapter will focus on the religion and nationalism of white Southerners.

Region and Religion

For many contemporary observers, religion is woven into the very fabric of social life in the South (e.g. Schaller 2006). According to Leib and Webster (2002, 232), religion "has played a more critical role in shaping the South's political culture than in any other region of the United States." Echoing the 'Southernization of America' thesis, some also argue that 'Southern religion' is conquering the rest of the country (Applebome 1996; Zeitz 2002). However, from a historical perspective, we can see that religion in the southeastern U.S. has not always played a dominant role, and that in fact other regions have experienced periods of religious development surpassing that which occurred in the South. From the beginning of English settlement in the 'New World,' the Southern (Virginia's Chesapeake) and Northern (New England) colonies were dramatically different, though not necessarily in the way that later generations would recognize. Between 1607 and 1660, Chesapeake society was "highly materialistic, infinitely more secular, competitive, exploitive," and focused on commercial agriculture for export (Greene 1988, 26–27). It also experienced slow family formation, marked individualism, and a weak sense of community. In contrast, the New England settlements were inspired by "powerful millennial and communal impulses," characterized by "strong patriarchal families, elaborate kinship networks, and visible and authoritative leaders," with strong social institutions and deeply rooted populations (Greene 1988, 27). Greene argues that overall, "the vast majority of early Southern colonists seem to have had a profoundly secular orientation from the beginning" (Greene 1988, 201).

Through the 1700s, white Southerners remained somewhat indifferent to organized religion. As late as the 1820s, it is estimated that under ten percent of white Southerners were church members (Cobb 2005, 48). Flynt writes that during the late 1700s Southern evangelicals "were mainly economically insecure frontier folk with a strong bias against the aristocracy" who "came to despise the established church and condemned the sins they associated with the social class it represented: horse racing, gambling, slavery. Southern folk religion became a volatile, defiant movement of alienated lower middle class and lower class whites" (1981, 24). The lower classes were attracted to evangelical Baptist and Methodist churches because of the element of egalitarianism that characterized their teachings (Tripp 1997). The white South, as a whole, experienced after 1800 "a prolonged religious awakening" which preceded the evangelical revival in the North and set the stage for broad acceptance of evangelicalism among the white population (Taylor 1969, 17).

It was the increasing tensions between North and South during the first half of the nineteenth century, sparked in part by the abolitionist movement, that would strengthen the role of churches in white Southern society (Wilson 1980a; Webster and Leib 2002). These churches supported and justified slavery, arguing that "superior peoples" had to care for the "inferiors" (Harvey 1997, 9) and that the Bible tolerated and even sanctioned slavery (Noll 1998). To the chagrin of abolitionists, Christian clergy often found it easier to use the Bible to defend slavery than to oppose it – with the important exception that the Bible actually seemed to condone *white* slavery and not black slavery (Noll 2008). An idea that we will come back to repeatedly in this chapter is that what is called (white) 'Southern religion' is continually created *through the interaction between Southerners and Northerners*, and as such the label 'Southern' elides the spatial dynamic that produced the 'Southernness' of Southern religion.

This is seen clearly in the experience of the Civil War. By 1860, in contrast to religious diversification in the North, "the southern churches remained orthodox in theology and, above all, evangelical in orientation" (Wilson 1980b, 220). The war itself strengthened religion among Confederate troops (Genovese 1998, 45), and in the larger white society of the South religious fervor grew such that the Confederacy was conceived as a Christian nation (Cobb 2005, 50). In fact, the Civil War can be understood as the event that forged a Christian, evangelical nationalism for the white South. This link between evangelical Christianity and Confederate nationalism is crucial for the understanding of contemporary Southern nationalism. As Smith (2003) argues, nationalism acquires its durability from its structural connections to religious beliefs and the discourses and practices that generate the sacred foundations of the nation. Nationalism can thus be understood as a "sacred communion of the people" (Smith 2003, 32). For Confederate nationalists, this sacred communion brought together white Southerners in defense of their homeland and their "Southern way of life," which involved an invocation of their divine duty. "Within the terms of the covenant theology that Confederates embraced as a central component of their emerging nationalism," notes Faust, "southerners stood in a relationship of benefit and obligation with God, who had anointed them with both a special status and special responsibilities" (1988, 82). The humanitarian and egalitarian aspects of evangelical Christianity facilitated the cultivation of popular consent required by nationalism and "reinforced the Confederacy's search for both popular and divine approval" (Faust 1988, 83). A religious nationalism thereby created and was sustained by the horizontal comradeship of the imagined community of the white Southern nation (Anderson 1991).

Furthermore, various religious leaders at the time advanced the notion that the war was at its root a theological war, and that the Confederacy was fighting for the survival of Christianity in North America (Sebesta and Hague 2002). Thus the "War Between the States" was also a war between different interpretations of the Bible and strategies for promoting moral norms in public life (Noll 2008). We can consider this an example of a religious geopolitics, in the sense that religion informed the worldviews of both parties, including their relations with each other and their standing in the global order. In fact, both the Union and the Confederacy saw their relationship to the rest of the world through the prism of Christianity, and Moorhead even argues that "it was from New England that the most articulate statements of America's millennial role were to come" (1978, 4). Both sides in the Civil War had their "geopolitical visions." Dijkink (1996, 11) defines geopolitical vision as "any idea concerning the relation between one's own and other places, involving feelings of (in)security or (dis)advantage (and/or) invoking ideas about a collective mission or foreign policy strategy" and which requires "a Them-and-Us distinction and emotional attachment to a place." We have seen how religion informed the us/them distinction between white Southerners and Northerners, and later in this chapter we will consider the role of Christianity in creating the emotional attachment to the South for League of the South members.

In spite of the 'fact' that they had God on their side, the Confederacy was defeated and the Southern states were gradually reintegrated into the Union. Having cast themselves as God's defenders, white Southerners were then faced with the problem of interpreting and understanding their loss in the war. Interestingly, many blamed themselves, "sadly acknowledging that they had lost the War because a persistent sinfulness had cost them God's favor," which presented a bitter irony. "By forfeiting God's favor, they had sentenced themselves to live under the very social system that they had condemned as un-Christian…Punished for their lapses, they now found themselves enmeshed in a materialistic, marketplace society that promoted competitive individualism and worshiped Mammon" (Genovese 1998, 102). The role of the evangelical churches in the South did not diminish after the war; if anything, they became even more entrenched. The white churches served as bulwarks against the influence of Northern religion and culture (Current 1983), and the postbellum years saw the evolution of a civil religion that stressed the conservative notions of moral virtue and an orderly society. This civil religion reinforced the wartime fusion of evangelical Christianity with white Southern identity, satisfying the need of white Southerners for reaffirmation in the aftermath of their defeat (Wilson 1980b). If the political dream was deferred, a cultural dream would take its place, and white Southerners could find their redemption by becoming better Christians and renewing the South as a Christian land.

This cultural/religious dream was embodied by the philosophy of the 'Lost Cause.' The Lost Cause "was the story of the use of the past as the basis for a Southern religious-moral identity, an identity as a chosen people" (Wilson 1980a, 1). Lost Cause advocates called for a war of ideas to reinvigorate white Southern identity, and "religious leaders and laymen defined this identity in terms of morality and religion…[white] Southerners were the chosen people, peculiarly blessed by God" (Wilson 1980a, 7). "The South was sacred to its citizens because they saw a sacred quality in it" (Wilson 1980a, 15), and part of the sacralization of the Southern nation during the Lost Cause era involved the memorializing of Confederate heroes, who represented "the transcendent truths about the redemptive power of Southern society" (Wilson 1980a, 10–11). Smith (2003) stresses the role of the "glorious dead" as an element of the religious nature of nationalism and the embodiment of sacrificial virtue, and Confederate soldiers were cast in the role of the fallen Christian hero. This developed to such an extent that Confederate figures such as Robert E. Lee were seen as Christlike. "So closely linked were Christianity and the Lost Cause," writes Richard Current (1983, 86), "that one can readily understand the mistake of the little girl who, when asked if she knew who had crucified her Lord, replied: 'O yes I know, the Yankees'."

This is a crucial moment in Southern history, as many of the contours of the social debate in the region during the twentieth century were sketched here. One aspect of this debate is the juxtaposition by white Southerners of an immoral North with a moral South, a morality based in a particular interpretation of Christianity. Many white Southerners embraced a religious moral cartography (Friedland 2001) that cast the North as betraying the country's Christian heritage while this heritage was preserved most faithfully in the South. As the little girl's comment about Yankees suggests, Northerners were constructed as the enemy not only of the South, but of Christ himself.

Throughout the Jim Crow era, well into the twentieth century, white Christian churches in the South played a largely conservative role. In many cases, white Southern preachers explicitly supported racial segregation, arguing that blacks were content with their position (Harrell 1971, 55). Southern white churches mostly sought to preserve the status quo against assaults such as the civil rights movement, though some marginal religious organizations supported whites who fought for civil rights. As history has shown, the opposition to desegregation was a losing battle, but the fact that the pro-segregation side lost would be a fateful event for evangelicalism in the United States. For "once legally enforced racism was gone, the great impediment that had restricted the influence of southern religion to only the South was also gone. Stripped of racist overtones, southern evangelical religion—the preaching, the piety, the sensibilities, and above all the music— became much easier to export throughout the country" (Noll 2008, 156–157). The successes of the civil rights movement had another, perhaps ironic, effect: the diminishing stigma of overt racism allowed southern whites to develop political coalitions with evangelicals in other parts of the U.S. based on ideologies of Christian morality and local control. Evangelicals throughout the country joined together to oppose federal mandates and regulations "as offensive intrusions attacking the family, gender, and sex" (Noll 2008, 157). This is the backdrop for the polemics attacking the political "power shift" to the South (Sale 1975) and the "Southernization" of the U.S. (Cochran 2001; Lind 2003; P. Taylor 2004).

The strength of interregional evangelical coalitions suggests, then, that the conditions are set for the contemporary abatement of Southern nationalism, itself historically grounded in evangelical Christianity. Yet this does not seem to be the case. If anything, based on the number of organizations, Southern nationalism seems to be increasing. Beyond the traditional groups such as the United Daughters of the Confederacy and the Sons of Confederate Veterans, which are still with us, recent years have seen the rise (and sometimes decline) of groups like the League of the South, the Southern Party, the Southern National Congress, and the Confederate Legion, among others.

Moral Geographies and the League of the South

The League of the South was founded in 1994 with the purpose of defending a particular notion of 'Southern' identity and culture, reclaiming a positive and largely racialized view of what it means to be Southern (Hague 2002) and pledging to defend Christianity in the South (Sebesta and Hague 2002, 256). They have also been labeled a hate group by the Southern Poverty Law Center (Webster and Kidd 2002; Beirich and Potok 2004) based on their linking "Southern" identity to a presumed Celtic heritage, their defense of Confederate symbols like the battle flag, and their opposition to immigration into the region. The League has made effective use of the internet to spread its message (McPherson 2000) through its website (www.dixienet.org), and it runs a summer school every year in which speakers give talks on a theme relevant to the group's interests. When I attended the 2003 summer school in Abbeville, South Carolina, the theme was "Our Confederate Heroes." I conducted sixteen interviews with League members, and seven additional people completed written questionnaires, resulting in 23 respondents.

The League does not keep data on the religious affiliation of its members and has no official religious requirement for membership, but as an organization it recognizes "the legacy of Christianity and the universal sovereignty of the triune God." According to the League, "Trinitarian Christianity cannot be separated or removed from Southern society without both ceasing to be Southern" (League of the South 2004, 160). Furthermore, its membership is drawn mostly from a region that is predominately evangelical Protestant. In the early 1960s, almost 90 percent of Southerners identified themselves as Protestants, compared with six of every ten non-Southerners (Reed 1975). According to polls done during the late 1950s, Southerners differed from non-Southerners most when answering the questions "Do you believe there is or is not a devil?" (86 percent of Southerners professed belief compared to 52 percent of non-Southerners) and "Do you think Jesus Christ will ever return to earth?" (81 percent of Southerners answered "yes" compared to 56 percent of non-Southerners) (Reed 1975, 60). More recently, an ABCNews/Beliefnet poll (2001) showed that 55 percent of Christians in the South identified as evangelical, compared to 21 percent in the North, 26 percent in the Midwest, and 31 percent in the West. League members are likely to be evangelical out of proportion to the rest of the Southern population, given their conservative orientation.

As with their Confederate ancestors, one could say that the League has its own geopolitical vision of a new Confederate nation-state, which involve secured borders, non-interventionism, and a rejection of international alliances and binding treaties. A part of this geopolitical vision is a certain understanding of the relationship between the South and the rest of the country. In addition to Christianity, this geopolitical vision is profoundly informed by the discourse of internal orientalism (Jansson 2003a, 2003b, 2005, 2007, forthcoming). Internal orientalism in the U.S. represents a material and representational relationship wherein the Southern states occupy a subordinate position in the national imagination. This discourse produces a binary that essentializes the imagined spaces of 'America' and 'the South' as it privileges the former at the expense of the latter. Internal orientalism holds that 'America' (or 'the North') and 'the South' are radically and fundamentally different socio-spatial entities. An exalted American identity is produced in part through the othering of the South as a place of racism, poverty, backwardness, xenophobia, and violence (for example). It is within the context of internal orientalism that the activities of the League of the South must be understood.

I argue that internal orientalism generates an imagined moral landscape that conceives of the South as the nation's moral nadir. As we shall see, the League members reject this characterization and seek to reverse the binary, asserting that in fact the South is the moral *apex* of the nation. The production of an alternative moral landscape by the League members is grounded in the assumption (consistent with internal orientalism) that the imagined spaces of 'the South' and 'the North'/ 'America' are fundamentally different, and in fact opposed. For example, Anne[1] wrote:

> We are a different people than those of the Northeast and actually all other parts of the country, and the nosy-Parkers of New England and other parts of the North cannot for the life of them leave us alone. And that is all we want. (questionnaire)

This way of thinking about the relationship between North and South was not challenged by any of the members I spoke to. Rather it was supported by all the speakers at the League annual meeting and is taken for granted by the League's statements on its website.

Part of what it means for the South to be different from the rest of the country is that the region is understood to embody a uniquely moral social order. The South is understood, in opposition to the internal orientalist representations, as the very seat of morality in the U.S. rather than as an immoral den of iniquity. This moral order is imperiled, however, by the encroaching American menace. In response to the question "what does the term 'Southern heritage' mean to you?" Bev wrote:

> The term 'Southern heritage' means a reverence for the values and ideas displayed by my forefathers and a desire to see those values preserved in today's hectic world. The morals and social structure today have crumbled and my Southern heritage is a desire to correct those corruptions. (questionnaire)

Bev does not place blame for the crumbling of "morals and social structure," but by describing 'Southern heritage' as the antidote to crumbling morality she highlights the understanding of the League members that the social problems afflicting the South are not inherently Southern but instead represent American/ Northern characteristics and processes. For example, I asked Tom what he gets out of his involvement with the League:

> It gives me a great sense of hope for the future. Like the status quo, I've opted out of that, I've opted out of the big government, unlimited immigration, unlimited abortion, the immorality and the decadence that's falling on this country, and the open borders that are allowing this country to be infiltrated. I feel like at least I am out of that and into an organization that is trying to stop that, or at least curtail it, and it's an organization of absolutely like minded people, you know with the same values that I have, the same values I was taught as a child…and these people show on a daily basis that we are strong and that we are a people,

---

1 All the names of the 23 participants involved in this study have been changed.

and that we are opting out of this decadence that the United States is seeping into. I guess, you know, it's a ray of hope in a dark world.

While the U.S. may be "seeping into" decadence and immorality, the League can act as a ray of hope lighting the path to the moral center of "the South," and this role provides a reason for hope for many of the League members as they seek to contest what they see as the degradation of their beloved region. There is an interesting parallel here with Christian fundamentalists' view of seeing themselves as being *in* the world but not *of* the world; Tom's comments suggest that he sees himself as *in* the U.S. but not *of* 'America.' Likewise, internal orientalism writes the South as being in the U.S. but not of 'America,' though from the other direction and with the opposite valorizations.

The League members come to their understanding of the South as the embodiment of a moral society through a particular interpretation of the historical geography of the region, an interpretation that is encouraged by programs such as the League's summer school. For example, I discussed with Kevin what he found valuable about the program.

> Based on what I've learned here at the Institute, I could argue comfortably that the racism in the South was nothing compared to the racism in the North. And that the agrarian success of the antebellum period, there was no hunger…The medical care of everyone to include the slaves, I mean to the poorest in the Southern society was better than the poorest in the North… Now, does that mean that Southerners are not prejudiced? No, we've had racism in the South, and the Jim Crow, and all the ugliness of that whole legacy, and the horrible issue of the lynching issue, but I've also learned about the starvation of the South in Reconstruction. There was no Marshall Plan for the South, but there was martial law – different word – disfranchisement of Confederate veterans, they had no government, and that, the Ku Klux Klan of Nathan Bedford Forrest was a completely different organization than what is perceived as the Ku Klux Klan today, and that, which makes sense to me, if I believe my Christian heritage and my ancestors were Christians, there's no way they could cooperate in lynchings and all these ugly things…and I just believe that if there was abuse of slaves in the South it was a small minority, as compared to when it came to the War the state legislatures of the North did not want any black migration, of where they even had laws saying that they wouldn't allow blacks as citizens in their state. So the cross that Southerners have had to bear on the race issue has been large as compared to no burdens Northerners have had to bear when it comes to having the history of their prejudice put in their face.

This passage is worth quoting at length because it concisely summarizes the historical interpretation that most of the League respondents embrace. First, Kevin's focus on racism suggests its primacy in representations of the South. But racism is clearly not the only part of the story; Kevin appeared to be sincere in his admissions of the sins of the South, but the summer school has taught him that these sins are not as overwhelming as he had previously believed and that there was much to praise in the Old South. Furthermore, he is quick to point out the racism of the North in defense of his region, so while he is rehabilitating the South as a moral landscape he is simultaneously questioning the North's own morality. The imagined moral landscape of the South thus begins to look much different because of the strategic intervention of the histories supported by the League. Such interventions are important to the future prospect of the League as the organization's growth and potential for success depends upon its ability to re-engineer the imagined moral landscape of the South in a very specific way. As an institution, the League seeks to create new social relations of knowledge and disseminate its 'true' vision of the South, past and present. The League needs potential sympathizers to conceive of the South as an inherently moral place so that Southerners will feel good about who they are; it simultaneously needs its supporters to believe that internal orientalism is a dominant force and a threat to the South so that they will be motivated to join and actively support an organization that seeks to defend an imperiled Southern heritage.

Christianity was also a major focus of discussion in the interviews, and in fact it would seem that for the League members, Southern identity is constituted fundamentally by its grounding in Christianity. A related element of Southern identity was agrarianism, which itself was also fundamentally linked to Christianity, and both deserve further reflection.

Christianity and the South

For most of the League members I interviewed, a central aspect of the moral landscape of the South is the region's relationship to Christianity. As Ben explained, "part of the reason why the Southern cause is so tantalizing to me and has so much importance is because I see it intermeshed with Christianity so highly." When I asked Kevin what being a "Southerner" meant to him, he replied, "the foundation of my Southern culture is my Christian tradition." To the same question, Michael responded:

> Southern heritage is deep in Christianity, I think that's the most important part of the Southern heritage. If you go back far enough in Southern history – well, let's just go back to the 1860s, since we're concentrating on that era this week, virtually everyone had a Christian worldview, and someone who didn't was considered abnormal, or something was wrong with that person, and so to me, the most important thing about the Southern heritage is that, is the belief that Jesus Christ is the son of God, that we have a book, the Bible, that is the word of God, that tells us about Christ, who he was, and his teachings, and the Old Testament, which he referred to and said was the word of God, and that God is omnipotent whether man wants to believe it or not. So to me, that's the solid foundation that Southern heritage is built upon.

"Being a Southerner," answered Rachel, "means being a hardcore Christian." And for most of the participants, 'hardcore' is the only kind of Christian to be. For them, if the South were allowed to manifest its true nature, the region would be a place where life is guided by the Bible.

The conceptualization of the South as inherently Christian is articulated in part through comparing the region with the rest of the country, mirroring the dynamic of internal orientalism. We see this idea in the response of Trevor to the question "what does being a Southerner mean to you?":

> ...the South...is the only Bible Belt in the country, it still is the most religiously oriented section in the country...So, and our ancestors who fought during the War Between the States and who thought that the South was the last really really last Christian civilization on earth...and so the remnants of that are still evident today, it is the most religious section of the country.

Thus the League members see a Southern nation united in Christianity as far back as Southerners have existed.

The association of the South with Christianity extends even to the symbols of the Confederacy. Many of the participants pointed out the link between the Confederate battle flag and the cross of St. Andrew. I asked Michael what the flag meant to him.

> It's, as far as I know, well it's certainly one of the few, maybe the only, national flag that has a Christian symbol in it, that's the cross of St. Andrew, St. Andrew one of the apostles of Christ. The legend goes, this is not in scripture, the legend goes St. Andrew was crucified on a cross that was shaped like that [forms an X with his arms]. So I understand it has a Christian background to it, which I think is appropriate, again because according to what I know about the history of the South, the source of Southern values, if you trace the ideas back as far as you can go was, the Bible. And so I think that's an appropriate symbol.

The Christianity of the South is thereby embodied in its most well known symbol. Webster and Leib (2002, 18) are thus right on the mark when they state that the "early intertwining of southern culture with religion aids in understanding the passionate conviction held by many southern whites over the sanctity of traditional southern culture and its associated Confederate era icons such as the Battle Flag." The many conflicts over the display of the battle flag in the southern states must be understood in this context.

The constant repetition of the idea that the South is the most religious region of the country has the effect of creating the assumption that the South is godly in a way that the North is not. But this does not necessarily imply the moral superiority of the South, because white Southern religion is frequently seen as eccentric, excessively emotional, racially prejudiced, and backward (Harrell 1971; Flynt 1981; Sweeney 2001). It is in part this internal orientalist notion of a religious

South and non- or less religious North that the League members are reacting to and reinforcing, as they see national and even some local institutions run by 'pagans' who have no interest in or respect for Christian values.

Agrarianism; or, Amish with Firearms, Computers, and Chainsaws

In addition to Christianity, agrarianism was understood as an essential aspect of Southern history and something that today's Southerners need to get back to. Smith (2003) points out that in fact, a 'return to nature' is a pervasive feature in nationalist ideologies around the world, and the League members are no exception in this regard. In response to the question "what does being a Southerner mean to you?" Ben saw the agrarian return to nature and Christianity two sides of the same coin:

> I would say that in some ways the agrarian aspect is probably one of the biggest, and in some ways the agrarian aspect has helped shape the focus of the religious nature of the Southerner. Agrarianism involves this period of work, of planting, of tending things, but there comes a time when one can do no more than sit back and wait and hope, and you know in the case of Southerners this develops into prayer. I mean you can actually see this effect in other lands, but where people have that period of time where nothing else physical can be done. In that period of time this spirit develops, and it may develop through the hopes and prayers, and that may be specifically to the crop but you're thinking also of the larger community.

Ben was not the only one to make the connection between religion and agrarianism. Indeed, the League's *Grey Book* states that "Farm life both requires and develops faith" (League of the South 2004, 129). Community is clearly an important part of this as well, as the Southern agrarian lifestyle that he describes occurs in a social context, which in theory is a community of people who share these values and embody the same spirit of which Ben speaks. The social network represented by the community is a crucial contributor to the ability of the community to formulate alternative truths that challenge the dominant discourse.

Some respondents linked agrarianism to family life. As part of his answer to "what does being Southern mean to you?" Steve discussed his experience with farming.

> Yeah, all my family had farms when I was growing up, my parents didn't, but there were a lot of farms in the family. I didn't have to be told to be an agrarian, I knew from that experience. When I started to learn about the Agrarians, I said 'well I've always been that'…So to divorce the South from agrarianism you don't have the South. So we've returned to that as a good way to live, and we've had our children live with us, and to learn things that are fast becoming unknown. Let me give you an example – I like to hold out the Amish…I'd like to

characterize ourselves as Amish with firearms, computers, and chainsaws...So we're trying, the Amish are a good example...we're trying to live like that with our own family and trying to form community, so that's the core part of being the South, and it's an extension of Christianity too.

Part of one's Christian duty, then, as a Southerner, is to get 'back to the land' and revive the agrarian tradition of the South, and by doing this through forming communities one can best ensure that the Southern Christian traditions live on.[2]

Finally, for some agrarianism is seen in the context of a tension between industrial and agricultural economies, or 'the city' versus 'the country:'

> I think a varied economy, entrepreneurial spirit, low taxes, independence, self-sufficiency, these are the Jeffersonian ideals. 'Course Jefferson didn't want any industry, except that which you had to have to stay free, you know, to fight a war. It came to that, but small farming, yeoman farming, that was the basis – and I think it ought to be finally the basis of our economy...And so, I don't like cities, but you tolerate them, you just put them in their place, they're too arrogant, they look down on the hinterlands and the countryside and the hayseeds...And that's been a theme in America for, from the beginning. City versus country...And I would say we need a better balance, we've gone too far to the urban and we need to protect our countryside. And also, I'm an environmentalist, without being a capital E, I'm a conservationist, meaning conserve. I love nature, I think we need to be better stewards, and I think we can be, if we put things in balance. And the Southern genius is to live close to nature, even if it means hunting and fishing – you know that. To live in the cycle, to live close to the cycle of the seasons in nature is the Southern way. (Charles)

One of the things this quote demonstrates is that the discourse of internal orientalism intersects with other discourses in the U.S., in this case the urban/rural discursive binary that mirrors internal orientalism in both its longevity and its ambivalence. Internal orientalism is a product of modernity (Jansson 2003a), and agrarianism scarcely has a significant place in a modern world focused on change, consumption, and accumulation. It is therefore no accident that Charles indicts the city as arrogant – in the modern imagination urbanity points to the way forward while rurality is backwardness and stagnation. The urban/rural (or city/country) discourse is not unique to internal orientalism, but the latter incorporates it in such a way as to conflate the urban/rural binary with the North/South binary such that rurality is associated with the South and urbanity with the North. If one is to shun the internal orientalist notion of America, there is no better way than to embrace agrarianism and strive to live close to nature, something that is part of the Southern heritage as seen by the League members.

---

2   Steve later made it clear that for him, the Southern community is a white community. Many of the other respondents, though, were more racially inclusive.

Charles's brand of environmentalism leads us to a closely related aspect of the moral landscape of the South as seen by the League, and that is as a place where a consumerist materialism is rejected. Materialism is the sine qua non of modernity, and particularly of the U.S. 'prime modernity' (P. Taylor 1999), a consumerist paradigm that is based, in part, on the assumption that a better life can be achieved by the purchase of *things*. Internal orientalism spatializes this prime modernity by associating it with the imagined space of 'America' in opposition to 'the South.'

The League members do not necessarily reject the association of America with material progress but instead revalorize the binary so that spiritual values are prized above material conditions. Southerners do not worship money, but rather take a different approach:

> One of the principle differences between the traditional Southern and the American attitudes is that Southerners feel that life is for enjoyment, not for continuous getting and spending. In other words, one acquires enough to enjoy himself and commences to do so whereas the American way of life seems to largely consist of a never ending cycle of acquisition and frenetic consumption. (Terry, questionnaire)

Patrick expressed similar views:

> Southerners value the human over the material and human bonds over the cash nexus men and women are subsumed into by corporate capitalism. (questionnaire).

These comments echo the findings of Nader (1989) that the Occidentalism of Muslims in the East juxtaposes a materialistic West with an anti-materialist and philosophical East. We see that the reaction to the social processes that are a part of internal orientalism does not disturb its fundamental assumptions but rather revalorizes the binary so that the 'Southern way of life' is understood as superior to the 'Northern' or 'American way of life.' In the next section, we will see how the fusion of Southern identity with Christianity teaches League members how to be faithful to this Southern way of life.

How to Live as a Southerner

A fundamentalist interpretation of the Bible and Christianity is also employed for the purposes of making judgments about the best way to live in a contemporary society that League members generally feel admits no public role for religious beliefs. One of the more interesting (and disturbing) exchanges I had in Abbeville was with 18 year-old Rachel, one of the most fervent and committed League members I met. At the end of the interview I asked her if there was something that she thought I should have asked her, and she suggested the question "Do you consider yourself a racist?" because "a lot of us around here have many different views on that, and many different reasons why." I asked her how she would answer that question.

> Am I a racist? Do I consider myself a racist? No. No, I'm not a racist, but I, I don't know how to use the word, not racist but, I do judge people: I see whiggers,[3] I see niggers, I see blacks, I see whites, and I see trailer trash. I see liars and thieves and sluts, I see them, and I *judge* them! But I am not racist, I am not black, I am not white, I am Biblically judgmental.

Earlier in the interview she confessed that "everything's so Biblical for me – I'm a black and white person," so such a sentiment is not surprising. And as we saw above, Rachel associates being a Southerner with being a 'hardcore Christian,' so this judgmental attitude is part of her Southernness.

Such attitudes did not sit well with at least one summer school participant. Morgan was bothered by much of what he heard during the week and was relieved to be able to discuss it.

> I'm glad to have the chance to talk to you about this stuff – I've been dying to let someone how uncomfortable I've been feeling. I think the League of the South attracts a judgmental personality, everything is very black and white for them. The League of the South people want to be seen as more complex and not like the stereotypes, yet they demonize and homogenize the Yankee themselves, so they're doing the same thing that gets done to them.

That "they are doing the same thing that gets done to them" expresses a form of 'Occidentalism,' where in the case of Orientalism, the othered group essentializes the oppressors (or the Orient others the Occident; Carrier 1995; Coronil 1996; Chen 2002). However, the othering of non-Southerners is a trap, a kind of 'derivative discourse' (Chatterjee 1986) that reinforces the essentialist binary of the discourse League members are trying to contest.

Morgan's was a minority view at the summer school. The others were less conflicted about applying their interpretation of Christianity to their daily lives. Steve, who happened to be a source of many of the comments that troubled Morgan, described the development of his Southern nationalist consciousness and how that related to his Christian perspective.

> But so it wasn't I think until I saw the attack on the last 30 years—the deconstruction that has come along, the rewriting of history—that I said "well what are we losing here?" And then more so if the South was wrong, then maybe it's correct we should lose it. If it's wrong Biblically—I try to base my worldview on that. And that caused me to do great study. And then I realized that the South

---

3 'White niggers,' or whites that adopt a stereotypical form of black culture.

was right, and it's worth honoring and emulating the transcendent truth of then. How do we live it out today?

For Steve and many other League members, the South was right (in their interpretation) in that social relations in the region were based on Christianity – the slaveowner had a Christian duty to provide for the well-being of the enslaved; women were biblically required to submit to the domination of their husbands; even Southern hospitality was "a response to the biblical mandate to provide for friend and stranger alike, with the possibility of 'entertaining angels unawares'" (League of the South 2004, 128). The answer to the question "how do we live it out today?" for Steve thus depends on a religious reading of Southern history and an attempt to recover the Christian heritage of the South.

Other League members could look to particular historical figures for life lessons. Michael Hill told the annual meeting that Nathan Bedford Forrest has long been his favorite Confederate general, in part because of the inspiration his strategies and tactics provide to the current struggles of the League. For Tom, that important figure was the most famous Confederate general, Robert E. Lee, whom he also referred to as "sacred":

> You know, the only thing I have to think about, and I think about it often when I make decisions in life, whether big or small—you know there's a saying, "What would Jesus do?" I use the saying "What would Lee do?" because in my eyes, Lee, whether it be as a soldier or as a man, as a father or as a husband, he exemplified what it is to be a good man. And you know not divine, he was just a man, which means that anyone could aspire to be at his level. And you know, he was Southern, he was absolutely Southern, he had his battle flags flying when he fought his battles.

Lee, the Christ figure, is thus the paragon of Southern masculinity and a fitting role model for one who seeks to fully embrace his Southernness. Tom elaborated on this in response to the question "What does the battle flag mean to you?":

> You know I live right here in South Carolina, I live in one of the states that actually flew the flag and was proud of it. And my ancestors fought for it, and I feel that I have a certain connection to it that other people might not have. And so it's a love and a reverence for my land, my people, my food, culture, and, and defiance—not defiance for no reason, not just blatant, like I don't want to be like you for no reason. I've got reasons, I've got ideals, and I've got morals and judgments and things that I uphold and that I hold dear, and, just because you say that this law is this or this is the way things should be it doesn't mean I'm gonna just conform.

Tom hints here at an association that was common and explicit for a majority of the respondents, and that is the way that the flag symbolizes resistance. In this case, the resistance involves a defense of a moral, Southern way of life against corrupting influences from outside (i.e., the North).

The common theme binding the quotes in this section is the notion that as League members go about their daily lives, they are guided every step of the way by their (Christian) Southernness, envisioning the South as a Christian and thereby inherently moral place. In looking to their Southern past, to their cherished Southern traditions or to specific Southern figures, League members at the same time assert their Christianity in the face of attacks from a secular North. When they look to their future, they see an independent Confederacy of Southern States that will take its place in the international community as a faithful representative of God's will on earth.

But how do they get there? Will secession require another apocalyptic event such as the Civil War, or will it come about in some other way? How will they achieve the restoration of a Christian Southern society? One might ask, to phrase these questions in line with the other chapters in this book, whether their geopolitical visions are postmillennial or premillennial. In the interviews, the League members allowed that secession was not a likely possibility in the near future, but they did stress that they saw it as a worthwhile goal. However, the strategies suggested to achieve this goal diverged in a way that borders on the contradictory. Rachel, for example, advocated the infiltration of local government positions with League supporters. Michael Hill told the members at the annual meeting to forego national elections in favor of engaging with the political process on the local and state level. These comprise reformist, not revolutionary, measures. This reflects postmillennial thinking because it reveals a belief in the efficacy of their agency, that they can make a difference in the South. This optimistic assessment of the possibilities of change *within* 'the system' rests side-by-side with a more fatalistic view of a Southern society that has been corrupted by non-Southern values and institutions. Expressing a kind of premillennial perspective, one of the prevailing themes of the interviews and the annual meeting was that Southerners should *culturally* secede at the local level within the South, separating themselves from the 'American' decadence and moral decay that has infected the region. Cultural secession was on the lips of the League members as well as the speakers at the annual meeting. The idea was that even if one cannot change the larger society, one can form local communities which nurture Christian beliefs and Southern traditions. The League defines cultural secession from the U.S. as (among other things)

> withholding our support from all institutions and objects of American mass culture that are antithetical to our beliefs and heritage...encouraging the formation of communities of like-minded Southerners...buying and reading Southern literature, poetry, and history...[and] encouraging homeschooling and private education. (League of the South 2004, 160)

But the League's discourse is not solely pre- or postmillennial. The *Grey Book* argues that the Southerner's problem is both easily solved and not quickly solved.

Easily solved because "If we repent and live our faith, sooner or later our culture and our government will reflect it" (League of the South 2004, 131). Not quickly solved because

> The fruits of faithful living require generations to fully ripen. It has taken more than one generation to decline; it will take more than one to strengthen. It would be great if there were an exception in the cosmic scheme of things for us baby-boomers, but there isn't, so we just have to get over it and make history the old-fashioned way—by the sweat of our brows, one generation at a time. We must be committed to Christ and to a Christian South over the long haul. (League of the South 2004, 131)

These thoughts suggest a postmillennial view that requires a slow and steady march toward secession, guided by the light of Christian faith and girded by Southern nationalism. This emphasis on the fruits of the Southerner's labor is somehow accommodated to a worldview that also incorporates a fatalistic call to withdraw from a corrupted society.

Conclusion

> What is vital for nationalism and the nation, therefore, is not some promise of imminent apocalypse, but the very core of traditional religions, their conception of the sacred and their rites of salvation. This is what the nationalists must rediscover and draw upon in fashioning their own ideals of community, history, and destiny. (Smith 2003, 15)

For the members of the League of the South, Christianity and Southern nationalism are essentially one and the same thing. The League and its members also assert the fundamental alienation of traditional Southern culture from the larger national society. The only way to preserve the Christian South (which is for them a redundant expression) is to secede from the foreign culture that has invaded the South and build parallel local, truly Southern, communities, from which Southerners can ground faithful Christian lives. This discourse leaves no room for compromise with national institutions. Earlier I wondered why Southern nationalism seems so invigorated (or at least visible) at a point in U.S. history where evangelicals have had considerable influence at the national level. I think part of the answer is that Southern nationalists inhabit to such an extreme degree the binary of internal orientalism (albeit with the valorizations reversed) that it is difficult to see the rest of the country as anything but the enemy, as an immoral and un-Christian force. This is not to say that League members do not have friendships that cross the Mason-Dixon Line. Anne mentioned that she has a few friends who "came from the North" and that they also happen to be Christians and that identity is as close or closer even than the Southern thing" (questionnaire). But to the extent that

Northerners share a Southern sense of Christianity (that is to say, evangelical), they would not be considered true representatives of the Northern/American population.

I do not wish to be understood as accepting the thesis of Southern distinctiveness, even when it comes to religion. Internal orientalism asserts this thesis, and the derivative discourse of Southern nationalism is more than happy to buy into it as well. But it is not a terribly useful way to think about the internal spatiality of the United States. As Moorhead (1978, 110) argues,

> Some historians still suggest that fundamental aspects of sectionalism may be traced to a prior intellectual cleavage—an industrious North disciplined in the Protestant ethic set against a "lazy South" unfamiliar with institutionalized Puritan nurture. The extent to which this theory can be sustained is a question for further research, but it is safe to warn that any simplistic portrayal of the North and South as opposed ideological entities runs aground on the hard fact that the two regions shared a similar heritage of evangelical religion and political rhetoric. Indeed, the confederate clergy drew upon many of the same motifs as the North to justify their cause. Southerners, too, believed their nation to be the new Israel, "the pillar of cloud by day and of fire by night" guiding the Army of Northern Virginia no less than the Army of the Potomac.

Elsewhere, Howard Zinn (1964) sees fundamentalism as a national trait, not unique to the southeastern corner of the country, and Noll (2008) stresses the national influence of religion on politics. But it remains the case that so much of the discourse on the South and its relation to the rest of the U.S. relies on the assumptions of internal orientalism. For the League of the South members, Christianity is an inherent part of the internal orientalist binary: the South is rejected by the rest of the country *because* of its Christianity, *because* it seeks to uphold biblical traditions, and *because* it dares to reject an unthinking embrace of consumerist modernity in favor of family, community, and agrarianism. These Southern nationalists see themselves as Christians above all else, but in order to *live* as Christians in the U.S. they must assert their Southern identities and fight for their Southern heritage. Thus will the loyal subjects of the coming Kingdom of God speak with a Southern accent.

References

ABCNews/Beliefnet Poll. (2001), 'Religious Affiliation in America: Many Answers, Mostly Christian' <http://abcnews.go.com/images/PollingUnit/855a3Religion.pdf>, accessed 16 April 2009.

Applebome, P. (1996), *Dixie Rising: How the South is Shaping American Values, Politics, and Culture* (New York: Times Books).

Anderson, B. (1991), *Imagined Communities: Reflections on the Origin and Spread of Nationalism*, revised edition. (London: Verso).
Beirich, Heidi and Mark Potok (2004), 'Little Men: Today's Neo-Confederate Ideologues Are the Latest in a Long Line of Highly Conservative Southern Intellectuals. Or Are They?' *Intelligence Report*, Winter <http://www.splcenter.org/intel/intelreport/article.jsp?aid=509&printable=1>, accessed 16 April 2009.
Boles, J.B. (1985), 'Evangelical Protestantism in the Old South: From Religious Dissent to Cultural Dominance', in Wilson (ed.) (1985), *Religion in the South* (Jackson: University Press of Mississippi).
Carrier, J.G. (ed.) (1995), *Occidentalism: Images of the West* (Oxford: Clarendon Press).
Chatterjee, P. (1986), *Nationalist Thought and the Colonial World: A Derivative Discourse* (Minneapolis: University of Minnesota Press).
Chen, X. (2002), *Occidentalism: A Theory of Counter-Discourse in Post-Mao China*, 2nd Edition (Lanham, MD: Rowman & Littlefield).
Cobb, J.C. (2005), *Away Down South: A History of Southern Identity* (New York: Oxford University Press).
Cochran, A.B., III. (2001), *Democracy Heading South: National Politics in the Shadow of Dixie* (Lawrence: University Press of Kansas).
Coronil, F. (1996), 'Beyond Occidentalism: Toward Nonimperial Geohistorical Categories', *Cultural Anthropology* 11: 1, 51–87.
Current, R.N. (1983), *Northernizing the South* (Athens: University of Georgia Press).
Dijkink, G. (1996), *National Identity and Geopolitical Visions: Maps of Pride and Pain* (London: Routledge).
Faust, D.G. (1988), *The Creation of Confederate Nationalism: Ideology and Identity in the Civil War South* (Baton Rouge: Louisiana State University Press).
Flynt, W. (1981), 'One in the Spirit, Many in the Flesh: Southern Evangelicals', in Harrell, Jr. (ed.) (1981), *Varieties of Southern Evangelicalism* (Macon, GA: Mercer University Press).
Frank, A.K. (1999), *The Routledge Historical Atlas of the American South* (New York: Routledge).
Friedland, R. (2001), 'Religious Nationalism and the Problem of Collective Representation', *Annual Review of Sociology* 27, 125–52.
Genovese, E.D. (1998), *A Consuming Fire: The Fall of the Confederacy in the Mind of the White Christian South* (Athens: The University of Georgia Press).
Greene, J.P. (1988), *Pursuits of Happiness: The Social Development of Early Modern British Colonies and the Formation of American Culture* (Chapel Hill: The University of North Carolina Press).
Hague, E. (2002), 'Texts as Flags: The League of the South and the Development of a Nationalist Intelligentsia in the United States 1975–2001', *HAGAR: International Social Science Review* 3:2, 299–339.
Harrell, D.E. Jr. (1971), *White Sects and Black Men in the Recent South* (Nashville: Vanderbilt University Press).

Harvey, P. (1997), *Redeeming the South: Religious Cultures and Racial Identities among Southern Baptists, 1865–1925* (Chapel Hill: The University of North Carolina Press).
Jansson, D.R. (2003a), 'Internal Orientalism in America: W.J. Cash's *The Mind of the South* and the Spatial Construction of American National Identity', *Political Geography* 22, 293–316.
—. (2003b), 'American National Identity and the Progress of the New South in *National Geographic Magazine*', *Geographical Review* 93:3, 350–69.
—. (2005), '"A Geography of Racism": Internal Orientalism and the Construction of American National Identity in the Film *Mississippi Burning*', *National Identities* 7:3, 265–85.
—. (2007), 'The Haunting of the South: American Geopolitical Identity and the Burden of Southern History', *Geopolitics* 12:3, 1–26.
—. (forthcoming), 'Racialization and 'Southern' Identities of Resistance: A Psycho-geography of Internal Orientalism in the U.S.', *Annals of the Association of American Geographers*.
League of the South (2004), *The Grey Book: Blueprint for Southern Independence* (College Station, TX: Traveller Press).
Leib, J. and Webster, G.R. (2002), 'The Confederate Flag Debate in the American South: Theoretical and Conceptual Perspectives', in Willingham, A. (ed.). *Beyond the Color Line? Race, Representation, and Community in the New Century* (New York: Brennan Center for Justice), 221–42.
Lind, M. (2003), *Made in Texas: George W. Bush and the Southern Takeover of American Politics* (New York: Basic Books).
Mathews, M.B.S. (2006), *Rethinking Zion: How the Print Media Placed Fundamentalism in the South* (Knoxville: The University of Tennessee Press).
Noll, M.A. (1998), 'The Bible and Slavery', in Miller, R. M., Stout, S. S., and Wilson, C.R. (eds) *Religion and the American Civil War* (New York: Oxford University Press).
—. (2008), *God and Race in American Politics: A Short History* (Princeton: Princeton University Press).
McPherson, T. (2000), 'I'll Take My Stand In Dixie-Net: White Guys, the South, and Cyberspace', in Kolko, B., Nakamura, L., Rodman, G. B. (eds) *Race in Cyberspace* (New York: Routledge).
Moorhead, J.H. (1978), *American Apocalypse: Yankee Protestants and the Civil War, 1860–1869* (New Haven: Yale University Press).
Nader, L. (1989), 'Orientalism, Occidentalism and the Control of Women', *Cultural Dynamics* 2:3, 323–55.
Prunty, M.C. (1977), 'Two American Souths: The Past and the Future', *Southeastern Geographer* 17:1, 1–24.
Reed, J.S. (1975), *The Enduring South: Subcultural Persistence in Mass Society* (Chapel Hill: The University of North Carolina Press).
—. (1993), *My Tears Spoiled My Aim and Other Reflections on Southern Culture* (San Diego: Harcourt Brace).

Sale, K. (1975), *Power Shift: The Rise of the Southern Rim and Its Challenge to the Eastern Establishment* (New York: Random House).

Schaller, T.F. (2006), *Whistling Past Dixie: How Democrats Can Win without the South* (New York: Simon & Schuster).

Sebesta, E.H. and Hague, E. (2002), 'The U.S. Civil War as a Theological War: Confederate Christian Nationalism and the League of the South', *Canadian Review of American Studies* 32:3, 253–83.

Smith, A.D. (2003), *Chosen Peoples* (New York: Oxford University Press).

Sweeney, G. (2001), 'The Trashing of White Trash: *Natural Born Killers* and the Appropriation of the White Trash Aesthetic', *Quarterly Review of Film & Video* 18:2, 143–55.

Taylor, P.J. (1999), *Modernities: A Geohistorical Interpretation* (Minneapolis: University of Minnesota Press).

—. (2004), 'God Invented War to Teach Americans Geography', *Political Geography* 23, 487–92.

Taylor, W.R. (1969), *Cavalier and Yankee: The Old South and American National Character* (New York: Harper Torchbooks).

Tripp, S.E. (1997), *Yankee Town, Southern City: Race and Class Relations in Civil War Lynchburg* (New York: New York University Press).

Webster, G.R. and Kidd, T. (2002), 'Globalization and the Balkanization of States: The Myth of American Exceptionalism', *Journal of Geography* 101, 73–80.

Webster, G.R. and Leib, J.I. (2002), 'Political Culture, Religion, and the Confederate Battle Flag Debate in Alabama', *Journal of Cultural Geography* 20:1, 1–26.

Willingham, A. (ed.) (2002), *Beyond the Color Line? Race, Representation, and Community in the New Century* (New York: Brennan Center for Justice).

Wilson, C.R. (1980a), *Baptized in Blood: The Religion of the Lost Cause, 1865–1920* (Athens: The University of Georgia Press).

—. (1980b), 'The Religion of the Lost Cause: Ritual and Organization of the Southern Civil Religion, 1865–1920', *The Journal of Southern History* 46:2, 219–238.

Zeitz, J. (2002), 'Dixie's Victory', *American Heritage* 53:4, 46.

Zinn, H. (1964), *The Southern Mystique* (New York: Knopf).

Chapter 2

Contests over Latter-day Space: Mormonism's Role within Evangelical Geopolitics as seen through Last-days Novels

Ethan Yorgason

Harry Reid and Mitt Romney represent Mormonism's deepest participation in the nation's electoral politics over the religion's nearly 180-year history. Both the U.S. Senate Majority Leader and the candidate for the 2008 Republican presidential nomination, respectively, are active members of the Church of Jesus Christ of Latter-day Saints.[1] Mormons are no more than one in 50 Americans, yet Romney and Reid's centrality within the nation's political conversation and LDS 'overrepresentation' in Congress suggest Mormonism's potential significance to U.S. geopolitics.[2] Over the past 25 years, many Mormons have arguably been part of the Christian Right movement, joining evangelicals and others to promote a traditionalist social agenda. Will Mormons similarly join evangelicals in promoting a coherent eschatological geopolitical program, one that attempts to influence U.S. foreign policy? This chapter attempts to answer that question.

The short answer is no. Mormons and evangelicals are unlikely to deliver a joint and singular geopolitical message stemming from a similar last-days vision. However, the full answer is more complex. Mormons and evangelicals may indeed promote similar geopolitical positions or even short-term ideological alliances of convenience. But anything beyond those is likely to founder. I explain by first discussing how Mormonism functions as a religion and social force, relating it to both Christianity generally and conservative/evangelical Christianity more

---

1   While other 'Mormonisms' exist, The Church of Jesus Christ of Latter-day Saints (headquartered in Salt Lake City) is overwhelmingly the largest one. My use here of 'Latter-day Saint,' 'LDS,' 'Mormon,' 'Mormonism,' 'Saints,' or similar constructions refers to it. 'Active' is LDS lingo for a believing, regularly participating relationship to the church.

2   The church reports just under six million American Mormons, or almost two percent of the U.S. population. National religious surveys report that only two-thirds of that number self-report as Mormons. If the latter is more accurate, then Congressional 'overrepresention' stands at double the percentage of members: nearly three percent of Congress is LDS. It should be noted, however, that a few religious groupings (Episcopalians, Presbyterians, Congregationalists, and Jews) are far more 'overrepresented' in Congress than Mormons are.

specifically. This chapter next characterizes Mormon eschatological beliefs in relation to prominent evangelical forms. Lastly, we compared contemporary Mormon last-days novels and evangelical novels. Popular geopolitical analysis is deployed to draw out assertions and assumptions about the geopolitical constitution of the world, with a particular thematic focus on Americanism, the family, and the Antichrist. Despite many similarities, Mormonism's social/cultural geography is among the decisive factors preventing eschatological geopolitical coalition with evangelicalism.

Mormonism as Religion and in Society

During the 1820s Joseph Smith claimed to receive visions/revelations leading to his discovery and translation of an ancient scriptural record—the Book of Mormon. Believers considered Smith's church, started in 1830, the re-establishment of Christ's ancient church. Smith argued a restoration was necessary because of an apostasy in the first few centuries of the church and Protestantism's failure to restore priesthood authority and prophetic gifts. Neither Catholic nor Protestant, the LDS church proclaimed itself the earth's "only true and living church." Mormonism's separation from other Christianities was greatly enhanced by its nearly immediate quest to create a City of Zion in Missouri, its persecution by non-Mormons, and its movements to (and expulsions from) various frontier locales, culminating in a semi-theocratic and polygamist society in the American West. The religion became as much society as doctrine, one as proud to claim distinctiveness as to emphasize Christian roots.

By the nineteenth century's end, American legal, political, and cultural pressure compelled abandonment of the church's most distinctive social practices: polygamy, theocracy, and communalism. While initially ambivalent, Mormons took just a few decades thereafter to fully rejoin America's gender, political, and economic norms. In their own minds, at least, they became normal Americans (Yorgason 2003). Their rhetorical battle with American Christianity also reached something of a détente during the twentieth century. Mormons still sought to stand apart doctrinally, but now squeezed their distinctive sociality into the 'proper' American sphere of religion. Mormons still gave obedience to the priesthood hierarchy and shared economic blessings between members. Yet those practices now, for the most part, did not preclude normal relations with their non-Mormon neighbors. Throughout the twentieth century, Mormonism walked the fine line between being just another American religious denomination and a singular religious system with uniquely strong internal sociality.

Evangelicals typically decry both Mormonism's ecclesiastical/social structure and its doctrinal grounding. It bears reminder that although prominent overlap exists, American evangelicalism does not equate with fundamentalism. Evangelicalism ranges from conservative to liberal, both theologically and socially. Nevertheless a fundamentalist impulse within it grew as the twentieth century progressed, pursuing a biblically literalist epistemology and a traditionalist social agenda. Mormonism of the early twentieth century flirted with progressivism. A few Mormons promoted unions and egalitarian economic ideologies, for example (Alexander 1986; Yorgason 2003). Some LDS leaders believed scientific and social progress would draw the world into ever tighter agreement with Mormonism. This important but never hegemonic hope within Mormonism lasted only through the 1930s, however, ending with those leaders' deaths. Thus unlike Catholicism, to which evangelicals sometimes compare Mormonism because of their similar ecclesiastical hierarchies, Mormonism experienced little liberal impulse after the early twentieth century. Mormonism has always contained liberals, but the larger trend was toward conservatism. For most of the past several decades, a traditionalist social and epistemological ethic has assumed dominance.

Evangelicalism's traditional *social ideals* virtually matched those of an increasingly conservative Mormonism during most of the twentieth century. But traditional *epistemologies* differed. Many Latter-day Saints increasingly made fairly literalist scriptural interpretations, to be sure (Barlow 1991). But Mormons did not regard the Bible as the single source of religious truth. They had additional scripture and believed their leaders possessed prophetic access to God's will. This access gave those leaders license not only to interpret scripture but also to give entirely new scripture and guidance. To evangelicals, Mormonism thus went beyond Catholicism, whose leaders' binding statements pertained only to scriptural interpretation (Mouw 2004). A traditionalist epistemology, to Mormons, meant that truth came primarily *through the church itself*, defined largely by its priesthood hierarchy.

Because Mormonism in America often sought conversions from among the same conservative whites that evangelicals targeted, evangelicals sometimes regarded Mormons not just as Others but as deceiving competitors (Shipps 2000). From very nearly the beginning of Mormonism, American Protestants had distrusted concerted LDS missionary efforts and during the late nineteenth century tried to limit LDS proselytizing in Europe. The political contest retreated during the early twentieth century, as Mormon growth was not especially rapid, hampered by two world wars and a depression. However, LDS growth rates picked up in the second half of the century, coinciding with evangelicalism's rapid rise to cultural/political/religious prominence and influence. Evangelicals and Mormons find each other's social message hard to argue against, but distrust each other theologically. Anti-Mormonism gained increased vitality within evangelicalism during the late twentieth century, with fundamentalism on the rise, and as Mormonism proved more durable than expected. Mormonism's continued claim to singularly hold the keys to salvation appalls evangelicals. Many evangelicals center on Mormonism in their warnings about 'counterfeit' Christian cults. Jerry Falwell's inclusion of Mormons in the Moral Majority during the 1980s thus met with considerable evangelical distrust. However, the coalition largely worked, even without full LDS institutional support, because many Mormons and evangelicals decided that the importance of issues such as abortion, pornography, family values, and national defense made compromise possible. The 2008 alliance between evangelicals, Catholics, and Mormons to proscribe gay marriage in California produced similar dynamics. Nevertheless the evangelical attacks on Mormon theology continue. Evangelicals fear that dignifying Mormon theology threatens the Christian gospel's integrity. Thus efforts by a few Mormon and evangelical scholars to more calmly compare the religions (for example, Blomberg and Robinson 1997) produced a slight rapprochement, but also many misgivings among evangelicals. As another example, some nationally prominent evangelical leaders publicly agreed to overlook Mitt Romney's Mormon theology during his recent presidential campaign. Other, especially 'grassroots', leaders and adherents insisted that, in spite of Romney's rapid lurch to the social Right, Mormonism was a dealbreaker.

LDS Eschatology

Mormonism began with eschatology in mind. The Latter-day Saints gathered into concentrated communities, expecting Christ's imminent second coming. Drawing upon extant ideas, but also fashioning a unique last-days configuration, Joseph Smith created both a new sacred time and new sacred geography, with profound consequences both theologically and in the realm of practical action (Barlow 2007). Mormons, Barlow argues, feel remarkable urgency to do God's work. In their early communities, Mormons busily sought to create the City of Zion, very literally intending a new Jerusalem on the American continent. It was to be a refuge from a latter-day world ripening in sin, a place the Saints would prepare to meet the Savior. Mormons believed God had inaugurated a new and final period prior to the Millennium, naming it the 'dispensation of the fullness of times.' Most commentators regard the Anglo-Irish priest John Nelson Darby, a contemporary of Joseph Smith, as Christian dispensational premillennialism's father, though the system was not popularized in the United States until the late nineteenth century (Prosser 2006). However Philip Barlow, an LDS scholar with expertise in the history of biblical interpretation, believes that 'dispensationalist' ideas circulated broadly enough during Darby's and Joseph Smith's key years (late 1820s-1830s) to question the singular label. In any case, Smith's LDS dispensationalism arose mostly independent of Darby's ideas.

Both LDS and evangelical dispensationalism are broadly biblical. Mormons refer to a dispensation as a period of time when God's inspiration and authority operates fully on earth. Evangelicals call it a period where God tests humans in particular ways (Prosser 2006; Lassiter 1992). Mormons worry less about naming the various dispensations than evangelicals do, but the final dispensation holds great LDS significance. In this last dispensation, all truths and keys of authority from prior dispensations will be fully revealed and reconciled. While evangelical dispensationalism regards the (current) Christian period as an interlude to God's focus on the House of Israel, LDS dispensationalism sees Mormonism as God's instrument to create the anchors of righteousness and religious authority that must precede Christ's return. Something of a cross between dispensational premillennialism and Christian covenant theology, Mormonism anticipates God's return of attention to the House of Israel—especially Jews (and thus is not fully supersessionist), but regards itself as part of that return of attention. As with evangelical dispensationalists, books such as Revelation, Daniel, Isaiah, and Ezekiel contribute heavily to Mormons' last-days expectations. Latter-day Saints generally predict greatly accelerated and devastating wars and natural disasters to precede Christ's return. God will redeem Israel only after much tribulation.[3] These tribulations will destroy the wicked. Israel's lost tribes will return and reclaim their covenant status. Mormons anticipate playing a leading role in supporting Christ during his Millennial reign, but it is not Latter-day Saints alone, but good people from all walks of life, who will be saved to live in that era.

The evangelical scholar Richard Mouw (2004; see also Sturm 2006) argues that evangelicals divide themselves eschatologically into premillennial, postmillennial, and amillennial positions (although they sometimes mix and match). These respectively suggest that: after the Tribulation, Christ's second coming ushers in a peaceful millennium (see Introductory chapter, this volume); a millennial period of peace and harmony precedes the second coming; and apocalyptic/millennial scriptures refer to the Christian period generally and are not literal expectations for the final days. Mormons, Mouw (2004) perceptively argues, are most essentially premillennial. Nevertheless, their Zion concept—the Church and its region become a peaceful and godly 'city on a hill'—has a postmillennial flavor (see also McGowan 1983; Introvigne 1997; De Pillis 2003). Mormons proudly claim their religion lacks a theological tradition. Perhaps for this reason, and much like less-theologically-aware evangelicals, Latter-day Saints do not worry much about tension between premillennial and postmillennial ideas. Thus while Mormonism's eschatology is unique in some details, large parts draw on well-known possibilities within broad Christian eschatology.

Other aspects are quite singular, however. Mormonism's prophets provided additional details. Most notably, Mormonism prioritizes the American continent and (in most accounts) the American nation. Other American religious traditions have had significant elements of Americanism, seeing America as a shining city on a hill, but Mormonism takes the ideas further, making literal what they hold more figuratively. Most significantly the Book of Mormon indicates that an American New Jerusalem will arise in tandem with Israel's redeemed old Jerusalem. Joseph Smith identified the location as Jackson County, Missouri, one of the Saints' 1830s gathering places. Just as Mormons expect a rebuilt Jewish temple in the last days, they assume they will return to Jackson County to build a temple and await Christ's second coming (De Pillis 2003; Campbell 2004).

---

3 Latter-day Saints use 'tribulation' eschatologically more as a generic rather than a proper noun, referring to a general period of societal decline rather than a specific seven-year period of turmoil.

Latter-day Saints' interest in eschatology has varied over time. Grant Underwood (2003) argues that Mormon (pre)millennialist fervor peaked during the nineteenth century, varying almost directly with persecution against the church. Despite occasional rhetorical flourishes, Underwood claims, the church itself has never been revolutionary, however; that is, it eschews violence (other than perhaps for self defense) and does not seek to bring the apocalypse to pass. Mormons see their eschatological work as preparatory rather than hastening. Once integrated into twentieth-century American culture, however, Mormonism paid very little attention to eschatology, Underwood suggests. I think he is correct at the level of the institution per se, but as he and Douglas Davies (1973) imply, the church still embodies its historical emphasis about last-days tribulation and the millennium in other ways. These range, for example, from its formal name to the regular, if usually unofficial, preaching to its teenagers that they have been reserved as God's valiant warriors in the final days (see Hamula 2008 more officially). Nevertheless, it is true that twentieth- and twenty-first-century Mormons have been somewhat less concerned with the current events/apocalypse relationship.

These days, Mormons vary considerably among themselves on how or even whether to read geopolitics eschatologically. For example, Mitt Romney, Harry Reid, and the religiously inactive Mormon Brent Scowcroft each hold very different philosophies of foreign-policy. No singular LDS *geopolitical* position exists, even though Mormonism has a general eschatological vision and despite many Mormons' recent individual support for George W. Bush's neo-imperialism.

However, this situation provides just a partial answer to this chapter's key question. Large numbers of Mormons could still, theoretically, seek common cause with evangelical 'geopoliticians'. In asserting official Mormonism's decreased concern with eschatology, Underwood underplays the urging of certain high church leaders and quasi-official doctrinal authorities (e.g. Brigham Young University scholars) *individually* for members to be aware of the 'signs of the times' around them (Stein 2007). This urging produces a more grassroots or folk level of last-days expectations, one with periods of greater and lesser activity. At times it unself-consciously accrues aspects of broadly evangelical eschatology from American culture (Barlow 1989; Introvigne 1997). Last-days forecasting was in fact a moderately popular, if never quite official, armchair endeavor for Mormons from about 1960–1990 (Yorgason and Robertson 2006). The Cold War and some LDS leaders' anti-communism (on Mormonism's historical dissociation from communalism, see Yorgason 2002) generated expectations that the United States would defeat the Soviet Union during World War III and Israel would triumph over its Arab neighbors. American society and politics would simultaneously deteriorate in the process, however. Latter-day Saints would 'rescue' important American principles. Consequently, and along with Christ's millennial reign, the LDS Kingdom of God would supersede the U.S. and other governments.

Dale Robertson and I (2006) have argued that the end of the Cold War greatly disrupted this settled expectation. While America and Israel remained chosen lands, LDS eschatologists looked in new places and at different groups for sources of trouble. Simultaneously Mormonism's overall enthusiasm for last-days forecasting greatly diminished for reasons more directly affecting Mormonism internally. Quasi-official attempts to map geopolitical events onto prophetic grids almost completely ended. Authors who attempt to eschatologically plot current places and events now typically peddle their wares with marginal presses. Everyday Mormonism—the Mormonism of church meetings, individual study, family gatherings, and social interaction with fellow members—became much less saturated with last-days emphasis. The theme did not disappear altogether. Most Latter-day Saints at some level still expect trouble in the end times and believe it may occur in their lifetimes. However, even among the grassroots, eschatology is no longer as central to 'performing' Mormonism. Millett's partially normative argument (2005) is thus probably correct analytically, too: most Mormons are currently prudent toward eschatology, avoiding overconcentration on signs of the times.

Mormon and Evangelical Christian Last-Days Novels

Yorgason and Robertson (2006) analyzed primarily 'non-fiction' monographs. We did not analyze LDS last-days novels. Such novels are important however. Their overtly fictional character and more complex scenarios reveal more about where Mormonism might or might not find common cause with evangelicals. In what follows, I compare LDS last-days novels to evangelical end times expectations, broadly following the methodology of popular geopolitics. Popular geopolitics asserts that relationships between political communities are constituted through popular culture just as significantly as they are through formal politics (Sharp 2000). Most novels, including those analyzed here, attempt to anchor imaginative stories to reality through believable and plausible descriptions of the settings. Authors hope readers find their initial description of places and communities realistic, and changes attributed to them within the realm of possibility. Popular geopolitical analysis thus places assertions and assumptions about places, communities, and their relationships at center stage. Much popular geopolitical analysis, both in this volume and elsewhere, exposes and questions geopolitical thinking within popular culture, but my use here is primarily comparative. Placing both the overtly argued and assumed geopolitics of evangelical and Mormon last-days novels side by side affords deeper evaluation of prospects for a Mormon/evangelical geopolitical movement.

I use Tim LaHaye and Jerry B. Jenkins's *Left Behind* series as the evangelical point of comparison, acknowledging that even many conservative evangelical dispensationalists disagree with its doctrine/geopolitics, but the series sold well over 60 million books and resonates culturally at some important level. It seems, in fact, that *Left Behind* inspired Mormon authors to imitate its form while relying on LDS eschatology.[4] I touch base with five LDS series here: Pam Blackwell's *The Millennial Series*, Kenneth Tarr's *Last Days*, Jessica and Richard Draper's *Seventh Seal*, Wendie Edwards's *Millennial Glory*, and Chris Stewart's *The Great and Terrible*.

The Evangelical Left Behind Series

Left Behind (LaHaye and Jenkins 1995+) begins with a pilot plotting to take the next step with a flight attendant to whom he is attracted. Before succeeding, the Rapture throws his flight into disarray; dozens of people suddenly disappear. Upon landing he encounters massive chaos; his wife and son, among millions of others, are gone. In short order he, his daughter, a crack magazine journalist, and a previously unsaved preacher become born again, devote themselves to learning the Bible's prophecies, and set off to convert others during the Tribulation God has initiated. Along the way, with both guerrilla media and violence, they fight the Antichrist, who is inflicting great suffering upon the world.

The series follows LaHaye's eschatological geopolitics. For example, even before the Rapture, Russia, aided by its allies in Iran, Ethiopia, and Libya, had attacked an unprepared Israel. A miraculous firestorm saved the latter nation, fulfilling prophecies commonly seen in Ezekiel 38 and 39. Additionally, Israel had perfected a method to grow vegetation in its sandy soil (the biblical desert blossoming as a rose), signed a peace treaty with its neighbors, and considered promoting a one-world government. Meanwhile, an international monetary conspiracy had also begun before the Rapture, preparing the way for LaHaye's prediction of a single world economy accompanying the Antichrist. When the Antichrist—a charming, deceiving politician preaching world peace—comes to power through the United Nations, he comes from Eastern Europe rather than the expected Rome.[5] However, the Christian Tribulation Force quickly realizes Romanians have Roman roots; this blond Antichrist seems untainted by his nation's historical admixture with Eastern European peoples. Along the way, the Antichrist advocates the UN, the World Bank, and the International Monetary Fund. He presents himself as the champion of tolerant, agnostic peacemaking, all the while ordering heinous murders and massacres. The Antichrist divides the world into 10 kingdoms under the globally unified economy and religion (with an exception for the Jews), again fulfilling biblical prophecy. Two Jews at the Wailing Wall preach of Christ before being killed, and a prophet-like Jewish convert to Christianity provides scriptural insight and leadership for the Tribulation Force.

---

4 Writing just over a decade ago, the only LDS-oriented 'millenarian fiction' Introvigne (1997) found worth analyzing was science fiction, such as that from the LDS writer Orson Scott Card. I found only one novel/series on the last days strictly speaking from an LDS perspective published before *Left Behind* (Newman 1989).

5 In a semi-ecumenical nod, the Lutheresque Pope is Raptured, though a cardinal becomes the Antichrist's spiritual leader.

The Mormon Series

The LDS series have not been nearly as influential as *Left Behind*. A prolific and respected reviewer for the Association for Mormon Letters, the leading organization concerned with LDS fiction, wrote of the "surprisingly few works of fiction in the LDS market promoting ... a sensationalistic view of eschatology." He attributed it to Mormons looking more backward in time (to their history) than forward (Needle 2003). This sentiment confirms my own observations. I have closely studied and written on LDS culture for more than 15 years. I taught Mormon Studies at an LDS university. I am an active Mormon. I think I have good access to major trends within grassroots Mormonism. Yet I did not know of the existence of these last-days novels until performing concerted research. Granted, I do not live in Utah, where grassroots Mormonism takes particularly vibrant (and idiosyncratic) form. I am not surprised these novels exist. Yet while the series are not hard to find, their position in Mormonism is very different than that which *Left Behind* occupies within evangelicalism. Mormons simply do not often discuss the LDS novels. Assessing cultural influence poses perhaps insoluble methodological challenges; my own opinion is that these novels have limited influence as guides to the last days for Latter-day Saints. Most active Mormons surely believe the general, long-standing LDS prophecies mentioned above. These novels' efforts to tie these (and more specific and obscure) prophecies into current events surely influence some minority of Latter-day Saints, especially among those closer to Mormonism's geographical core. But their larger significance is their collective expression of how (especially conservative American) Mormons envision the world getting 'from here to there.' The novels do not express official LDS positions, though they try to be faithful to LDS prophecies. But if Mormons thought Christ would arrive within 25 years (a decreasingly valid assumption, I believe), I think the novels fairly represent the general trajectory many Mormons expect the world would take.

One semi-solid indicator of LDS cultural influence is sale at Deseret Book bookstores. This church-owned chain's offerings very nearly define normative LDS culture.[6] Out of scores of LDS-themed novels, just two last-days series (*Millennial Glory* and *The Great and Terrible*) were available there during a set of recent visits (July 2008); only the first volume of a third series (*Seventh Seal*) was additionally available on the store's website. I use these three series, all published after 2000, as my major Mormon points of comparison to *Left Behind*. The two other, slightly earlier beginning, series identified previously also receive mention.

---

6 Deseret was the original name assigned for what (in geographically whittled-down form) became the territory and then state of Utah. The Book of Mormon indicates the word means honey bee; early Utah Mormons used it to connote communal industry. Various nineteenth-century church enterprises during took the name. Deseret Book is one of a few church-owned businesses that retain this label pointing back to Mormonism's communal, separatist social geography.

Analysis refers especially to early storylines, since novelists there fashion their most direct links to the world as currently constituted. To help illustrate my final point about the LDS novels, I draw on a Mormon novel that admittedly does not quite fit the last-days genre: Gerald Lund's *One in Thine Hand*. Although this novel is set in the recent past, before the 'end of days,' it clearly anticipates that coming period.

This chapter cannot adequately summarize each LDS series. The Mormon novels vary almost as much from one another as they do from *Left Behind*. Collectively, they partially challenge stereotypes of a homogeneous, unwaveringly structured LDS society: women and divorced men are sometimes central characters, ethnic diversity receives more than just a token nod in a few novels, the most interesting events do not always occur in Utah. Nevertheless, some common elements characterize the LDS stories. For example, since Mormons do not believe in the Rapture, the novels begin with a world gradually descended (or gradually descending) into chaos. No single overwhelming sign tips off believers. Instead, evidence gradually accumulates that the very last days have arrived. A few strong, common geopolitical themes emerge. I discuss how Antichrist, Americanism, and family relate to key plot elements in this section. These themes are neither unique nor treated wholly differently within LDS novels compared to *Left Behind*. But understanding their manifestation lays the groundwork for the argument that follows.

Antichrist Mormon novels do not insist that the Antichrist has literal ties to Rome (for further discussion of Antichrist figures, see Chapter 3, this volume). Nevertheless, that figure usually has clear connections to Europe. The *Last Days* (Tarr 1999+) imagines the mastermind of a conspiracy to take over the world as a wealthy man of finance living in Britain. Hating Arabs almost as much as he hates Jews, he schemes to set them at permanent war against each another. He aids the cause by sending the United States into chaos. Using connections within the former Soviet nuclear industry to 'accidentally' shoot off nuclear warheads—since he could not successfully entice/threaten the governments of Russia and Kazakhstan to attack the United States preemptively—he destroys Chicago and Detroit. In two other series, an Islamic reference is added. For example, Stewart's *The Great and Terrible* (2003+) series's key evil character is Abdullah, crown prince of Saudi Arabia. He takes over from his father to end his father's attempts to democratize and create stronger relations with the United States, pursuing and murdering much of his own family in the process. The new king is mentored by Lucifer himself, who appears as an old man Abdullah met one day on a European beach.

These Antichrist figures resemble *Left Behind*'s Antichrist in their combining of military power rapidly gathered into a single pair of hands and a pathological willingness to use it, on the one hand, with an intense hatred of particular peoples, particularly the people of God, on the other. In fact, this type of Antichrist results as much from borrowing popular Christian eschatology as from presenting mainstream LDS understandings. Clearly a personified Antichrist makes for a powerful representation of evil. And, to be sure, a singular last-days figure is not unknown within Mormon thought. But Latter-day Saints typically give more emphasis to Antichrist as a type of condition characterizing many persons or societies at any time or place than as a singular last-days figure (Frandsen 1992). Despite the novels' similarities, this Mormon tendency may result in a subtle difference: *Left Behind* focuses more, especially early in the series, on who the Antichrist is and what he believes; the LDS series gives relatively more emphasis to how the Antichrist's agenda confronts forces of goodness in the world.

Millennial Glory (Edwards 2001+) typifies this point by showing how the Antichrist's actions create a geopolitical scenario for the LDS church itself. A deceiver named Imam Mahdi who promises world peace and unity jeopardizes Edwards's central family (whose key character is the mother, Corrynne).[7] Readers encounter this character when Utah teenagers and young adults get caught up in his rave parties and mark themselves with the sign of the beast. This Antichrist emerged out the Himalayan mountains and now resides in London's Pakistani/Indian community. Unsurprisingly, Edwards figures Europe as the center of much last-days evil. The United Nations has moved its headquarters there. The UN *per se* is actually not as much of an evil to Edwards as it is to many authors; one of Corrynne's sons is spiritually drawn to work with and try to save the organization. Nevertheless, it is easily manipulated by one-world government/economy evil. These world-government developments directly affect Mormonism's geopolitics. Ironically, LDS proselytizing has now opened in both China and Israel, helping to fulfill prophecy. The UN took over in Israel because of the lack of peace there, thereby creating enough stability to allow the Tribe of Ephraim (as most Mormons consider themselves) to fulfill its last-days responsibilities: bringing Judah back to the truth. In later volumes, many Jews convert to Mormonism. In the U.S., pockets of virtuous people, particularly Mormons, set up tent cities amid the chaos. There, in shades of *Left Behind*, they resist the world government's order to be implanted with a computer chip. Their refusal eventually saves them, and they reinstall a society (along with the Lost Ten Tribes from the north) built on the original values of American freedom.

Americanism Edwards finds in the United States the remnants of goodness needed to fight the Antichrist. For Tarr the world only spirals into chaos after the nuclear bombings of that country. Similarly, other LDS authors regard the American nation-state as a particular focus of the Antichrist's attention. Like *Left Behind*, Mormon novels portray the U.S. as vulnerable to takeover by the UN or some other world government, but there is an important difference. In *Left Behind*, the fight against evil by some in the United States seems the result of contingency: many Christians are there, and the nation has an independent streak. It is powerful

---

7 Edwards deliberately patterned him after a real self-proclaimed prophet of the same name (who presumably borrowed the name from the Mahdi redeemer figure of Islamic, particularly Shiite, tradition).

and important, yet it is still just another nation the Antichrist will conquer. In the Mormon novels, the Antichrist's attempts to destroy America are prophesied necessity. The U.S. is clearly not just another nation. It does not just happen to be the Antichrist's most significant national opponent. In fact, America does not just happen to be. It will face trials and may succumb temporarily, but God has willed that its values and principles will emerge victorious.

Chris Stewart, author of *The Great and Terrible* (2003+), is a former Air Force pilot. Unsurprisingly, Stewart pens the most militaristic series, from, for example, having the central characters come from a military family to the portrayal of the U.S. military as an unambiguous force for good. The family's father, a military man, sacrifices himself. He warns the U.S. government rather than fleeing Abdullah's nuclear attack, to which he alone is alerted. The bomb, which follows a nuclear blast in Israel, destroys Washington DC. Shortly thereafter coordinated nuclear explosions high in the atmosphere above the U.S. destroy virtually everything electronic in America. Famine and desperation ensue. Lucifer had helped Abdullah obtain nuclear capability from Pakistan and now insists that the United States must be destroyed. He gives three reasons: it protects freedom for peoples around the world, God's people (Mormons) are most strongly concentrated there, and Israel cannot be destroyed when a powerful United States exists. In stoking Abdullah's desire for power, Lucifer uses ethnically appealing lies: America is the Great Satan and Arabs are the chosen people. The novel's central family, along with many other Mormons, survive and subsequently perform heroic deeds—such as preserving a semblance of the U.S. government.

The Family Like *Left Behind* each of the Mormon novels centers on key characters' family relationships. Themes of finding, saving, anguishing over, and loving family members dominate. Even more than *Left Behind*, however, each LDS series uses the (reified) family not simply to signal a geographically imprecise condition of moral breakdown, but also to explain geopolitical trouble.

In *Seventh Seal* (Draper and Draper 2003+), wife-husband authorial team, Jessica and Richard Draper, refer in first volume's epigraph to an (appropriately Mormon theologically) personified earth:

> Only the spiritually sensitive heard the cry of pain. It wasn't quite a voice, more a feeling that translated itself into words coming from deep within the Earth: "Wo, wo is me, the mother of men: I am pained, I am weary, because of the wickedness of my children. When shall I rest, and be cleansed from the filthiness which is gone forth out of me?"[8]

---

8 The quotation is from Moses 7:48 in the LDS scripture, Pearl of Great Price. The passage hints toward a critique of ecological violence, a point that (along with a critique of imperial violence) one critic of *Left Behind* regards as the Book of Revelation's most profound message (Rossing 2007). However, though they are not as anti-ecological as *Left Behind*, none of the Mormon novels can be regarded as pro-ecological.

The story opens with three simultaneous earthquakes around the world and a gigantic tsunami devastating Southern California, among other locales. The LDS Church has already moved its headquarters to Missouri, even while facing sporadic attacks throughout the world, for example an LDS apostle is abducted and then killed in the Philippines. What animates the story, however, is the identification of evil. This identification points to Mormonism's and evangelicalism's somewhat different theologies. While worldly evil in *Left Behind* signifies that God will commence the last-days drama, in *Seventh Seal* evil seems more directly the cause. The Drapers focus particularly on U.S. society, pinpointing many evils Mormons regard as family matters: pornography, abortions, homosexuals and their supposed agenda, feminism, consequence-less notions of sexuality. The Drapers link these rather directly to the world's emerging geopolitical chaos. Even the characters themselves embody judgments about family-related evils. The book's most important character, Merry Galen, is an LDS convert. Before she fights the corporate conspiracy she and her husband discover, she has had to overcome her upbringing: her mother is a lesbian who does not know Merry's sperm-donor father and had wanted Merry to live a feminist, lesbian lifestyle. Most of all, Merry's mother had not wanted her daughter to marry into a patriarchal relationship within the LDS Church.

Millennial Glory's (Edwards 2001+) message is similar. Her prime target is family breakdown, manifest through a transformed geography of love and care.

> Corrynne's thoughts turned to the ironic isolation of the world's inhabitants, despite the growing number of people who lived on its surface. She marveled at Satan's cunning. What a great way to destroy the souls of mankind! Ruin natural affection. Divide the family, and conquer the souls. Make everyone isolated and defensive. Create so much dishonesty and selfishness that trust is abolished. Cut the bands that bind. This was the recipe for the breakdown of the family worldwide. (2001, 116)

With people seeking companionship elsewhere, Edwards argues, much of the evil tearing society apart results from misapplied efforts to be part of a group. She imagined future concrete manifestations:

> Corrynne reflected on the gay community's recent successes in the courts to redefine their relationships as familial. Then she thought about the rampant divorce rates. ... The failure of the family to serve the basic needs of companionship was a testimony of the breaking down of the world's primary unit. The family was the one organization that had the power and influence to solve most of the evils of society, but the world denied the traditional family, in the name of "modernization." A transient society could never take the place of the dependable background that a solid family could give. (2001, 117)

Comparison

We should be wary of overly literally postulating the novels' effects on readers. Many *Left Behind* enthusiasts do not feel bound to LaHaye's exact scenario (Dittmer 2008). Some evangelicals are more embarrassed than enthused by the series (Mouw 2004). Others surely read *Left Behind* a good deal more loosely than LaHaye might wish (Coulter 2005; Dart 2002). While acknowledging geopolitical literalism among some readers, I believe last-days predictions (whether evangelical or Mormon) more effectively perform other cultural work: confirming a sense that the world cannot continue on its present path; validating the religion's epistemology (ie, biblical or prophetic); expressing concern with or alienation from society's trends; and encouraging commitment to the faith (Rossing 2007; Anderson 2005; Peterson 1976). In fact, Mormon authors typically ask readers to regard their stories more as parables than accurate accountings of last-days events. Yet even acknowledging a variety of readings, the critical geopolitical point is that the novels' scenarios create and reinforce particular ways of seeing the world and its constituent parts, many of which are quite problematic. Like *Left Behind*, the Mormon series grossly simplify and brazenly stereotype people, places, and political motivations. Everything important in life reduces to a battle between good and evil. A strident right-wing ideology saturates the novels, the LDS series even more than *Left Behind*. Violence and dishonesty are tolerated and even required in support of righteous causes. Similar to *Left Behind* and *contra* contemporary LDS trends elsewhere (McAlister 2003; Green 2001), the LDS novels essentially erase Palestine/Palestinians as last-days actors of any moral standing in an uncritical valorization of Israel/Israelis. More theoretically, the Mormon novels combine problematic regimes of vision, knowledge, and theatricality (cf, Sturm 2006). Anything beyond superficial historical and geographical knowledge appears irrelevant, trumped by detailed knowledge of prophecy.

Common Ground   Nevertheless, my goal here is comparison more than critique. Much in *Left Behind* and the LDS series suggests that Mormons and evangelicals could find common ground. Leaving aside the Rapture, both expect a quick last-days descent into chaos. Once John the Revelator's seventh seal is open, there is no turning back. Both *Left Behind* and the LDS novels strongly imply that most of the world will be unsalvageable. Agreeing that the Antichrist must be fought, both find little room for sentimentality toward things lost. Better to look forward with joy and hope to the earth's cleansing and Christ's second coming. Causes become more important than people. Despite a few nods to the contrary, fighting sin takes priority over reclaiming sinners. Each LDS series can, perhaps with similar issues for *Left Behind*, be considered a reaction against the recent LDS trend of downplaying signs-of-the-times discussion. The novelists push back against embarrassment over prophecy, striking claim for a literalist interpretations by invigorating prophecy with current events.

The two types of novels imply partially similar eschatological maps. Both expect Israel to be purged through violence, despite many miracles of protection along the way. They agree Arab societies will fight Israel and effectively destroy themselves in the process. Both expect two prophets to preach on the streets of Jerusalem. Each series deeply distrusts the United Nations and believes it is destined for evil. Most of the books imply, and some explicitly justify, a conspiratorial view of the world. The Antichrist will rise through secret combinations. This nearly identical prism for viewing, fearing, and identifying evil in world politics is the most likely focal point of any Mormon/evangelical coalition. Readers strongly influenced by the novels will likely interpret the day's news similarly. Barack Obama must frighten many in both camps. Even if they resist associating him with the Antichrist, the right-wing tune about Obama as a smooth, perhaps partially foreign, multi-lateralist who preaches peace—someone willing to apologize for America's flaws and beloved by Europeans—strikes all the wrong chords (again, see Dittmer, this volume).[9]

Evangelical Obstacles As worrisome as these shared interpretive dispositions may be, they fall far short an organized partnership would that persist for long. The novels hold hints that any nascent affiliation would almost immediately run aground. Some evangelicals will have trouble with Mormons' expectation of no Rapture or with other small differences from LaHaye's template. For example, *Left Behind* anticipates the two preachers in Jerusalem will be resurrected Israelite prophets while Mormons usually assume they will be either young LDS missionaries or LDS apostles. However, small details will be the least of the problems. After all, evangelicals differ among themselves on such matters.

For other evangelicals, the real problems are Mormonism's additions to biblical prophecy. Evangelicalism simply encompasses little tolerance for those. Evangelicals might overlook the LDS claim that the Book of Mormon offers special insight into last-days conspiracies. *The Great and Terrible*'s (Stewart 2003+) initial story setting in the (uniquely LDS conceived) pre-mortal battle of all humanity against Lucifer may cause a few more problems, but evangelicals almost certainly will not disregard Mormonism's spatial shift in geopolitical attention, with Missouri and Utah joining Jerusalem as key foci of last-days action. Virtually every LDS series uses the formal return of Mormons to Missouri to mark the last days' arrival. More generally, the differences in Americanism might cause problems. Granted, some evangelicals write the United States into their expectations. But they do so on a non-scriptural basis. To them, the U.S. role contingently exists since the United States is Israel's strongest ally and presumably the world's most genuinely Christian nation (see Mouw 2004, though he puts these

---

9 Though the connection to the last-days novels addressed here cannot be known, see the mild furor surrounding Mormon and neo-conservative CNN talk show host Glenn Beck jokingly asking evangelical pastor John Hagee about the likelihood that Obama is the Antichrist in March, 2008 ("Cameron" 2008). Hagee's answer: "no chance."

ideas to a different conclusion). LDS doctrine, on the other hand, brings the United States into almost direct theological equivalence with Israel. Mormon novels are as much about saving the United States as worrying about Israel. LDS authors insist that America must prevail, even if in humbled and purified form. In *Left Behind*, the American characters care about their country. But once it fails, the novels imply, it is best to ride out the storm and look ahead to Christ's Kingdom. *Left Behind* may universalize U.S. identity implicitly as the type of culture toward which true believers gravitate, but most LDS novels universalize it theologically as the necessary cultural/political stepping stone into Christ's kingdom. Mormons have a (not fully verified) tradition that Joseph Smith prophesied the United States Constitution would "hang by a thread" in the last days; in the end, Mormon elders' heroics will save it. One of Edwards's titles is even *Hanging by a Thread*.[10]

Mormons' insistence on calling themselves the Ephraimic leaders of the House of Israel—and occasionally calling particular racial groups (Native Americans and Polynesians) House of Israel warriors who will help establish the American New Jerusalem or referring to them as having special spiritual insight—will also strike evangelicals as crazy talk. The themes are admittedly not very overt in the novels, given LDS statements in other forums, though they are implied in most of the novels. The racial argument is strongest in Blackwell's *The Millennial Series* (1996+). There, a key character is a Maori woman who, along with her mother, serve as unofficial prophetesses. She miraculously delivers a son after a troubled pregnancy. Prophecies accompany the baby: he will perform an important mission during his lifetime, fulfilling a charge that (racially) belongs to members of the House of Israel.[11] Extra-biblical notions led evangelicals to regard Mormonism as a non-Christian cult in the first place. It is hard to imagine evangelicals overlooking them to forge a meaningful geopolitical coalition.

Mormon Obstacles Perhaps most importantly, the novels hint at the Mormon impulse that will ultimately preclude alliance with evangelicals.[12] Basically, eschatological thinking invites a turning inward for Mormons, toward the church's

---

10   Utah Senator (and Mormon) Orrin Hatch was criticized during his ill-fated 2000 presidential campaign when, in a Salt Lake City radio interview, he argued the Constitution was hanging by a thread. Whether he had visions of being the Mormon elder who saved the Constitution, or, more likely, was speaking to the 'home crowd' about 'activist' judges in language resonant for them, the nearly immediate critical reaction shows the resistance in America toward any hint of heightened last-day importance for Latter-day Saints.

11   This episode links to House-of-Israel prophecies for Book of Mormon remnant peoples. The Book of Mormon explains that its people were a branch of ancient Israel that traveled across the sea to a promised land (on the American continent). Many LDS leaders have declared that Polynesians descended from a Book of Mormon group that itself left on a voyage, never to be heard from again.

12   While LDS eschatology envisions last-days alliance with good people of any religion, Mormons envision these allies as individuals rather than corporate units. Goodness of the heart will trump particular belief. A similar allowance for individuals who trust

corporate body. A virtual litmus test for Latter-day Saints showing whether the final days have truly arrived is renewed persecution of Mormons and Mormonism.[13] In this regard Mormons expect the last days to mirror the days of Joseph Smith and Brigham Young. They anticipate unique and concerted attention aiming at both the church itself and individual members. For example, Blackwell's world government targets Utah for martial law because of that state's success and independence. In addition to having the Antichrist's False Prophet come from a Mormon offshoot that hates the LDS church, Tarr writes of born-again Christians and deceptively 'tolerant' Methodists trying to destroy the church, as well as a Missouri militia attack on Mormons in Iowa (shades of Joseph Smith's day). For Edwards, the Imam Mahdi appears quite early in Utah, targeting especially LDS youth. The Drapers conjure multiple depredations against Mormons and Mormon property, as well as the refusal of the "United Baptist Churches" to cooperate with Mormons within the Red Cross. And Stewart's Lucifer simultaneously singled out Israel and the Mormons.

Additionally, the LDS series portray Mormons coming together as a fundamental last-days event. Each emphasizes obeying the LDS prophet, for example. While the novels anticipate a great spiritual outpouring within Mormonism, not only among church leaders but also individuals, these revelations are expected to reconcile with the prophet's instructions. Interestingly, even the novels' descriptions of last-days evil themselves enact obedience to Mormon prophets. In 1995, the church's hierarchy issued a Proclamation on the Family, urging the world to give more attention to family ties, love, and responsibilities. The Proclamation argued that wives and husbands should be companionate helpmates to one another within a system of clearly differentiated gender roles. Many evangelicals would not object the message. *Left Behind* also clearly has a gendered message. However in comparison to the Mormon novels, *Left Behind* offers a generic and implicit reinforcement of patriarchal norms. By contrast, the LDS novels strongly imply that Mormons *own* the key to successful family life. Thus obeying the prophet is vital for last-days happiness. Family messages are not directly tied to the eschatology in *Left Behind* (though they are in other evangelical publications). The Mormon novels place such messages squarely at their eschatological and didactic center (a phenomenon inconceivable in Mormonism before 1995). Mormons thus define the world's political geography of evil in response to not only changes in

---

Christ seems available within *Left Behind*, but again, particular religious bodies, such as Mormonism, are not expected allies.

13 I am aware of informal attempts within the LDS grassroots to read the animus directed against Mormons after the 2008 passage of California's Proposition 8 (banning gay marriage) as part of the fulfillment of this expectation. The joining of Mormons with evangelicals (and Catholics, among others) raises the possibility that Mormons could shift expectations of persecution to include other Christians standing alongside them rather than as persecuting them. But while this state of affairs bears watching, I expect movement toward this that inclination will be temporary and ambivalent, at best.

world politics, but also changes in LDS prophetic emphasis. Mormons differentiate themselves through the Proclamation. Messages on the family reinforce biblical epistemology for evangelicals; however the very similar messages for Mormons strengthen the prophetic epistemology that divides Mormons from evangelicals.

The LDS novels' portrayals of Mormons coming together also have social and even geographical components. Mormon congregations (or sometimes loose coalitions of Latter-day Saints on the run) become mini-societies unto themselves. In some stories, the return to Missouri begins. Tarr anticipates a prior gathering to Utah by non-American Mormons; Utahns open their homes to accommodate. Almost all novels utilize the LDS notion of the Rocky Mountains as a place of refuge from a chaotic world. The novelists clearly expect the LDS ethnic impulse to strongly re-emerge in the last days. They imply that, in addition to saving America and Israel, Mormons properly engage with the last days by closing ranks and strengthening their own community (Peterson 1976; McGowan 1983).

Finally, Mormonism's sense of geo-ethnic possession will hinder inter-faith eschatological coalition. Put quite simply, the novels suggest that Mormons anticipate a greater sense of ownership toward the events, places, and even peoples significant in the last days than evangelicals do. Yet the difference between *Left Behind* and the LDS novels in this regard is quite subtle, as the senses of ownership are much more implicit than overt. In fact, these senses are subtle enough that teasing them out through summary is difficult. I therefore turn to a final LDS novel to help illustrate. I use this novel not directly in comparison with *Left Behind*, but rather to better define this notion of geo-ethnic possession.

One in Thine Hand (Lund 1982) is not strictly speaking a last-days novel, although implicit last-days expectations saturate the story. It takes place primarily in Israel of 1973, nearly 10 years before the story's publication. The main character, an American named Brad, served a two-year LDS mission immediately prior to two more years as a soldier in Vietnam. Returning home restless and agitated, Brad decides to visit Israel to find himself. On the flight he meets a recent Palestinian convert to Mormonism. Ali had tried terrorism some years before, but decided it was not for him. Now, as a Mormon, he is returning to his homeland to start up a school. Ali sets Brad up with an Israeli family that owns a hotel, a family Ali knows well. The family's daughter, Miri, is beautiful and headstrong. She and Brad, also stubborn, get along badly at first, but a clear attraction soon emerges. The plot centers on whether and how she will accept the Mormon gospel, and whether she will marry and live with him in Salt Lake City (both eventually occur). Along the way, Ali loses his life trying to save his friends during a particularly dangerous incident. Ali's death allows Miri to understand and accept Jesus's atonement.

While many readings/critiques of this novel are possible (see for example, Abunuwara 1989), a fairly straightforward one has Miri and the land of Israel representing one another. Precisely as Brad grew fascinated with Israel (he intends to become a professor of Jewish/Israeli history by story's end), he became captivated by her. She opened herself up to him emotionally just as she guided him through Jerusalem's sites.[14] Her conversion to Mormonism and willingness to marry and live in Utah symbolizes not only Brad's taking possession of Israel within his own life but also Mormonism's prophesied marriage to the Jews in LDS eschatology. Brad (American Mormonism—Ephraim) will take the lead. From an American LDS point of view, the union is entirely natural, desirable, and necessary in a last-days context. Reading Ali similarly, Arabs, as joint heirs to Abraham's covenant, need not fail in their struggle against Israel. If they learn of their true (though secondary) role in the last days, Lund intimates, they can be redeemed with Israel through Mormonism. The novel thus implies that Mormonism carries not only the key to last-days meaning and individual salvation, but also last-days rights and responsibilities to conceptually possess lands and peoples.[15]

One in Thine Hand is still sold at Deseret Book, despite its age, perhaps because its author became part the church's second tier of general leadership (though he is also a highly popular author of other LDS novels). Nevertheless, the inclination to possess Israel has moderated within Mormon culture since initial publication. And, of course, some parts of evangelicalism also yearn to possess Israel in certain senses. Yet while the difference is subtle, I believe that even the recent LDS last-days novels exhibit a deeper sense of this type of geo-eschatological possession than does the *Left Behind* series. Mormon novels more strongly suggest that the LDS church will play a proprietary or managerial role in directing last-days events. For example, LDS leaders arrange a seminal meeting between a Tibetan holy man

---

14 This reading reveals a couple of surely unintended sexual innuendos, though ones that illustrate well the overlap between desires for land and desires for women in Western thought. Upon showing him a sunrise view of the Old City: "[Miri] nodded, pleased in spite of herself, 'Each time of day has its own mood,' she said softly, 'but I love this one the best. The city opens herself to view, innocently, fully, as though she had nothing to hide.' 'It is beautiful," Brad murmured. 'This is what I came for.'" Later, she said, "'Let us stop looking at Jerusalem from a distance and go to meet her. She has seen much sorrow, but she has also seen much joy.' [Brad answered] "I am anxious to come to know her well'" (pp. 68, 72).

15 *One in Thine Hand* was first published as the LDS Church prepared to build a Brigham Young University Jerusalem Center. The role of Jerusalem in LDS eschatology was surely in church leaders' minds at the time, as it probably continues to be today. But local Israeli worries over LDS proselytizing during the mid-1980s construction of the center led the church to be quite circumspect, promising no proselytizing by those participating in the center's programs, a promise the center takes great pride in upholding even at present. The center thus functions almost exclusively as a study-abroad site for Brigham Young University students, a place to soak in Old and New Testaments as well as the various civilizations/languages of the Near East. Perhaps ironically, while the 1980s were a time of relatively outspoken geopolitical support for Israel by some Mormons, my personal observation is that those who have participated in the center's programs almost universally report gaining a great deal more respect for the Arab side of the Arab-Israeli contest over Palestine/Israel than they possessed before. So while eschatology is surely not absent from the center's function in LDS life, the center has also acted for a generation as a brake on the momentum of imminent eschatological expectations for some of Mormonism's soon-to-be leaders.

who possesses newly discovered ancient scripture and the Jewish LDS convert who will translate them in Blackwell's *The Millennial Series*. Few such notions exist in *Left Behind*; God may have his Tribulation Force, but God himself directly runs the geopolitical show.

Conclusion

Problems inhere in reading world geopolitics through eschatological lenses, as other chapters in this volume attest. The difficulties are as relevant for Mormons as for evangelicals. Nevertheless, I find little reason to expect convergence between Mormonism and evangelicalism on eschatological geopolitics. Many elements from the two systems are certainly similar. Both tend toward literalist readings of eschatological scripture, for example. In both, those who most insistently read geopolitics through eschatology share similar right-wing politics. Agreement between evangelicals and Mormons should not be surprising on certain U.S. foreign policy issues.

However several factors mitigate against deeper alliance. Perhaps most important, neither of the two faiths shows internal unity about the eschatology-geopolitics relationship. Mormonism's emphasis on unity under prophetic leadership, in particular, suggests that lack of unity on eschatology may mean lack of importance. Mormons have collectively proposed many, sometimes mutually contradictory, scenarios matching eschatology with geopolitics. The majority of Mormons, however pay the relationship very little mind. Second, Mormons are less prone than evangelicals to act directly on their eschatological inclinations. The relatively few Mormons who desire to do so are not in a political position to achieve much. No LDS campaign is analogous to the pressure campaign on American politicians from evangelical dispensationalists (McAlister 2003). Mormonism's current institutional effort to build strong relationships with national governments around the world also suggests that the Mormon hierarchy is not interested in promoting any blatantly partial geopolitical objective (Underwood 2003; for something similar in evangelicalism, see Mouw 2004). As a social phenomenon then, eschatology motivates little direct, foreign policy-related action among Mormons.

Finally, the last-days novels themselves, which in the LDS case originate from somewhere toward a fringe in Mormon culture, seem to suggest the improbability of anything beyond a diffuse, temporary, and reactionary coalition of evangelicals and Mormons eschatologically/geopolitically. Despite many similar interpretive tendencies, important differences exist. Evangelicals are unlikely to compromise on eschatology stemming from LDS prophetic additions to the Bible. In addition, the Latter-day Saints' eschatological impulse leads them to turn inward and emphasize the religion's distinctiveness. Finally, Mormonism's sense of possession of last-days events, peoples and places will not be easily reconciled by those who believe God's purposes do not run through the LDS church. These are hardly the ingredients that make for a solid geopolitical coalition.

References

Abunuwara, E. (1989), 'Nothing Holy: A Different Perspective of Israel,' *Dialogue: A Journal of Mormon Thought* 22:3, 92–101.
Alexander T.G. (1986), *Mormonism in Transition: A History of the Latter-day Saints, 1890–1930* (Urbana, IL: University of Illinois Press).
Anderson, C.C. (2005), 'A Catholic and Ecumenical Response to the *Left Behind* Series,' *Journal of Pentecostal Theology* 13:2, 209–30.
Barlow, P.L. (1989), 'Why the King James Version? From the Common to the Official Bible of Mormonism,' *Dialogue: A Journal of Mormon Thought* 22:2, 19–42.
—. (1991), *Mormons and the Bible: The Place of the Latter-day Saints in American Religion* (New York: Oxford University Press).
—. (2007), 'Toward a Mormon Sense of Time,' *Journal of Mormon History* 33:1, 1–37.
Blackwell, P. (1996+), *The Millennial Series* (Houston: BF Publishing).
Blomberg, C.L. and Robinson S.E. (1997), *How Wide the Divide?: A Mormon and an Evangelical in Conversation* (Downer's Grove, IL: InterVarsity Press).
"Cameron," (2008), 'Glenn Beck, *Think Progress*, Barack Obama & the Antichrist' (Blog posted on 10 March), Magic Valley Mormon [website], <http://magicvalleymormon.blogspot.com/2008/03/glenn-beck-think-progress-barack-obama.html>, accessed 20 January 2009.
Campbell, C.S. (2004), *Images of the New Jerusalem: Latter Day Saint Faction Interpretations of Independence, Missouri* (Knoxville: University of Tennessee Press).
Coulter, D.M. (2005), 'Pentecostal Visions of the End: Eschatology, Ecclesiology and the Fascination of the *Left Behind* Series,' *Journal of Pentecostal Theology* 14:1, 81–98.
Dart, J. (2002), '"Beam Me Up" Theology,' *Christian Century* 25 September–8 October, 8–9.
Davies, D.J. (1973), 'Aspects of Latter Day Saint Eschatology,' in Hill, M. (ed.), *A Sociological Yearbook of Religion in Britain · 6* (London: SCM Press), 122–35.
De Pillis, M.S. (2003), 'Christ Comes to Jackson County: The Mormon City of Zion and its Consequences,' *John Whitmer Historical Association Journal* 23, 21–44.
Dittmer, J. (2008), 'The Geographical Pivot of (the End of) History: Evangelical Geopolitical Imaginations and Audience Interpretation of *Left Behind*,' *Political Geography* 27, 280–300.

Draper, J. and Draper R.D. (2003+), *Seventh Seal* Series (American Fork, UT: Covenant Communications).

Edwards, W.L. (2001+), *Millennial Glory* Series (Provo: Seventh Seal Publishing).

Frandsen, R.M. (1992), 'Antichrists,' in Ludlow, D.H. (ed.), *Encyclopedia of Mormonism* (New York: Macmillan), 44–45.

Green, A.H. (2001), 'Mormonism and Islam: From Polemics to Mutual Respect and Cooperation,' *BYU Studies* 40:4, 199–220.

Hamula, J.J. (2008), 'Winning the War Against Evil' (Address to the 178th Semiannual General Conference of the Church of Jesus Christ of Latter-day Saints, 4 October), The Church of Jesus Christ of Latter-day Saints [website], <http://www.lds.org/conference/talk/display/0,5232,49-01-947-17,00.html>, accessed 27 October 2008.

Introvigne, M. (1997), 'Latter Day Revisited: Contemporary Mormon Millenarianism,' in Robbins, T. and Palmer, S.J. (eds), *Millennium, Messiahs, and Mayhem: Contemporary Apocalyptic Movements* (New York: Routledge), 229–44.

LaHaye, T. and Jenkins, J. (1995+), *Left Behind* Series (Wheaton, IL: Tyndale House).

Lassiter, C.J. (1992), 'Dispensations of the Gospel,' in Ludlow, D.H. (ed.), *Encyclopedia of Mormonism* (New York: Macmillan), 388–90.

Lund, G.N. (1982), *One in Thine Hand* (Salt Lake City: Bookcraft).

McAlister, M. (2003), 'Prophecy, Politics, and the Popular: The *Left Behind* Series and Christian Fundamentalism's New World Order,' *The South Atlantic Quarterly* 102:4, 773–98.

McGowan, T. (1983), 'Mormon Millenarianism,' in Bryant, M.D. and Dayton, D.W. (eds), *The Coming Kingdom: Essays in Millennialism & Eschatology* (Barrytown, NY: New Era Books), 149–68.

Millett, R.L. (2005), *A Different Jesus? The Christ of the Latter-day Saints* (Grand Rapids, MI: Eerdmans).

Mouw, R.J. (2004), 'What does God Think about America? Some Challenges for Evangelicals and Mormons,' *BYU Studies* 43:4, 4–21.

Needle, J. 'Review of *The Gathering Storm*, no. 1 in *The Last Days* series, by Kenneth R. Tarr,' *The Association for Mormon Letters-List Review Archive* [website], (updated 19 September 2003) <http://www.aml-online.org/reviews/b/B200031.html>, accessed 24 July 2008.

Newman, M. (1989), *Fire and Glory: The Millennial Story: A Novel* (Salt Lake City: Wellspring).

Peterson, S. (1976), 'The Great and Dreadful Day: Mormon Folklore of the Apocalypse,' *Utah Historical Quarterly* 44: Fall, 365–78.

Prosser, P.E. (2006), 'Dispensationalism,' in Burgess, S.M. (ed.), *Encyclopedia of Pentecostal and Charismatic Christianity* (New York: Routledge), 137–41.

Rossing, B.R. (2007), 'Prophecy, End times, and American Apocalypse: Reclaiming our Hope for the World,' *Anglican Theological Review* 89:4, 549–63.

Sharp, J. (2000), *Condensing the Cold War:* Reader's Digest *and American Identity* (Minneapolis: University of Minnesota Press).

Shipps, J. (2000), *Sojourner in the Promised Land: Forty Years among the Mormons.* (Urbana, IL: University of Illinois Press).

Stein, S.J. (2007), 'Historical Reflections on Mormon Futures,' *Journal of Mormon History* 33:1, 39–64.

Stewart, C. (2003+), *The Great and Terrible* Series (Salt Lake City: Deseret Book).

Sturm, T. (2006), 'Prophetic Eyes: The Theatricality of Mark Hitchcock's Premillennial Geopolitics,' *Geopolitics* 11:2, 231–55.

Tarr, K.R. (1999+), *Last Days* Series (Springville, UT: Bonneville Books).

Underwood, G. (2003), 'Millennialism, Persecution, and Violence: The Mormons,' in Wessinger, C. (ed.), *Millennialism, Persecution, and Violence: Historical Cases* (Syracuse: Syracuse University Press), 43–61.

Yorgason, E. (2002), 'No Grounds for Conversation: The Regional Construction of Fundamental Difference between Mormonism and Socialism,' *Antipode: A Radical Journal of Geography* 24:4, 707–29.

—. (2003), *Transformation of the Mormon Culture Region* (Urbana, IL: University of Illinois Press).

Yorgason, E. and Robertson, D.B. (2006), 'Mormonism's Raveling and Unraveling of a Geopolitical Thread,' *Geopolitics* 11:2, 256–79.

Chapter 3
Obama, Son of Perdition?: Narrative Rationality and the Role of the 44th President of the United States in the End-of-Days

Jason Dittmer

> Let no man deceive you by any means: for that day shall not come, except there come a falling away first, and that man of sin be revealed, the son of perdition; Who opposeth and exalteth himself above all that is called God, or that is worshipped; so that he as God sitteth in the temple of God, shewing himself that he is God.
>
> (2 Thessalonians 2: 3–4, KJV)

> The anti-Christ will be a man, in his 40s, of MUSLIM descent, who will deceive the nations with persuasive language, and have a MASSIVE Christ-like appeal.... the prophecy says that people will flock to him and he will promise false hope and world peace, and when he is in power, will destroy everything. Is it OBAMA?
>
> (email 'forward' circulating in 2008)[1]

In November 2008, millions of American citizens cast their ballots in the most exciting (and longest) election campaign in American history. The election had several highlights – the first major party African American nominee for president (Barack Obama), the strongest showing by a female candidate for president (Hillary Clinton), the first nomination of a woman to a Republican ticket (Sarah Palin), and the most money raised (and spent) in a campaign (Barack Obama's $750 million). Further, while religion has long been an issue in presidential politics, never before has one candidate been so closely associated with Satan.

Origins of Conspiracy and Rumor

Barack Obama's quest to secure his identity to the voting public began in 2004, when he spoke before the Democratic National Convention. Obama emerged as an 'inspiring' figure with the potential to transcend racial and partisan divides in the U.S., but this was conjoined by the initiation of whisper campaigns against him. The most persistent rumor about Obama originated with Andy Martin, a man

---

1 http://www.snopes.com/politics/obama/antichrist.asp (accessed 11/18/08).

who was barred from practicing law in Illinois because of a diagnosis of paranoia and associated character defects, and who later ran for president twice, in 1988 and 2000 (Rutenberg 2008). Notwithstanding his notable failures, he nevertheless successfully originated a viral internet campaign against Obama.

Martin's 2004 press release, stating that Obama was a secret Muslim, found its way into the hands of Ted Sampley in 2006, shortly before Obama declared his candidacy for the presidency. Martin's argument was then expanded to include the now well-known claim that Obama was schooled in Wahhabist schools during his childhood in Indonesia, as well as the following (Sampley 2006, n.p.):

> Debbie Schlussel, a conservative political commentator, columnist, was more pointedly critical about Obama's Islamic connections, she wrote on her blog[:] "while Obama may not identify as a Muslim, that's not how the Arab and Muslim Streets see it."
>
> "In Arab culture and under Islamic law, if your father is a Muslim, so are you. And once a Muslim, always a Muslim. You cannot go back. In Islamic eyes, Obama is certainly a Muslim. He may think he's a Christian, but they do not."
>
> Is a man Muslims think is a Muslim, Schlussel asked on her blog, "a man we want as President when we are fighting the war of our lives against Islam? Where will his loyalties be?"

Sampley's writing, originally published on his personal website, rapidly circulated via email and blog postings.

This chapter begins and ends by analyzing the second permutation of the viral whispering campaign: that Barack Obama is the Antichrist whom many evangelicals believe is prophesied in the Bible. This rumor even broke out of the blogosphere to the 'mainstream' media, with Nicholas Kristof repeatedly addressing the issue (with disgust, it should be noted) in his *New York Times* column (Kristof 2008a, 2008b) and with CNN airing a three minute segment in August 2008 on the question of Obama's role in the end of days after the McCain campaign launched its internet-based advertisement titled 'The One'.[2] This advertisement played off the messianic qualities that Obama's followers purportedly projected onto him (and which, the McCain campaign argued, Obama fostered).

How did the claim that a major party presidential candidate is the vicar of Satan arise? Several interconnected themes merit attention: 1) the importance of narratives and metanarratives to evangelical identities and geopolitical visions; 2) The role of the United States in some evangelical narratives of the end of days; and 3) the role of the 'new' media, especially the Internet, in mediating this controversy. This chapter will demonstrate the ways in which competing geopolitical narratives of both evangelicalism and Americanism are intertwined.

---

2   http://uk.youtube.com/watch?v=rUpM42X-DCs (accessed 1/29/09).

This chapter continues with a substantive overview of the literature on narrativity and evangelical eschatology before critically engaging with contemporary Internet discussion threads that debate the possibility of Obama being the Antichrist.

Empirical evidence from several websites on which debates took place during the campaign is deployed in this chapter. In particular, quotes are drawn from two lengthy discussions found on www.anti-christ.com (as of Nov. 2008, this thread had been ongoing for nine months and had 1,759 postings) and on www.godlikeproductions.com (as of Nov. 2008 this thread had been ongoing for 11 months and had 59 pages of postings). Neither of these sites are associated with particular theologians, who often police the content of their website. Thus, these threads became key loci for evangelical debate about the taboo subject of Obama's Antichrist potential. The ensuing quotes are taken 'as-is', including many spelling and grammatical errors. Readers should remember that these posts were made in an informal online setting and therefore their casualness should not be judged to be indicative of intellectual rigor or political (in)significance. Given the subject matter of this chapter, it is perhaps not surprising that several of the posts quoted have racist or otherwise objectionable overtones. They are also provided 'as-is' to enhance critical engagement, and in no way reflect this author's beliefs.

Narratives, the Internet, and Conspiracy Theory

Narratives and Metanarratives in Dispensational Thought

Narrativity emerged as a major focus within the humanities and social sciences when the central role of narratives within processes of identity formation became apparent (White 1973). Walter Fisher (1987) has argued that people utilize two different forms of reason: *Logos* and *Mythos*. Logos refers to reason, or empirically-based decision making. However, Mythos refers to story. The narrative paradigm argues that people are more inclined to make decisions based on concepts of narrativity than is often publicly acceptable in traditional Western modes of thought. Narrative rationality involves evaluating the coherence of a story to see how well it fits with a person's own experiences, rather than comparing it to external, empirical benchmarks. The division between Logos and Mythos is an artificial distinction – as subjectivities (and the narratives that compose them) are always in flux as a result of new experiences, and these are always filtered through our subject positions.

The key to narrative rationality is an appreciation of individual events as occurring, not in isolation, but rather as episodes within a larger internally coherent plot. A plot can be understood as a series of relationships between people and things which are rooted in temporality and space. As one critic notes, "in the face of a potentially limitless array of social experiences deriving from social contact with events, institutions, and people, the evaluative capacity of emplotment demands and enables *selective appropriation* in constructing narratives" (Somers 1994, 617, emphasis in original).

There are, for the purpose of this analysis, three scales at which narrativity can be observed (Somers 1994). Ontological narrativity refers to the narratives of individuals who use these stories to define themselves both to themselves and others. These definitions are important as justifications for action and are thus open-ended and always in process as both new actions and experiences contribute to the 'leading edge' of who we believe ourselves to be and as old behavior is re-interpreted. The second scale at which identities can be understood is at the collective level. Public narratives are shared among a number of people, from small groups (such as a family narrative) to large ones (such as the American experience of the 9/11 attacks). Subjects view themselves as embedded in multiple public narratives, all of which purport to tell the story of groups of people. Powerful institutions such as governments and corporations often try to shape these narratives through mechanisms such as history books, memorials, and advertising.

The third scale of narrativity is metanarrative, which includes any narrative that claims the whole world as its subject. For example, the notion of human history as a never-ending conflict between Orient and Occident (e.g., Gress 1998) attempts to put all of human history into a universal narrative. Metanarratives are often the most unexamined form of narrativity because they are so fundamental to a subject's worldview. All three of these scalar narratives provide plots through which events are evaluated and through which subjects can understand their 'role': "Narratives help to construct personal and social identity, provide sense and order to experience, and frame and structure action" (Roberts 2006, 710). These narratives can be incredibly resilient in the face of contradiction.

Narrative rationality is not solely a feature of any one community. For example, the inauguration of Barack Obama as the first 'black' President of the United States was interpreted by newspapers and individuals as a significant episode in the American narrative. However, the role of narrativity in constructing dispensationalist identities should be readily apparent. Dispensationalism is, strictly speaking, a teleological metanarrative that purports to emplot individual humans in a story of iterative covenant-making and covenant-breaking. This plot is derived from various ancient source texts which have been collected together and bound as the Bible. Some of these texts purport to tell of the time during which the Earth was created and peopled; others tell the ancient history of the Jewish people, including the life of Jesus of Nazareth; yet others purport to be prophetic visions, some of which have been characterized as descriptions of the end of the world. Thus, the entire history (and future) of humanity on earth is framed as a series of dispensations, or covenants, between humanity and God, each of which humanity breaks.

The moral of the Dispensationalist metanarrative (and Christianity more broadly) is a constitutive binary – humans are flawed but an eternally forgiving God is not. This metanarrative is obviously important to other public and ontological narratives; evangelicals' personal and collective identities are rooted in this emplotment. However, missing from the source texts (it is usually believed) is a clear description of the current moment, leaving this open to a wide array of interpretations. Religion is, after all, an unstable signifier, meaning different things in different contexts (Ivakhiv 2006). This incredible flexibility of evangelical narratives has enabled a never-ending series of tactical shifts by those attempting to read the end times into current events:

> The participants in this discourse all keep the end near, but it seems never to quite arrive. There are always new possibilities for who will be the Antichrist or how the technology of the mark of the beast will be implemented. When confronted with new information or even a new historical event such as the attacks of 9/11, the participants in this discourse immediately can assimilate this new information into their worldview without confronting any challenges to the narrative structure in which they exist. (Howard 2006, 30)

It is critical to keep in mind then that individual subjects will ascribe different meanings and degrees of importance to elements of the metanarrative. Nevertheless, this metanarrative carries with it a variety of geographic consequences, as several of the authors in this volume have argued. Certain places figure quite prominently, such as Israel (Dittmer 2007), and others are less obvious in the source texts, such as the United States (Dittmer 2008). However, just because a place is not named in the source texts (as it is obviously impossible for the United States to be in a text written thousands of years ago), does not mean that it cannot be interpreted as part of the narrative, especially in texts that are understood to be prophetic and opaque to unschooled readers, such as the Books of Daniel, Ezekiel, and Revelation.

These narratives are thus highly contingent, varying in time and space even as elements remain incredibly resilient. One of the factors that influences narrative is medium, which brings its own particularities to the narrative; how has the Internet impacted the dispensational narrative? To answer this we must turn to a discussion of the role of the Internet in conspiracy theory.

Conspiracy Theory and the Internet

Scholars (Jameson 1988; Knight 2000) have pointed to globalization as a driving force in the construction of conspiracy theory, as existential insecurity caused by a democratic deficit in governance and a sense of instability in border practices (from the scale of the body to the international). Thus, as geographies are reworked at all scales by social processes often beyond the immediate experience of everyday, disempowered citizens, conspiracy theories emerge as ways to try to understand the world.

In an alternative to this structural understanding of conspiracy theory, postmodern scholars have elevated conspiracy theory's legitimacy via academia's continued interrogation of 'Truth' and the processes through which it is produced (e.g., Foucault 1977). Thus, conspiracy theory can be understood as one of many master narratives that are in competition at all times for hegemony; being labeled a conspiracy theory is therefore the mark of de-legitimation. Jones (2008) even identifies an argument for conspiracy theory as a privileged, ironic form of knowledge that takes as a starting point its own (and others') ultimate inadequacy.

It is not as an act of delegitimation that I bring conspiracy theory into a chapter about evangelical beliefs; rather it is because of certain similarities between conspiracy theory and certain forms of prophecy interpretation:

> [C]onspiracy discourse establishes causality and intentionality amongst circumstances and events that might ordinarily be considered random, coincidental or accidental; where everything is significant of something else and therefore taken as evidence of a larger whole. [...] Conspiracy is situated, contextual, and highly subjective, as individual theorists weave narratives from these variety of influences and sources of information to create an explanatory chronology of events. (Jones 2008, 43)

Much the same can be witnessed in the web fora analyzed in this chapter; posters are constructing a narrative of the end times from news media, the Bible, and each others' web presence in ways that draw on extant cultural resources, enabling them to reach the conclusion their faith predisposes them towards – that what they see around them can be tied together with a coherent apocalyptic narrative. For example, here is part of a post that argues for Obama's potential to be the Antichrist:

> According to Moody News (Moody Adams is a christian pastor I heard speak at my church, and have since followed his website), Chris Matthews (Hardball) speaking about Obama, "This is the new testament, I am confident we can create a kingdom right here on earth". A Chicago Art Gallery displays a sculpture depicting Obama crowned with a neon halo. Nation of Islam Minister Louis Faraakhan said "Presidential candidate Barack Obama is the hope of the entire world". The *Dallas Morning News* 2-26-08 "He is running a theological campaign- at some point he took off his arms and grew wings". His wife Michelle said " We have to fix our souls - our souls are broken in this nation." Per the *Jamacia Gleaner* 3-1-08, "The similarity between Obamian hope and biblical hope are extraordinary". Charlotte, SC 3-31-08, "He communicates God-like energy".[3]

As Jones (2008) points out, these are individual narratives that can be produced via any variety of paths through people, places, and events. Thus, depending on each narrators' subject position(s), race, religion, national identity, and any number of other possible elements might be a focal point of their reason to emplot Obama as

---

3 http://www.anti-christ.com/?p=25&cp=all#comment-1190 (accessed 11/21/08).

the Antichrist. As Tristan Sturm (2006) notes, this same narrative flexibility has existed historically regarding evangelicals' eschatological geographies.

Michael Barkun (2003, 18) argues that the relentless intertextuality visible in the above is characteristic of "improvisational millennialism". This form of theorizing draws on multiple, often contradictory sources of inspiration, such as the links between Nostradamus and the Bible found in posts such as this one: "Nostradamus prophesized the name of the antichrist is 'Mabus.' This can relate to why Obama is the antichrist. If you write obama+bush, you end up with obamabush. Do you spot 'mabus' in the middle? Bingo. This is a strong indication that Obama will succeed Bush as president in 2008, further confirming the prohpecy."[4] This is but one of a myriad of posts that involved the occult seer Nostradamus in what is ostensibly a Christian apocalyptic narrative.

Barkun goes on to attribute the recent rise of improvisational millennialism to a politico-technological assemblage that incorporates the decline of traditional authority and the technical capacity to bring a variety of texts into relationship with each other. Similarly, Laura Jones (2008, 43) argues that "the Internet has now provided individuals with instant access to an overwhelming network of information, opinion, rumour and knowledge through which to 'evidence' their conspiratorial claims, as well as fuelling the formation of new ones". Indeed, the hyperlinked nature of the Internet, combined with the growth of broadband capability and the Internet's consequent flowering as a multimedia platform, have resulted in a digital environment that allows these disparate elements to be interwoven in the construction of conspiratorial narrative.

Emplotting the United States in Dispensationalist Eschatology

The United States has long been the subject of prophetic scrutiny. The founding of the American colonies was accompanied by postmillennial emplotment of the 'New World' as a new Jerusalem, one that would have a special relationship with God similar to the one claimed by ancient Israel, if only sin could be exterminated. Christopher Collins (2007, 117) quotes Francis Blake's 1796 Independence Day speech:

> For behold, ye who have been exalted up to heaven, shall, ere long, be cast down to hell! The final period of your crimes is rapidly approaching. The grand POLITICAL MILLENNIUM is at hand, when tyranny shall be buried in ruins; when all the nations shall be united in ONE MIGHTY REPUBLIC!

This early belief in the United States as a hearth of righteousness unto the world remained strong until it began to decline following the bloodshed and trauma of the Civil War. President Woodrow Wilson's debilitating stroke and removal from public life marks perhaps its final gasp of hegemony in American foreign policy

---

4 http://www.godlikeproductions.com/forum1/message486743/pg1 (accessed 11/21/08).

– his view of the First World War as a final, apocalyptic battle to end war on earth remains the last strong expression of this narrative in public life, although more secular versions of it continue into the present in the form of technological and other forms of utopian modernity.

Subsequently, the experience of the Depression, the Second World War, and the onset of a more consumerist domestic culture began to undermine the narrative coherence of the older prophecies, convincing believers that the United States was moving further away from its potential as a New Jerusalem (Boyer 1992). Instead of a New Jerusalem, the United States was more likely to be compared to a New Babylon, the villainous nation of the Jewish prophetic books. The sexual revolution, feminism, and the rise of homosexual culture in the United States all played their part in the rising sense of alarm and pessimism among evangelicals and their leaders, as illustrated in this recent quote from an online forum: "America brought this on itself. Since we are all high and mighty and belive that we are so saved. [...] God gave us a promisland filled with milk and honey and we turned it into a whore house for the world..".[5]

However, Boyer, drawing on the work of O'Leary and McFarland (1989), argues that the fusion of secular belief in American exceptionalism and religious narrative offered a way out from the country's present sinfulness in a way that harkened back to Puritan jeremiads: "[Pat] Robertson in [his writings from the mid-1980s] offered a *Heilgeschichte*, or salvation-history, of America, in which discovery and settlement, Declaration of Independence, Revolution, and Constitution all function as sacred events and texts," (Boyer 1992, 240). The notion of the United States as a chosen country was also perpetuated by a contrast during the Cold War with the 'godless' Soviet Union.

The contest between these two narratives, the jeremiad-based postmillennialism and the more pessimistic premillennialism, continues to this day (on jeremiads and American identity, see also Campbell 1992). It could be argued that this is maintained by a disconnect between the optimistic religio-political rhetoric of politicians, which often taps into Americanism as a civil religion, and the more pessimistic discourse of premillennialism that is more dynamic theologically. As alluded to earlier, the notion of the United States as a redeemer of the world has, in many circles, become secularized. Despite being stripped of its overt postmillennialism, this civil religion nevertheless is dependent on postmillennial rhetoric. Jewett and Lawrence (2003, 4) cite several speeches by presidents (or men who would become president) that do so: "America had the infinite privilege of fulfilling her destiny and saving the world," (Woodrow Wilson); "[O]ur beliefs must be combined with a crusading zeal, not just to hold our own but to change the world...and to win the battle for freedom," (Richard Nixon); "This will be a monumental struggle of good versus evil. But good will prevail," (George W. Bush). All of these quotes hint at the role of the United States as a redeemer for

---

5   http://www.anti-christ.com/?p=25&cp=all#comment-1023 (accessed 11/19/08).

the world through the universal export of Americanism, rather than a symptom of the world's decadence.

Given all these ambiguous understandings of the role of the United States in the end of days, it was possible for the country to be found in the plethora of passages in the prophetic books of the Bible. The United States has alternately been described as one of the "young lions" of Ezekiel 38:13 who fight against Gog and Magog on behalf of Israel (the other lions being the 'white' dominions of the British Empire) or as the "isles" of Ezekiel 39:6 whom God will cast fire upon in his war against Gog and Magog, among other interpretations. Most prophetic authors, however, maintained that neither the United States, nor any other country in the Western Hemisphere, was referred to in Bible prophecy except in common with other corrupt and hedonistic countries (Boyer 1992).

While strands of postmillennial thought that maintain American exceptionalism have remained alive in contemporary political discourse, by far the stronger dimension of American popular eschatology in recent decades has been the pessimism of premillennialism. Given the assumed proximity of the Second Coming, based on readily-apparent 'signs of the times', most prophetic writers have scripted a narrative in which American hegemony is the final stage of Church Age geopolitics, only to be ended by the rise of the Antichrist and his renewed Roman Empire (Sidaway 2006). Alternatively, the doctrine of the Rapture has been used in online discourse to open up the possibility of Christian America being excised from the world, thus enabling the rise of the Antichrist: "'We' Americans are not even around in 'end times' [having been Raptured], so therefore it may not really matter who wins the election."[6] This perspective both enables an optimistic view of Christianity as a force for good in the present, but a country that will be unmentioned in the future narrative of the Antichrist's rise to power. It is to this Antichrist figure that this review now turns.

Temporality and the Role of Antichrist in the Apocalyptic Narrative

The Antichrist is a figure with deep roots historically, and it is important to examine these roots to understand the contemporary context in which his identity is sought. The Antichrist is not only an eschatological figure, but also an apocalyptic one. The eschatological and the apocalyptic are not entirely congruent concepts – while eschatology focuses on the end of days, the apocalyptic tradition is about all the secrets that God has unveiled. These secrets are generally understood to include the prophecies of the end times, but also include the historical narrative of creation – both looking back to our divine origins and also forward to our earthly finale (Emmerson 1981).

This linear narrative trajectory is punctuated by occasional divine interventions, such as in the incarnation of Jesus Christ (still today the fundamental pivot of chronology, constituted by BC and AD). A divine order is understood to be behind

---

6   http://www.anti-christ.com/?p=25&cp=all#comment-813 (accessed 11/19/08).

all of history, and in the apocalyptic tradition numerical patterns are often seen in events. For example, the dispensational system of seven covenants between God and humanity can be traced to apocalyptic numerology from the medieval era: "The six days of creation delineate historical epochs – periods of time within history – whereas the seventh day, the Sabbath when the creator rested, symbolizes human rest beyond historical time" (Emmerson 1981, 16). The delineations of history have varied from Augustine, to Isadore, to the Venerable Bede, and so on until John Nelson Darby, but it was the Venerable Bede who added an oscillation within each era, from joy to lamentation; from good to evil. Thus, the previous age began with the rebuilding of the Jewish temple after the Babylonian Captivity and ended with Antiochus Epiphanes, the Seleucid king who sacked Jerusalem and attempted to Hellenize the Jews. The current age began with the goodness of Christ and must conclude with the evil of Antichrist. The Antichrist then, is not just a figure resulting from the 'literal' interpretation of scripture as contemporary evangelicalism generally holds, but is instead a holdover from older, medieval understandings of Biblical exegesis (Fuller 1995).

Acting as a narrative foil to the perfect Jesus, the Antichrist has long fascinated believers. However, there is little actual scripture from which to draw for understanding. In the New Testament, the term is only used in the epistles of John (Fuller 1995). The first reference introduces the idea of the Antichrist-type, those whose existence indicates the nearing of the end times: "Little children, it is the last time: and as ye have heard that Antichrist shall come, even now are there many Antichrists; whereby we know that it is the last time" (1 John 2:18). Nevertheless, as the quote indicates, beyond the Antichrist-types is an actual Antichrist who will exceed all others. Verse 22 marks the addition to the apocalyptic narrative that the Antichrist will deny the divinity of Jesus: "Who is a liar but he that denieth that Jesus is the Christ? He is antichrist, that denieth the Father and the Son." This narrative role of the Antichrist as a deceiver who will weaken the Church is emphasized in 1 John 4:3 ("And every spirit that confesseth not that Jesus Christ is come in the flesh is not of God: and this is that spirit of antichrist, whereof ye have heard that it should come; and even now already is it in the world") and 2 John 1:7 ("For many deceivers are entered into the world, who confess not that Jesus Christ is come in the flesh. This is a deceiver and an antichrist"). These passages refer to itinerant preachers who claimed to have a direct connection to the Holy Spirit in the same way that Jesus did – indeed, these 'antichrists' are generally understood to have been Christians themselves who taught 'heretical' beliefs, namely, that Jesus was not special (Fuller 1995). Collectively, these quotes are the only references to the Antichrist (and the Antichrist-type) by name in the New Testament (Emmerson 1981).

The belief reflected in those passages that the Antichrist was afoot and the end times were near has been true from the earliest believers until the present. However, the passages have always been understood intertextually, with current events being read through the lens of the Bible and the Bible being understood in relation to current events. For example, the Emperor Nero was interpreted by many Christians as an (or the) Antichrist, in part because some historiography links him to the initial persecution of Christians and the martyrdom of both Peter and Paul. Indeed, an oral tradition lasted for some time throughout the Roman Empire that Nero would return from the dead, as a mirror image of Jesus would have to in order to maintain the narrative (Fuller 1995).

However, the narrative understanding of the Antichrist has not been limited to the overt textual references. Biblical scholars have long expanded their characterization of the Antichrist to include other passages, such as those referring to the "son of perdition" quoted in this chapter's epigraph. 2 Thessalonians 2:3–11 has been interpreted to mean that the Antichrist will rebuild the temple in Jerusalem and anoint himself inside it as a god to be worshipped. The only thing restraining him has been Roman influence (variously interpreted in different times and places to be the Roman Empire, the Roman Catholic Church, or various extensions of those institutions, such as the Holy Roman Empire), but when the end times are nigh he will be empowered by Satan to perform false miracles, all as part of God's plan to test humanity. However, he will be defeated by the return of Jesus, who will then usher in the new age (continuing Bede's apocalyptic pattern by oscillating from the darkness of Antichrist to the brightness of Christ).

The apocalyptic narrative has been further constructed from scripture by those who have interpreted parts of the Book of Revelation as referring to the Antichrist, albeit even more obliquely than in 2 Thessalonians and the epistles of John. Here, the Antichrist is seen in the character of the seven-headed beast. From Chapter 13 of the Book of Revelation are taken a number of elements of Antichrist's role in the apocalyptic narrative: he will rule for forty-two months, he will have a prophet (the two-horned beast of verse 11), he will give his mark to those who are deceived (and those lacking it will be unable to buy or sell), and he can be recognized by his number, 666 (alternatively, the number of the Beast is sometimes identified as 616, see Emmerson 1981). This requirement of the mark in order to buy or sell has intersected throughout European history with anti-Semitism, as Jews were resented for their emergent economic niche in urban commerce. This re-emerged at the beginning of the twentieth century with the forgery that is the *Protocols of the Elders of Zion*.

Biblical scholars have been able to 'snowball' their understanding of the Antichrist to recognize him in other books and passages. For instance, Genesis 49:16–17 ("Dan shall judge his people, as one of the tribes of Israel. Dan shall be a serpent by the way, an adder in the path, that biteth the horse heels, so that his rider shall fall backward.") has been used to claim that the Antichrist will be a Jew from the Tribe of Dan, and John 5:43 ("I am come in my Father's name, and ye receive me not: if another shall come in his own name, him ye will receive.") has been used to claim that the Jews will generally reject Christ in favor of Antichrist when the time comes. Perhaps most critical to the construction of eschatological narrative is the Book of Daniel, which has been much debated by Bible scholars who have sought to use these passages to stake out a narrative timeline and ordering in which end times events will take place (Emmerson 1981). This has led to a wide variety of interpretations, as recognizable in the premillennial, postmillennial, amillennial, and other theological positions described elsewhere in this volume.

The American Search for the Antichrist

Over the last two thousand years many have been identified as the Antichrist, including the aforementioned Nero, the later Caligula, various popes, Napoleon, Hitler, Henry Kissinger, Vladimir Putin, Javier Solana, Prince Charles, and Saddam Hussein (Fuller 1995; Dittmer 2008). In the past three hundred years or so, however, much of this speculation has originated from the area now known as the United States. The original nonconformist settlers in the English colonies of the east coast had much stronger belief in the apocalyptic than the Anglicans they left behind in Great Britain. Wary of the ecclesiastical hierarchy in the Anglican Church, they sought a new place from which to start over, and they saw themselves as playing a key role in the apocalyptic narrative by rejecting the 'papist' habits of the Anglicans. The providence which God showed them was contingent on continued faithfulness but nevertheless could be counted on in that context. Thus, the first colonists equated the Antichrist with the hierarchy and ornamentation of the established Church of England, and more obviously the Roman Catholic Church that epitomized (to them) the corruption of God's word.

Later, Native Americans would be identified as Antichrist-types themselves, because the colonists were soon attracted to the wilderness and lifestyle of the nearby tribes, forsaking their covenant with God and each other. Colonial leaders believed that Satan had lured the Native Americans there to ruin their plan for a New Jerusalem. Initial attempts to Christianize the local tribes soon gave way to holy war. The same struggle between good and evil was directed domestically during the Salem Witch Trials, in which women who were associated with the more commercial elements in town were being driven out by the more theocratic ones (Boyer and Nissenbaum 1974).

The American Revolution provided a new external target—King George III, the nominal head of the Anglican Church and the functional head of the empire that was oppressing the believers. The Stamp Act of 1765, in which all legal documents and contracts in the Colonies were required to pay a tax in order to receive a stamp, was quickly identified by colonists as the Mark of the Beast (which is supposedly required to buy or sell in the end times), thus strengthening the anti-British fervor. Following the successful Revolution, "[t]he millennium awaited the building of a Christian Commonwealth in the new republic. The only task that remained was to identify the lingering pockets of Antichrist's resistance" (Fuller 1995, 73).

A period of relative quiet in Antichrist-spotting masks the rise of premillennialism during the nineteenth century as a rival for the previously common postmillennial perspective. However, the ebb tide of Antichrist-spotting began to turn in the 1880s as the United States began to go through the changes associated with modernity. Anti-immigrant and anti-Catholic sentiment went hand-in-hand, especially in an atmosphere drenched in the fear of industrial society. The rise of megacities, ethnic diversity, Darwinist thought, and liberal Bible scholarship all fueled an angry rejection of American society by most premillennialists, even as many postmillennial groups turned towards the social gospel and the amelioration of modern inequality. The Antichrist was deemed to be afoot, and embodied by these forces that sought to deny the new premillennial orthodoxy.

During this period figures symbolic of the industrial and diverse America, were labeled Antichrist: immigrant Jews, who brought with them union demands and also quickly integrated into the 'polluting' entertainment industries; Franklin D. Roosevelt, who established diplomatic relations with the U.S.S.R. and was thought by some to be Jewish; John F. Kennedy was suspicious because of his 'papist' faith, as were several popes of the last century, including Paul VI and John Paul II; during the Cold War various Soviet premiers were suspected, including Stalin and Khrushchev; Henry Kissinger was suspected not only for advocating detente but also for being Jewish (Fuller 1995).

Of particular interest to the subject of this paper is the more recent focus on Islam as a source for the Antichrist. While Muslims have been identified as Antichrist-types since the inception of Islam because of the religion's disavowal of the divinity of Jesus, this perspective began to intensify in the United States in the 1970s with the onset of the oil crisis. This also dovetailed with Arab opposition to Israel, which also marked Israel's hostile neighbors as nations likely to be involved in the war described in Ezekiel 38. Of course, peace efforts in the region were equally a mark of Antichrist, as it is generally understood by premillennialists that the Antichrist will make a peace treaty with Israel and eventually break it. Hence, at varying points the Ayatollah Khomeini, Yasser Arafat, Anwar Sadat, Yitzhak Rabin, and Saddam Hussein have all been tagged 'Antichrist' (Boyer 1992; Fuller 1995). Since the 11 September 2001 attacks, Islam has been demonized within apocalyptic discourse, most overtly by Jerry Vines and Jerry Falwell, who over the summer of 2002 announced variously on CNN's *Crossfire* and CBS's *60 Minutes* that Muhammad was a pedophile, genocidal anti-Semite, and a terrorist (Collins 2007).

With that, the claims described at the beginning of this chapter, that Barack Obama is a Muslim and the Antichrist, has come into view. Throughout history, the naming of Antichrist has served as a deflection of anxiety and a bulwark for threatened identities. The very first "anti-christs" described in the epistles of John were deemed outside the 'real' Christian community, as were their followers. Similarly, the postmillennial belief in the newly founded United States as a 'New Jerusalem' and a 'shining city on a hill' intertwined American identity with evangelical hopes of redeeming the world for Jesus. Those who questioned this notion were cast out, as in the case of the Salem Witch Trials, which staved off (for a time) the redefinition of Americanism along capitalist rather than theological lines. Thus, American apocalyptic and geopolitical narratives become mutually imbricated and difficult to isolate from one another. Antichrists in the more recent past, during which dispensationalism has become hegemonic among the evangelical community, have tended to be either domestic threats to fundamentalist identity (e.g., feminists and secularists) or foreign threats to American geopolitical power (such as the 'godless' Soviets or, more recently, Islamist movements).

Barack Obama and the 2008 U.S. Presidential Election

It is key here to reiterate the diversity of evangelicals and evangelical thought. In particular, a major divide can be seen to exist between the white evangelical tradition and the African American evangelical tradition. Ehrenhaus and Owen (2004) argue that the white evangelical tradition is founded primarily on the primacy of personal salvation from sin among the various possible morals drawn from the apocalyptic narrative, located specifically in the Old Testament's calls to sacrifice and the New Testament's blood sacrifice of Jesus Christ. However, African American evangelicals are historically more likely to emplot themselves with the character Jesus Christ as both persecuted and favored by God, than to focus on the actual plot of expiation. These different emplotments come together, they argue, in the Jim Crow-era performance of lynching, in which African Americans accused of crimes were sacrificed for the 'good' of the community, as Jesus had been. This finding points towards the complex intersection of race and religion in American narratives.

Race, Religion, and American Identity

Let us consider the contributions of race, religion, and American identity to the evangelical case for Barack Obama to be the Antichrist. Race is a much-discussed element of Barack Obama's election, whether through the narrative of the first 'black' president or whether through the narrative of his mixed-race background as a curative for the wounds of slavery and prejudice. Race in evangelical thought is very complex, but in the apocalyptic narrative modern categories of race and ethnicity are descended from particular individuals. For example, Arabs have been described as the descendents of Abraham's son Ishmael, and Africans the descendents of Noah's son Ham. By linking whole races, ethnic groups, and religions to one person, especially one alternately favored by God or cursed, it becomes possible to justify the political and economic dominance of one ethnic group over another as simply the effects of God's will (Collins 2007). In the case of Africans, their blackness was seen as an outward manifestation of their curse (in their case, Ham's refusal to cover his father's drunken nakedness). The same was done in online fora in regards to Muslims:

> all muslim's are decendents of ISHMAEL son of the slave hagar. ISHMAEL whom GOD hated,now you know why those people are evil and act as they do.obama is the antichrist,mohammad was a child molester had sex with

a 6 year old child then married a 9 year old girl.what type of person follows someone(something) like this? evil followers of the devil that's what.[7]

Thus, in white evangelical thought there is a tradition of essentializing races and other identities. Needless to say, white skin has been associated with purity – where Africans were believed to be descended from Ham, Europeans (and hence Euro-Americans) were believed to be descended from another of Noah's sons, Japheth, whose name was translated in the American South as 'fair' or 'beautiful' (Collins 2007). America's racial hierarchy was thus rooted in God's will, with whites destined to rule over the cursed Hamitic blacks. Of course, it has been not only evangelicals, but indeed most whites, who have historically believed in, and reinforced, America's racial hierarchy, for a variety of religious and secular reasons, most often rooted in the perpetuation of politico-economic hegemony.

Therefore, the possible rise of a prominent African American politician to the presidency threatened not only the racial order, but also the apocalyptic racial narrative in which many evangelicals have personally invested. Alongside the need to reinforce the racial order is a more anxious, inward-looking concern that white America's 'original sin' (as slavery has been commonly referred to) is yet to be paid for, and this might be the moment. The essentialization of Obama as 'black man' left him no role to play in the narrative except the vengeful force demanding blood sacrifice from (white) America.

Given this, it should not be surprising that an initial post beginning a long discussion thread devoted to the question of Obama's Antichrist status focused on his racial characteristics, playing off of both his 'blackness' but also his existence as the result of 'miscegenation':

> the NOI [Nation of Islam] prophesied the final muhommed (a muslim) would come into the world from the union of a white female and black male based on their interpretation of the story of Mary Magdelin and Jesus Christ; being white and black respective. Obama fulfills this. Obama hails from chicago whose zipcode is **60606** (do you see the three sixes ?) Obama would be a "black" president in the "white" house (satan is described as black in attribute and who seeks to take over the white mansion known as heaven).[8]

Later in the same thread (many posts of which mocked the original post), another author posted support for the original post:

> The ANTI CHRIST will come into power over a nation out of the tribe of Dan. This nation will be the nation that clenses the holy lands of evil and rebuilds the new Isreal. The tribe of dan was the last tribe to be awarded land in the book of

---

7 http://www.anti-christ.com/?p=25&cp=all#comment-941 (accessed 11/19/08).
8 http://www.godlikeproductions.com/forum1/message483746/pg1 (accessed 11/18/08).

> Numbers from the Bible. This tribe was given land which is now considdered [sic] AFRICA... The tribe of Dan were black people. The ANTI CHRIST will come to a new land to give its people HOPE!!!!![9]

This focus on the race of Obama has had well-known consequences regarding his personal security. The reason for this concern was obvious in the message board: "Well if the asswipe is elected president I just pray for him to go the way of Lincoln and Kennedy".[10] The possibility of Obama being assassinated was acknowledged by many (although outright advocacy as in the last quote was rare), but was seen as yet more confirmation that he might be the Antichrist because the Antichrist is expected to be killed but then rise again three days later in order to serve as the narrative mirror of Jesus: "i beleive that if obama gets shot than its him the anti christ and if he reincarnates himself lets just keep shooting the bastard and see how many times he can come back to life and make him mad."[11]

At times the discussion ranged away from theological issues and instead became a clearinghouse for racist anger at the Obama candidacy:

> If we end up with a president whose child has dreds, as obama's does, I'm moving out of the fucking country. Yhe [sic] last thing we need is a couple of little niglets running around our WHITEHOUSE. It would never work anyway, because there is no way the secret service is going to escort the obama family to walmart at 3 am everytime he gets paid.[12]

Certainly these views were not universally shared ("Good! Move away as quickly as possible! The country is better off without biggots like you! Wrap your self in your Confederate flag and get out!"[13]) and indeed the racist element in the thread was often marked as doing the work of Satan, as in this post:

> The Devil is tricking good Christians into thinking Obama is the Antichrist in an attempt to further divide the people. The result of this fear will put a wedge in the unity America has strived to achieve between races, religions, and ways of life. Just because he isn't a White individual doesn't mean he doesn't deserve a chance to make a difference by utilizing the political system of this great country we call America.[14]

---

9 http://www.godlikeproductions.com/forum1/message483746/pg4 (accessed 11/18/08).
10 http://www.godlikeproductions.com/forum1/message483746/pg7 (accessed 11/18/08).
11 http://www.godlikeproductions.com/forum1/message483746/pg9 (accessed 11/19/08).
12 http://www.godlikeproductions.com/forum1/message483746/pg10 (accessed 11/19/08).
13 http://www.godlikeproductions.com/forum1/message483746/pg11 (accessed 11/19/08).
14 http://www.godlikeproductions.com/forum1/message483746/pg12 (accessed 11/19/08).

Thus, race formed an important component of the online debates about Obama's potential to be the Antichrist. This occurred either because his skin was seen as a mark of affiliation with evil and otherness, or alternatively, because his race was seen as tangential but nevertheless related to the eschatological narrative, for instance as a reason for an assassination attempt. Many posters argued that many 'white' candidates for the Antichrist have been offered up over time, and that therefore race was of minimal importance. However, in the case of Obama, race seemed to be implicated in the debate as a replacement for characteristics traditionally attributed to the Antichrist, such as arising from the former Roman Empire (typically used to denote a European or someone of European or Jewish descent).

Another long discussion thread about Obama's Antichrist possibilities highlighted the role of religion in emplotting Obama in the eschatological narrative. This connection between religion and American identity should not be surprising given the above discussion of America in premillennial apocalypticism, and as mentioned in the introduction to this chapter much of this is dependent on the belief that Obama is a secret Muslim. This was a taken-for-granted fact in the discussion thread, and was often combined with a focus on the exotic 'Arab' sound of his name to indicate an essentially un-American nature: "He is of middle eastern and african descent. He is a muslim, his names are close/same as two previous antichrists, Osama and Saddam Hussein. Is this all a conincidence? [sic]"[15] In Internet claims such as these, historical claims that the Antichrist will arise from the Jewish tribe of Dan are replaced by a new focus on Islam.

Obama's name also featured in Gnostic interpretations of scripture. These forms of prophecy interpretation imply, in contradiction to most mainstream evangelical thought, that the Bible cannot always be read 'literally', as it has truths encoded within it that can only be unveiled through a process of code-breaking. Often these codes revolve around connecting the Number of the Beast, which is commonly understood to be 666, to various things associated with Obama:

> what is the matter with all these people that are voting for obama does'nt the name tell us something are they all been brain wash by this anti christ man that came out of nowhere to run for president of our country his name Barack has 6 letters 2008 the month of november when the elections come has 8 letters here is the scary part if you add 2–8 for our year it becomes 10 now november has 8 letters if you add 10 and 8 it.s 18 now divide it by 3 you get 666 scary huh well all those supporting obama are not in there right mind.[16]

Other formulations that connected Obama to the Number of the Beast were the number of letters in his name (Barack Hussein Obama = 6+7+5 = 18, and hence 666) and also a ZIP code for Chicago (60606).

---

15 http://www.anti-christ.com/?p=25&cp=all#comment-800 (accessed 11/19/08).
16 http://www.anti-christ.com/?p=25&cp=all#comment-817 (accessed 11/19/08).

While these numerological games have been played before with many potential Antichrists, Obama's name proved to be something of an obsession for many online Antichrist-spotters, who in particular elevated his middle name to the level of a 'sign of the times' all on its own, as in this post by a European evangelical:

> When I first heard on TV, somebody with a name similiar to "Hussein Osama" is running for American president I thought it is a joke. You know... those independant candidates without any chance to win, but with some crazy agenda. I was shocked when I realized that he is a serious candidate. I would have never thought Americans vote for somebody with such a name.[17]

The disbelief that someone with the middle name "Hussein" could potentially achieve the presidency was rooted in a geopolitical imagination that emphasized the fundamental differences between identities. Obama's Muslim heritage was not only seen as cause for suspicion but also as an unalterable element of his identity:

> Muslims are also able to deny their faith to accomplish the goal for their faith, but Christians are taught to die for theirs never deny Christ no matter what. Obama's stepfather was a Muslim extremist, and his father from Africa but, he was Muslim. Obama's mother although Atheist had a strong attraction for these sort of men. We are by nature a product of our enviroment. Obama attended a Muslim school in his early childhood, these are the most important years of influence.

Thus, Obama's name served as a signifier of otherness that was easily connected to his skin color and other personal attributes, such as his international upbringing, his 'blackness', and even the religion attributed to him, Islam.

However, it is important to point out that these were an always-shifting set of claims regarding Obama. Most posters would criticize any racial basis for his Antichrist-hood, while others would admit that they did not "know" whether he was Muslim or not. This allowed for a variety of different arguments to co-exist rather comfortably in what Howard (2006) has referred to as a plastic narrative. For example, here is one African American poster who rejects the racist basis for concerns about Obama but nevertheless remains concerned:

> God made me as black as they come. But aint none of us black or white or any color on the inside. Still I feel what you saying. [...] But one thing is for sure. He is different, and his promises are different. Yes he is black. But that isn't the different part i'm talking about. He might be the first U.S. president who went

---

17 http://www.anti-christ.com/?p=25&cp=all#comment-925 (accessed 11/19/08).

to a Muslim school and has a Muslim name in a time when radical Muslims are calling for the destruction of america.[18]

However, the plasticity of the narrative is rooted fundamentally in an anxiety of the other, as evidenced by the litany of groups in the beginning of the following post – note that the litany includes both Islam and African Americans twice while trying to describe Obama's popularity:

> It is not about color. It is about acceptance. Obama is accepted by many. Democrats, blacks, whites, muslims, catholics, protestants, black supremacy groups, Islam and ... HE IS THE ONE just ask Oprah, the highest paid entertainer in the U.S.. A black woman. Again, it is not about race. It is about position. Positioning himself to take over ... the country ... the continent, the world!![19]

The centrality of race and religion in the collective formulations of Obama as Antichrist is obvious, even as individual posters were able to deny one or the other element. Thus, even as the posters reflected widely divergent perspectives and subject positions, the overarching (tentative) emplotment of Obama within the eschatological narrative remains intact.

Contesting Orthodoxy, Constructing Narratives

This chapter has shown how a temporary crystallization of populist evangelical narrative around the notion of presidential candidate Barack Obama being the Antichrist emerged through a variety of 'signs of the times' dovetailing with various overdetermined cultural signifiers, such as race and religion, in a pastiche that represents what was described earlier as improvisational millennialism. Obama's seeming 'otherness' for many evangelicals led them to slot him into a preexisting apocalyptic and eschatological narrative as the right hand of Satan. However, this chapter's empirical focus on Internet discourse has occluded the views of evangelical elites, who fiercely contested the notion that Obama might be the Antichrist. On 4 March 2008, CNN host Glenn Beck asked evangelical pastor John Hagee if Obama could be the Antichrist. The answer: "No chance." Similarly, in the pages of the *Washington Post* (Boorstein 2008, n.p.),

> The debate over the [McCain ad, 'The One'] has gotten so intense that Tim LaHaye and Jerry Jenkins, the authors of the "Left Behind" series, issued a statement clarifying that Obama is likely not the Antichrist in the Book of Revelation. "I've gotten a lot of questions the last few weeks asking if Obama

---

18 http://www.anti-christ.com/?p=25&cp=all#comment-1129 (accessed 11/19/08).
19 http://www.anti-christ.com/?p=25&cp=all#comment-1130 (accessed 11/19/08).

is the Antichrist," Jenkins told Christian Newswire. "I tell everyone that I don't think the Antichrist will come out of politics, especially American politics."

Of importance here is the notion that the United States is not a place from which the Antichrist will rise. Most dispensational elites' geopolitical narratives of the end times, as described above, have at most a tangential role for the United States. Thus, internet-based emplotments of Obama in the role of Antichrist puts these institutional elites in the precarious position of de-legitimizing these populist theories of the end-of-days without alienating their free-thinking followers (a personal relationship with the Bible being a central Protestant tenet). Consequently, the debate over Obama is not simply one about the geography of the Antichrist, but is also one over power, authority, and systems of knowledge legitimation.

This is not simply a binary division between elites and *hoi polloi*; somewhere in the middle is a vast segment of 'populist elites' who serve as leaders and occasional spokespeople for the evangelical masses. For example, Todd Strandberg, who runs the popular *Rapture Ready* website for evangelical prophecy-watchers, was recently quoted in *Newsweek* on the Obama phenomenon: "Strandberg says Obama probably isn't the Antichrist, but he's watching the president-elect carefully. On his Web site, he has something called the Rapture Index, a calculation based on signs and prophecy of the proximity of the end. According to Strandberg, any number over 160 means 'fasten your seat belts.' Obama's win pushed the index to 161" (Miller 2008). Despite this, the *Rapture Ready* website does not allow supposition about the identity of the Antichrist in its fora, viewing this as not only unfaithful (as a pre-tribulational premillennialist website, Strandberg believes that no real Christian will know who the Antichrist is because they will be Raptured before he is known) but also embarrassing when read by unbelievers. In the case of Obama, a special reminder[20] had to be posted to warn off would-be Antichrist-spotters.

Implications

This case study has implications for the study of evangelical geopolitics beyond simply the case of Barack Obama. First, it is a reminder of the role that power, both institutional and technological, plays in the construction of narrative hegemony. The politico-technological assemblage brought into existence by the evangelical community on the Internet enabled a contestation of elite knowledge that would have been unthinkable prior, or that would have played out in different, less immediate ways. Similarly, Strandberg's institutional role as webmaster enables him to control the acceptable discourse on his website. While other, more traditional evangelical elites also have a web presence (such as Tim LaHaye's *Left Behind Prophecy Club*, see Dittmer 2008) they no longer dictate the boundaries of acceptable discourse; instead we see a fragmenting of the evangelical community as those who seek more freedom to speak their mind find common cause with

---

20 http://www.rr-bb.com/showthread.php?t=67300 (accessed 11/21/08).

each other on particular websites (like www.godlikeproductions.com and www.anti-christ.com).

A second implication is that the evangelical community is far more diverse than is commonly understood. In this case study we have seen a variety of different subject positions all united by their interest in Obama as the Antichrist. Further, posters drew on a variety of cultural resources that would typically be seen as beyond the belief system of Christianity, such as Nostradamus and numerology. Categorizing evangelicals through a particular theological definition will certainly provide an artificial boundary and fixity to the group, but will miss out on the 'fringe' that participates in Barkun's improvisational millennialism.

Finally, the role of race, religion, and national identity in American identity has been confirmed, although not in a clear causal fashion. Instead what is visible in the case study is an overdetermined link between these factors, in which some elements can be missing and the conclusion can still be reached. More traditional, modernist histories and geographies of identities often lose sight of the overdetermined processes that maintain identities in favor of simpler causal description. The multiple paths through a wide array of cultural resources enable a diversity of personal narratives that nevertheless coalesce into a massive Internet-based phenomenon that surely had influence among a great many of voters in the 2008 election. Evangelical geopolitics is therefore not just about the relationship between biblical texts and foreign policy; it is instead embedded at a variety of scales, from the grand strategy of Christian Zionism, to the scale of the nation and its public narrative, to the scale of the individual in the ballot booth. Of course, it is not only the evangelicals found on the websites featured in this chapter who are negotiating these cultural resources; a variety of different subjectivities can be brought to bear on these decisions, enabling new and previously unthinkable political possibilities – such as Obama's landslide election.

References

Barkun, M. (2003), *A Culture of Conspiracy: Apocalyptic Visions in Contemporary America* (Berkeley: University of California Press).
Boorstein, M. (2008), '"Left Behind" Authors: Obama Not Antichrist', *Washington Post*, 14 August. <http://voices.washingtonpost.com/the-trail/2008/08/14/left_behind_authors_obama_not.html>, accessed 21 Nov. 2008.
Boyer, P. (1992), *When Time Shall Be No More: Prophecy Belief in Modern America* (Cambridge, MA: Harvard University Press).
Boyer, P. and Nissenbaum, S. (1974), *Salem Possessed: The Social Origins of Witchcraft* (Cambridge, MA: Harvard University Press).
Campbell, D. (1992), *Writing Security: United States Foreign Policy and the Politics of Identity* (Minneapolis: University of Minnesota Press).
Collins, C. (2007), *Homeland Mythology: Biblical Narratives in American Culture* (University Park, PA: Pennsylvania State University Press).

Dittmer, J. (2007), 'Of Gog and Magog: The Geopolitical Visions of Jack Chick and Premillennial Dispensationalism', *ACME: An International E-Journal for Critical Geographies* 6:2, 278–303.

—. (2008), 'The Geographical Pivot of (the End of) History: Evangelical Geopolitical Imaginations and Audience Interpretation of *Left Behind*', *Political Geography* 27(3): 280–300.

Ehrenhaus, P. and Owen, S. (2004), 'Race Lynching and Christian Evangelicalism: Performances of Faith', *Text and Performance Quarterly* 24:3, 276–301.

Emmerson, R. (1981), *Antichrist in the Middle Ages: a Study of Medieval Apocalypticism, Art, and Literature* (Manchester: Manchester University Press).

Fisher, W. (1987), *Human Communication as Narration: Toward a Philosophy of Reason, Value, and Action* (Columbia, SC: University of South Carolina Press).

Foucault, M. (1977), *Discipline and Punish: the Birth of the Prison* (London: Allen Lane).

Fuller, R. (1995), *Naming the Antichrist: The History of an American Obsession* (Oxford: Oxford University Press).

Gress, D. (1998), *From Plato to NATO: The Idea of the West and its Opponents* (New York: The Free Press).

Howard, R. (2006). 'Sustainability and Narrative Plasticity in Online Apocalyptic Discourse After September 11, 2001', *Journal of Media and Religion* 5:1, 25–47.

Ivakhiv, A. (2006), 'Toward a Geography of "Religion": Mapping the Distribution of an Unstable Signifier', *Annals of the Association of American Geographers* 96:1, 169–75.

Jameson, F. (1988), 'Cognitive Mapping', in *Marxism and Interpretation of Culture*, Nelson, C. and Grossberg, L. (eds), (Basingstoke: Macmillan), 347–58.

Jewett, R. and Lawrence, J.S. (2003), *Captain America and the Crusade against Evil: The Dilemma of Zealous Nationalism* (Grand Rapids, MI: Wm. B. Eerdmans).

Jones, L. (2008), 'A Geopolitical Mapping of the Post-9/11 World: Exploring Conspiratorial Knowledge through Fahrenheit 9/11 and The Manchurian Candidate', *Aether: The Journal of Media Geography* 1:3, 37–57.

Knight, P. (2000), *Conspiracy Culture: From Kennedy to the X-Files* (London: Routledge).

Kristof, N. (2008a), 'Obama and the Bigots', *New York Times*, 9 Mar. <http://www.nytimes.com/2008/03/09/opinion/09kristof.html>, accessed 4 Nov 2008.

—. (2008b), 'The Push to 'Otherize' Obama', *New York Times*, 20 Sept. <http://www.nytimes.com/2008/09/21/opinion/21kristof.html>, accessed 4 Nov 2008.

Miller, L. (2008), 'Is Obama the Antichrist?' *Newsweek*, 24 Nov. <http://www.newsweek.com/id/169192>, accessed 24 Nov. 2008.

O'Leary, S. and McFarland, M. (1989), 'The Political Use of Mythic Discourse: Prophetic Interpretation in Pat Robertson's Presidential Campaign', *Quarterly Journal of Speech* 75:4, 433–52.

Roberts, G. (2006), 'History, Theory, and the Narrative Turn in IR', *Review of International Studies* 32:4, 703–14.

Rutenberg, J. (2008), 'The Man Behind the Whispers About Obama.' *New York Times*, 13 Oct. <http://www.nytimes.com/2008/10/13/us/politics/13martin.html?hp,> accessed 4 Nov. 2008.

Sampley, T. (2006), 'Barak Hussein Obama - who is he?' <http://www.usvetdsp.com/dec06/obama_muslim.htm>, accessed 4 Nov. 2008.

Sidaway, J. (2006), 'On the Nature of the Beast: Re-charting Political Geographies of the European Union', *Geografiska Annaler. Series B, Human Geography* 88:1, 1–14.

Somers, M. (1994), 'The Narrative Constitution of Identity: A Relational and Network Approach', *Theory and Society* 23:5, 605–49.

Sturm, T. (2006), 'Prophetic Eyes: The Theatricality of Mark Hitchcock's Premillennial Geopolitics', *Geopolitics* 11:2, 231–55.

White, H. (1973), *Metahistory: The Historical Imagination in Nineteenth Century Europe* (Baltimore, MD: The Johns Hopkins University Press).

PART II
American Evangelical Exceptionalism

Chapter 4
Apocalyptic Exceptionalism: Rosenberg, Clancy and the Prophecy of Americanism

Simon Dalby

Exceptional Geopolitics

American exceptionalism runs through American political culture and is evoked in numerous tropes in geopolitical discourse. David Campbell (1998) traces its themes from early Puritan texts through much more contemporary articulations of national security. Hofstadter's (1967) theme of a paranoid style in American politics has long informed geographers concerned to tease out the geographical dimensions of American identity (Agnew 1983). The Cold War invoked numerous threats to America, and given the specification of the putative enemy in the machinations of Soviet power as godless communists, a quasi-religious justification emerged for the prosecution of Cold War (Knelman 1985). These themes permeated American political thinking widely although the exact influence of assumptions of end times and imminent rapture is not easy to identify precisely (Boyer 1992). More recently these themes appear again in critical geopolitical writing where American identity has been examined in terms of the construction of places and popular identities (Sharp 2000; Dittmer 2005) and more specifically in terms of the modes of conduct implicit in how the War on Terror is conducted (Hannah 2006), and the larger geopolitical contextualizations of that war (Dalby 2007, 2008a).

The popular legitimation of American military conduct in the world is tied into understandings of America as different, exceptional and morally superior; it is also frequently contested precisely where it apparently violates the moral codes that should follow from such claims to superiority (Hannah 2006). From the dominant neoconservative view of the Bush Administration this was a world of danger, of threats to political order and Christianity itself, of decadent liberal politics in danger of appeasing threats, Islamic and otherwise. It is a matter of insecurities and struggles to overcome numerous evils, perpetual geographies of danger requiring vigilance. With the War on Terror as a backdrop, these themes have been rearticulated into further justifications of American violence in a world that clearly needs it if justice and development are to be extended across the planet and threats and dangers to modern prosperity removed in the process.

While this is the theme of many texts on American foreign policy and endless political commentary it was also the logic structuring official national security policy in the administrations of George W. Bush (Dalby 2006, 2009) and as this chapter will suggest more clearly, a theme which links the literary genre of techno-thrillers with some interpretations of biblical prophecy that claim America is in some senses the new Israel, the promised land and the hope for Christianity, and in some interpretations that the present circumstances are clear indication that we live in the end times (Sturm 2006). Given the sheer popularity of religious bestsellers, and the invocation of particular versions of truth to justify many things in their pages, American religious geopolitics is clearly in need of more attention than it has received from critical geopolitics until recently (Dittmer 2007). This is the case not least because as Wallace (2006) clearly shows, such invocations of religious sanction are much more obviously American than they are Christian. Themes of end times, prophecy and the violence made possible by religious contextualizations of life in these times are not new in American thinking (Boyer 1992). But they now spill over into the technothriller genre too, reinforcing the articulations of extreme nationalism that have been part of Amerian political discourse since 9/11 (Lieven 2004).

Religious interpretations of geopolitics thus take place in the larger cultural articulations of identity and danger, although the practical attempts to set up theological interpretations as sufficient unto themselves do not necessarily encourage engagement with wider culture. Nonetheless these discussions do take place in a world where politics and identity are contested interminably in a mediated culture where security narratives invoke endangerments while simultaneously portraying the responses, frequently military ones, as justified to restore the moral order whose insecurity was originally invoked (Debrix 2008). These themes are a matter of popular geopolitics, of culture providing the repertoire of discursive interpretation for political events. These appear in movies and novels just as much in the news, and have been doing so for a long time (Gogwilt 2000). Putting fictional representations into the cultural context within which they will be read is important to understand both the fiction as a cultural production and also how intertexts between genres are mutually reinforcing in at least some important senses (Dittmer and Dodds 2008).

Most importantly for the argument in this chapter is the simple fact that the genre of techno-thrillers has now spilled over into religious fiction and spies, violence and political mayhem now have religious sanction too. All this works through the War on Terror, in the revenge scripts of American neoconservative rationalizations of America's place in the world and the necessity of violence in a dangerous world where "America" is under threat (Faludi 2007). Security is very directly a matter of culture and danger, as well as imperial geopolitical framings wherein actors are called upon to do things in very particular places, frequently far from home, as part of grand global narratives (Dalby 2008b). As such the cultural intertexts between fiction and practical geopolitics are a matter of concern for contemporary scholarship in a number of disciplines (Martin and

Petro 2006; Power and Crampton 2007). Tabloid geopolitics is widespread in American culture; its analysis in geopolitics, international relations and security studies requires scholars to put it in this context if the larger intertexts are to be examined.

What follows in this chapter is but a small contribution to this task. First all this needs to be situated in terms of the larger cultural context of politics in contemporary American life. To do this the chapter turns first to Michael Williams's (2007) analysis of culture and security, and then to Francois Debrix's (2008) formulation of "tabloid geopolitics". These theoretical lenses then lead to a reading of two novels, first Tom Clancy's post 9/11 *Teeth of the Tiger*, and then Joel Rosenberg's *The Ezekiel Option*.[1] Both rearticulate American exceptionalism but do so in line with the larger contemporary cultural themes of insecurity and the war on terror. In these novels the knowledgeable subject and the bearer of virtue is the hero in these times, one possessed of superior detailed information, and in the genre of the techno-thriller, the common sense understanding of the world that can interpret intelligence information to put the technological tools of the secret agent to good use fighting evil and saving the virtuous victims in a dangerous world. The invocation of an explicitly Christian ethos in Rosenberg raises questions of the geopolitical context invoked to structure contemporary narratives of Americanism. The chapter closes with some reflections on other possible religious contextualizations which might structure very different political possibilities. Such a contrast emphasizes how important geopolitical framings are to the articulation of danger and the mobilization of justifications for violent forms of security.

Culture and Security

In his critique of the failures of constructivist international relations scholars to effectively link culture and strategy, Michael Williams (2007) suggests the necessity of linking culture directly to matters of security if security is to be appropriately understood. More specifically in his sophisticated engagement with international relations thinking, he argues that what it is that is secured is not simply a matter of territorial borders and national frontiers. Security is not just about armies, air forces, navies and the practices of diplomacy; its also about what it is that is threatened, the social and political order that invokes threats to its existence is also a necessary part of the analysis as are the various tele-visual representations of contemporary danger which link popular culture to national geopolitical imaginaries (Williams 2003).

---

1 For novel readers it is appropriate to issue the standard "spoiler warning" here! Details of the plots in both Clancy's and Rosenberg's novels follow. Likewise the ending of Frederick Forsyth's *The Afghan* is revealed too in the paragraphs below.

Williams (2007) goes on to show how the expansion of NATO after the Cold War was as much about the extension of a civilizational project as it was about military strategy. Liberal democracy, capitalist economies, human rights and technical practices of economic regulation were key to all this; the boundary of the external antagonist was understood as societies that did not adhere to these norms. The expansion of NATO suggested clearly that if states remade their political cultures and modes of rule they too might participate in the virtuous peaceful community of liberal democratic states, although in practice this process has been much more contested than the simple conventional narrative that expansion suggests (Kuus 2007).

Looking to the debate about security and culture in the United States more specifically Williams (2007) examines the so called culture wars that have dominated political discourse in the United States since the 1970s. More specifically he looks to the political logic that propelled the neo-conservative cause through the last few decades. In particular he suggests that American conservatives faced a profound challenge after the Cold War. While they did attribute the end of the superpower standoff to a victory brought about by the Reagan administration, and its determination to challenge the Soviet Union militarily, without that threat the purpose of American power was much less easy to articulate (Dalby 2008a). Neoconservatives needed an elevated patriotism, a sense of purpose for the American people now that the Cold War had apparently been won; they also needed new threats to justify the rebuilding of military power and its use on the global stage, which was a necessity, they argued, to maintain American preeminence (Project for a New American Century 2000).

This was however related to the domestic neo-conservative project which reworked American politics in terms of culture and identity: "Neoconservativism has been instrumental in promoting a political culture increasingly defined and divided by questions of culture and a politics of virtue and identity. In this domain, one of neoconservatism's most powerful elements resides in its adoption of a specific representational strategy and a cultural politics of authenticity" (Williams 2007, 107). The question of authenticity invokes a common sense understanding of the world, one superior to the academic abstractions and obfuscations of the liberal intelligentsia, and one that is an intrinsic claim to epistemological superiority. The neoconservatives do know the truth, and with this common sense practical understanding of the real world they can claim to speak for America.

As such it is a political and social strategy to reject the sophistication of intellectual debate in favor of a populist rhetoric that simultaneously is true and supposedly believed by most Americans, at least those who are not confused by too much exposure to the liberal media and academics whose loyalties are obviously suspect given their preference for what can easily be dismissed as moral relativism, a failure to believe in American virtue and the necessity of vigorous action to protect "the American way of life" from dangers within and without. Within such cultural framings "virtuous wars" can then be fought (Der Derian 2001). But in the American politics over the last few years such claims to simple truths link up fairly directly with the mobilization of religious organizations to support the Republican party; neo-conservatives may not be explicitly religious in their politics but the links to conservative religious political campaigns have long worried those who advocate a separation of state and church (Linker 2007; Connolly 2008).

The supposed loss of virtue on the part of liberals and their media, and the lack of values, is then linked to the larger problematic of security and a political strategy of mobilizing a political coalition to oppose these threats to a very specific vision of America. "Representing itself as standing against those who disparage these values and people, who deny their virtues, and who in doing so imperil America both at home and abroad, has provided neoconservatives, and the political right in the United States in general, with powerful forms of cultural and symbolic capital and power" (Williams 2007, 112). This allows a dismissal of debates about what the national interest is and a claim that such debates actually weaken the United States and reduce its willingness to act in what is apparently a dangerous world. Linked to a critique of the enfeeblement implicit in the decadence of modernity this allows for an assertion of a muscular foreign policy to assert American values in a world apparently much in need of such guidance (Kagan and Kristol 2000). Virtue is challenged by liberals, and Americans can be portrayed as victims of intellectuals, cultural critics and vaguely defined left wingers who do not understand America and its virtues. Neoconservative rhetoric attempts to reclaim domestic virtue and claim to speak for the majority of Americans denied an authentic voice by decadent liberal culture.

These political claims were powerfully reinforced in the aftermath of September 11, 2001 when key neoconservative intellectuals were in power in Washington and quickly shaped the discourse in response to 9/11's events into war, and more specifically a global war (Dalby 2003). The revenge narrative that resulted followed many neo-conservative themes, despite the rather obvious point that prior to 9/11 most of the neoconservative concerns about international threats to America related to states and not to terrorists (Dalby 2006). The subsequent fate of the ill-considered invasion of Iraq, powerfully justified by the simplistic neoconservative understandings of the political world follow on from this invocation of "a few grand ideas" (Kaplan 2008). But these failures are usually not a matter of any inadequacy on the part of the initial formulation; endless twists and turns, and the attribution to either incompetence on the part of Americans, or ungratefulness on the part of those who have been liberated, maintain the narrative as essentially correct (Gray 2008).

These themes are not just a matter of intellectuals in neo-conservative think tanks in Washington. This cultural field of virtue, certainty and the need for violence to protect the victimized ordinary American is also present in precisely the media that the neo-conservatives have repeatedly labeled as liberal and incapable of standing up for American values. The imposition of meaning on the complexity of contemporary events and the television images that saturate public discussion of politics, is a matter of culture and the invocation of danger. It is a practical everyday repetitive procedure repeated endlessly on television and in newspaper and now online blogs and news feeds. The scripts of political meaning are structured to give intelligibility to the world, but do so frequently in manners that have precious little to do with anything except the relentless reproduction of mediated meaning.

In short this is, to borrow Francois Debrix's (2008) terms a matter of "tabloid geopolitics". "Tabloid geopolitics is the result of mediatized discursive formations that take advantage of contemporary fears, anxieties, and insecurities to produce certain political and cultural realities and meanings that are presented as commonsensical popular truths about the present condition" (Debrix 2008, 5). Tabloid geopolitics frequently ends up employing vengeful and violent frames to interpret the events it reports, but likewise cannot adequately prepare audiences for tragedy when it breaks through the endless repetition of the security discourse. It moralises while remaining powerless. Evil figures and desperate deeds reappear frequently, and spill over into fictional representations, video games, novels, movies and numerous cultural genres. Hence in trying to see the practical effects of the link between culture and security looking to popular culture is entirely appropriate in examining spaces of insecurity and the generation of fear (Pain and Smith 2008; Ingram and Dodds 2009).

Tom Clancy has been publicly praised by many leading American politicians, and indeed Ronald Reagan's endorsement of his first novel *The Hunt for Red October* was apparently key to its initial success. In this novel Russians were the threat that was thwarted by American technological superiority and the brilliance of CIA operative Jack Ryan; more recently these novels portray Arabs and Muslims as the external danger (Malinen 2006). Endorsements from public figures are used for publicity in the book publishing business so this is not surprising, but nonetheless these comments add an air of factual authenticity to the genre. Rush Limbaugh's comment on the front cover of Joel Rosenberg's (2008) *Dead Heat* which states "Rosenberg has become one of the most entertaining and thought-provoking novelists of our day. Regardless of your political views, you've got to read his stuff" may be an extreme example of celebrity endorsement, but the suggestion of the plots as essentially credible interpretations of contemporary or near contemporary events emphasizes the point that tabloid geopolitics blurs any claims that fiction and fact can easily be distinguished. The simple moral universe of the tabloid genre is reproduced in contemporary technothrillers.

Tom Clancy's *Teeth of the Tiger*

Techno-thrillers came of age, as it were, in the 1980s when Tom Clancy's action packed novels, starting with *The Hunt for Red October*, linked American machismo and technical acumen in a simultaneous celebration of middle class manliness and violent destruction. Their shifting plots mirror the larger political themes of security discussion, and as the Cold War ended they variously encompassed matters of war in the Middle East, drug wars, terrorism of various sorts, potential clashes between Japan and the U.S., China and Russia, and various other speculative matters at the heart of American security discussions (Dalby 2008a). The theme of American exceptionalism in a harsh world, of the reluctant warrior disgusted by the necessity of the violence he was called upon to perform, was meshed into tropes of superior knowledge, intelligence analyst cum moralist; the CIA analyst and sometime action hero Jack Ryan eventually rising to the heights of American power as a reluctant president in Clancy's later novels.

Once more the geopolitical map has shifted, apparently requiring the reinvention of the old tropes of American identity affirmation in new post-modern technological guises. In Tom Clancy's 2003 novel *The Teeth of the Tiger*, all this literally involves a new generation as now retired president and former CIA analyst Jack Ryan's son, Jack Ryan, Jr., makes his appearance as a rookie analyst. Now, after 9/11, the "Ryanverse" world of Clancy's geopolitical imagination, as it is sometimes known to his fans, has to face a complicated world of potential terrorist attacks by jihadists whose apparent motivation is solely to kill Americans: as many of them as possible with the limited means available. In *Teeth of the Tiger* an unlikely alliance is hatched with some drug dealers in Mexico and the Islamic terrorists make their way across the border, heading to American shopping malls to shoot as many shoppers as possible and to bring terror to Middle America.

In one of the attacks they meet up with the Caruso brothers, cousins of Jack Junior, on a training exercise run by "the Campus", a secretive organization set up in part by Ryan, Sr. to tackle terrorist threats. The Campus is unofficial, unsanctioned by the authorities and funded by its own financial dealings; the cover story for the organization is that it is a financial investment company. The source of its technical prowess and superior information about the world, essential to the heroic figures at the heart of the plot, is a sophisticated intercept system where the American CIA and NSA communications systems are hijacked to read the raw intelligence as well as the assessments made about it by their analysts and feed it into the Campus's own analytical system.[2]

Doubts about secret assassination missions evaporate as the brothers tackle and kill the terrorists in the mall. Subsequently they are sent to Europe to assassinate recruiters and organizers of the terrorist organizations, and do so with an especially lethal but undetectable poison. They operate beyond the rule of law, which does not really matter because the constraints on necessary action, and the obstacles to taking revenge for obvious crimes, purportedly have to be circumvented to deal with extraordinary dangers presented by terrorists. All this is legitimate because those who operate outside the law use the common sense knowledge of Americans who really do understand the world, its dangers, and how to use lethal

---

2 The Campus building is conveniently located in between Langley and Fort Meade allowing rooftop communications gear to read the encrypted traffic between the headquarters of the two agencies, a huge security breach which neither of these agencies actually discover!

force effectively in America's defense. In short, Clancy's assassination squad is precisely the neoconservative vision of virtuous American masculinity using violence to deal with the bad guys despite the enfeebled efforts of liberals whose moral qualms interfere with necessary forthright action.

The intertexts between American exceptionalism and the license to kill evade the remaining claims to universal liberalism and the virtues of due process; Oliver North's 1980s fantasy of Nicaraguan Contra-style "off the shelf" stand-alone military entities is now realized. All American boys rise to the challenge to be a gunslinger in new guise (one holster now holds a Blackberry not a Colt, but now the six shooter has laser sights ... or, in the case of the Caruso/Ryan hit team, has morphed into a lethal syringe) in the wild spaces that are the political geography of contemporary American fear. While they may have qualms of conscience, in the end they do the right thing; vigilante violence subdues the threats on the frontier in distant lands and the bad guys learn the hard way not to mess with American masculinity.

There are numerous oddities in this novel and some on-line critics have excoriated it for historical anachronisms and too much talk in contrast to the expected action that is the hallmark of techno-thrillers. The role of analyst cum action hero is crossed once again in this plot, Jack Junior stepping up to do one of the assassinations at a crucial moment when it appears that the Caruso brothers might have been identified by their putative victims. The novel appears to be a set up for at least one sequel as the Caruso/Ryan team will presumably be ready for more missions in forthcoming volumes, perhaps even in competition with the elite Rainbow international counter-terrorism team, lead by John Clark, that appears in earlier novels in the Ryan series. But six years after its publication in 2003, a sequel has yet to appear.

Joel Rosenberg's *The Ezekiel Option*

While the Ryans and Carusos of Tom Clancy's imagination are portrayed as having vaguely Catholic theological inclinations on the very rare occasions religion enters the discussion of the morality of killing bad guys, various strains of evangelical Christianity are now also showing up in the plots of thrillers. The necessity of dirty deeds in the cause of the larger moral cause of American civilization has been expanded upon at length in Joel Rosenberg's series of novels *The Last Jihad* (2002), *The Last Days* (2003), *The Ezekiel Option* (2005), *The Copper Scroll* (2006), *Dead Heat* (2008), and his non-fiction *Epicenter* (2006), all published by Tyndale House in Illinois. Borrowing closely from Clancy the hero, Jon Bennett, and in a rather un-Clancy innovation, heroine Erin McCoy, are political and intelligence operatives working at the heart of the Washington power structure, traveling the world on diplomatic missions. With dangers everywhere they struggle to work out a peace agreement between Israel and Palestine, fueled, quite literally, by newfound oil wealth that will soon be discovered in the Eastern

Mediterranean sea, which offers the promise of prosperity if the peoples of the troubled region cooperate.

Rosenberg's novels are in one sense not so new at all. As Davis (2006) notes, the contemporary technothriller is a reinvention of the genre of the "future war" tale, a popular venue over the last couple of centuries for journalists, military officers and novelists with a concern about geopolitics and the possibilities of wars of one sort or another, to express their concerns about, and hopes for, the future in fictional mode. The theme of prediction of events likely to occur in the near future thus fits both this theme in the technothriller genre and the larger audience in search of "biblical" interpretations of contemporary events where prophesy is supposedly interpreted in light of the unfolding of themes from the Books of Ezekiel, Daniel, and most notably Revelation.

Davis (2006) also emphasizes the point that the technothriller genre requires imminent catastrophe as the backdrop for heroic action to fend off the danger, whether it is from nuclear attack, biological or chemical warfare, or most recently the prospect of further 9/11 style devastation launched by terrorists. This mirrors directly the narrative structure of the Bush Doctrine and the various articulations of national security that appeared in the immediate aftermath of the attacks on the Pentagon and the World Trade Center. Both an earlier Tom Clancy novel *Debt of Honor*, and the first of Joel Rosenberg's series, *The Last Jihad,* involve suicide flyers attacking American buildings so this theme resonated with the genre.

Prediction of imminent disaster and the moral choices about how to respond to warnings also play well with narrative themes in religious prophesy and discussions of end times. True believers are of course most likely to triumph in the face of disaster, because either their moral superiority or superior knowledge, or both, gives them interpretative resources that others lack. This adds all sorts of dilemmas and difficulties about action, predestination, and the limits of human agency, but this too is not a new discussion in theological terms either. Nonetheless American virtue as the land of true believers, the contemporary technology of warfare, and all manner of technological and political anxiety in a fast changing world all come together in the religious technothriller. Rosenberg's plots thus link to the "end times" interpretations of contemporary events quite directly (see Sturm 2006).

In the case of *The Ezekiel Option* a coup in Moscow and an unlikely alliance between Islamic forces and the new government in the Kremlin threatens the world with war. Caught in the middle of the coup while in government offices in Moscow, Bennett and McCoy enthusiastically join in the gunfight, firing at the soldiers storming the building. Quite what they were doing carrying weapons into the middle of a foreign government office is not clear; it is just assumed that they will be armed. McCoy is shot and captured, Bennett is sent out of Russia, and of course a plot of rescue and action is set in motion. The geopolitical scenario sees gullible Europeans wooed by the Russian regime as part of a plot to isolate the United States. In search of a way to rescue McCoy, Bennett meets up once again with a retired Israeli intelligence officer Elizier Mordechai, who has quietly converted to Christianity (a very courageous, or stupid, thing for a Mossad officer to do given the likely penalties should such an obvious shift of allegiance be discovered) and learned to believe prophesy concerning the end times and imminent divine intervention to stop the destruction of Israel by, of course, that new regime in Moscow.

The novel provides a detailed discussion of the prophesy, and knowing of the imminent destruction of Russia, Bennett sets in motion a distinctly unlikely, even by the standards of contemporary thrillers, single-handed rescue plan. An obligatory parachute jump, in this case into Turkey for some reason that is not entirely clear apart from the fact that half the world is supposedly looking for a now former American government official gone missing and Christian groups conveniently located to help him on his way despite the death such assistance brings to them. This adds a new twist to the usual mayhem wrought in techno-thrillers by linking up with a theme of missionaries abroad that will obviously play well with the American evangelical audience used to tales of missionary daring by smuggling Bibles into the Soviet Union and more recently Iran.

Buoyed by his faith, and the clear understanding that God would obviously not wish him to live without his true love, he gets to Moscow and of course finds McCoy, who has already escaped captivity, conveniently found another group of Christians willing to shelter her at their own peril, and has learnt by radio intercept of the Moscow police that Bennett is driving to her rescue. To leave Moscow in a hurry while the police are chasing them (in the obligatory car chase scene), Bennett and McCoy have to kill yet more Russians. While they may regret this, knowing that divine intervention is imminent whatever moral scruples they may have are overcome by the knowledge that those standing in their way are due to meet their end imminently, so a few hours or minutes will not matter. Divine intervention does indeed occur in the last part of the novel and the evil Russian plot to annihilate Israel with nuclear missiles is thwarted by a spectacular divine massacre on a grand scale coupled with a completely effective ballistic missile defense system!

Virtue is triumphant as Bennett's God allows him to rescue his true love while simultaneously seeing at least some of his foes eradicated in fiery vengeance (Only some die, after all some of them have to survive to appear in future novels!). Here religious knowledge of God's various plans for the planet now complements the merely geopolitical analysis available to CIA agents and other functionaries of the War on Terror. Coupled to the superiority of the born again who know the truth, and with the promise of Rapture and salvation as an additional sanction for violent illegality, the techno-thriller now embodies violent technological geopolitical mayhem as the apotheosis of the genre in which the saved are triumphant.

Rosenberg's *The Ezekiel Option* is noteworthy for its explicit attempt to educate the reader about the prophecies of the end times and that time's divine intervention to save a threatened Israel. In places the text, through the discussions between Mordechai and Bennett, and Bennett and some other characters such as his long suffering mother, becomes a primer on the literature that describes the prophesies. A couple of key pages in the novel explicitly review the books, most notably

Hal Lindsey's bestselling *Late Great Planet Earth* (1970), in which it is claimed that the events of the novel are foretold. Not that all of them actually unfold in the novel; large Chinese armies marching Westward and other events are notably absent. But as Sturm (2006) has pointed out, the geographical interpretations of contemporary events in the interpretations of prophecy are amazingly flexible while invoking supposed geographical veracity.

What is interesting here is Rosenberg's invocation of theo-political imaginations as an adjunct to the technical knowledge from spies, satellites, signals intercepts and so on that normally populate the pages of techno-thrillers. Rosenberg explicitly links this discussion with Ronald Reagan's discussion of living in end times and the invocations of prophecy that appeared in Reagan's statements both before and during his presidency (see Halsall 1986; Knelman 1985). In doing so the authority of prior presidential belief is used as a device to legitimize another fictional president's interest in this argument in the plot of the novel. But it also acts to acquaint any reader not familiar with this discussion with the key sources. Despite the disclaimer at the front of these novels, Rosenberg clearly has pedagogic intent in addition to pecuniary motive in publishing best sellers. Regardless of the political economy of thriller publication, clearly American virtue now is supported by a very particular interpretation of biblical prophesy to legitimate further mayhem and murder across the world as virtuous agents deal with distant dangerous people and places with the application of force. Except in Rosenberg's novel, it is God who has the really sophisticated technological instruments of destruction, mere human technical acumen just does not measure up.

Reading Novels

Clearly in these various articulations of tribulations, signs and claims that we live in end times, American Christians have a part to play in the great geopolitical drama that is being played out. Many of the arguments that support the religious contentions invoke geographical interpretations to "prove" the veracity of their claims (Sturm 2006). But the question of violence is not unrelated to the invocation of geographic settings to justify actions of many kinds. In a world understood in theological terms "all interpretive claims have geopolitical implications. They speak of a God who is not a philosophical abstraction but an agent historically and geographically engaged with humankind" (Wallace 2006, 215). Those who invoke prophesy in various forms, or divine design as the appropriate code for behaviour situate these in geopolitical contexts. As Agnew (2006, 186) writes: "They are providing nothing less than a Bible-based geopolitics for U.S.-policy on a wide range of issues, from taking sides in the Israel Palestine conflict and doing nothing about global warming to the obviously diabolical meaning of the terror attacks of 11 September 2001." In Rosenberg's novel these themes reappear again too where his interpretations draw on American culture and in particular in the period of the

War on Terror, on claims to virtue in the face of decadence at home and threats abroad.

Many of the themes of *The Ezekiel Option* and *The Teeth of the Tiger* also appear in other contemporary discussions of secret agents, geopolitical conflict and the war between Al Qaeda and the West. Notable is Frederick Forsyth's bestseller *The Afghan* which parallels both Clancy and Rosenberg's themes but does so without the simplistic invocation of American superiority, the masculinist themes found particularly in Clancy, or the religious justifications for murder and mayhem that run through Rosenberg's novels. In *The Afghan* the British do it the traditional way with regional specialists and old fashioned secret agents actually penetrating the inner operations of a global terrorist network and thwarting evil deeds from within. There is a whiff of James Bond in Forsyth's writing, but the plot is much more intriguing, not least because of Forsyth's sophisticated knowledge of international economics and his much sharper eye for the telling detail.

Huge ironies run through these novels and their attempts to tell stories that simultaneously entertain and reassert American superiority to their avid readers. Not least is the invocation of the phrase "no greater love has a man than that he lay down his life for his friends" by Rosenberg in contrast to Frederick Forsyth's discussion of a mission by a Western agent, heading for certain death, who infiltrates Al Qaeda in an attempt to stop another major attack on the political leadership of the Western world. This is actually the motif that Forsyth uses to justify the reinvention of a warrior ethos among Western agents, although he has a secular version of this ethos in his narrative (Samet 2005). While Rosenberg invokes the theme, it is Forsyth's hero who actually follows through and ends up being forced to sacrifice himself to prevent carnage on a spectacular scale. The irony here is that it is Forsyth's secular hero who sacrifices himself; Rosenberg's hero and heroine kill apparently with impunity, but seem reluctant to forego the delights of their love affair in the cause of doing God's work.

The question of sacrifice, and the necessity for a warrior to give their life for a cause *in extremis*, adds all sorts of difficulties to a simple moral tale. As critics of Al Qaeda's attacks on 9/11 quickly discovered, it is hard to frame a suicide attack as cowardice; madness works better as it effectively evades discussions of morality (Faludi 2007). Dying for a cause has to be interpreted as delusional, or evil in these circumstances; because to do otherwise is to investigate motivations and many aspects of international politics closely, which precludes the simple moralising of tabloid mode. On the other hand the heroes of many American tales end up giving their lives for a cause, a point that dramatically contrasts with the notion of suicide bombers as evil (see Diken and Laustsen 2004). It seems that our warriors give their lives for a noble cause; others' causes are simply evil.

Likewise Rosenberg, in attempting to mimic the precise technological details which have long been a particular rhetorical device to appeal to readers of techno-thrillers, is actually short on technical credibility while long on religious justification. In that sense Forsyth is true to the genre in a way that Rosenberg is not; not a surprising literary fact, but one which does make Forsyth's prose much more compelling to a techno-thriller reader, although probably not to a born again believer relieved to find their version of the universe portrayed in the pages of a novel they can purchase at the local Christian bookstore!

This is of course the point; these novels are popular entertainment while simultaneously providing support for at least some of the prior geopolitical presuppositions that the readers bring to the texts. They are part of the cultural practices that articulate security, threat and the need for virtuous war. American practical knowledge of guns and the need to act in a morally corrupt world with a clear vision of right and wrong is powerfully reinforced as the heroes make their way through a decadent world incapable of facing up to the dangers that await it, ones known best to the neoconservatives and their friends whose patriotism and moral clarity allow the clouds of uncertainty to be penetrated by courage and determination. Political virtue and a superior knowledge combine in this patriotic act of self sacrifice; if God gives blessing to all this then so much the better. It is a wild world out there, full of threats and dangers that need the grand pan-opticon of the intelligence community to be combined with the practical technical savvy of Americans, a combination that will ensure all turns out well in the end, at least for those who share the patriotic truth of this vision of the world.

Writing Geopolitics

But this is an imperial view of the world, one of dangerous others inhabiting wild places that threaten violence to the civilized center (Gregory 2004). It is one of a perpetual insecurity in the face of challenges to American power, and hence it is a view of the world in which violence is necessary, and American power is hence legitimate. It is a theme that runs through much of American thinking (Campbell 1998), and draws on larger European fears of the decline of empire and threats from barbarians which run through modernity's geopolitical imagination. Linking this to American vulnerability, and the divine blessing that America as the new Jerusalem brings with it, ties the geopolitical anxieties of national security to the tropes of religious protection; God is apparently on America's side, or at least in Rosenberg's view on the side of those who are saved in America. As Jason Dittmer points out in his chapter in this volume, in the 2008 presidential election campaign there were suggestions that Barrack Obama was the antichrist; not all Americans are saved and the cultural struggles that the neo-conservatives have partaken in, that Williams (2007) emphasizes, have their parallels in the religious struggles within the United States. All this also seems to suggest the inevitability of war as a permanent situation; the national security strategy of the United States during the Bush Administrations after 9/11 explicitly said that only the expansion of economic liberty and American power to all the parts of the planet that could potentially threaten America would provide ultimate security (Dalby 2008a, 2009). In the fictional representations in Clancy and Rosenberg's novels, this geopolitical imagination clearly fits Debrix's (2007) formulations of tabloid imperialism. This is a vision of violence and conquest, not just one of a defensive mode of realism in the face of supposedly external dangers.

Simultaneously popular discussions of America as an empire, whether it is one, and if so what this means for American policy, and how lessons might be learned from the British and Roman empires, are also in circulation. The presence of American troops in many parts of the world, the ubiquitous use of the dollar and the widespread emulation of American popular culture all suggest imperial themes. Building bases that look remarkably like garrison towns from earlier imperial episodes suggest an imperial culture imposed on the landscape as well (Gillem 2007). Popular discussions of American culture and contemporary history frequently invoke the analogies with Rome, both in historical comparison and the re-emergence of popular culture (Murphy 2007), only most obviously in the movie *Gladiator* (Dalby 2008b). Google now has a virtual Rome in which online viewers can review its glories, if not the violence and destruction wrought by its military adventures and the extraction of commodities and slaves from the provinces to enrich the imperial rulers in Rome.

While many of the supposedly "factual" accounts of end times and contemporary geopolitics read current events in terms of what Francois Debrix (2008) calls "Tabloid Realism" and in terms of "Tabloid Imperialism" (Debrix 2007) they remain stuck in the arguments about the inevitability of war and violence. But such arguments have been challenged by a very different geopolitical imagination that resists the pre-millennial interpretations of imminent or nearly imminent destruction and the resultant need for virtuous violence (Wallace 2006). These theological arguments frequently suggest that Jesus of Nazareth was not on the side of empire and certainly would not respond to current circumstances in terms of invoking divine sanction to prosecute a War on Terror that abandons the rule of law, international norms and diplomacy and assumes that violence is the appropriate response to contemporary political difficulties (Northcott 2004). The debate within the American evangelical community raises some of these questions, the violence of warfare sitting uneasily with Christian messages of a God of love and the possibilities of building some version of the Kingdom of God on earth (Fitzgerald 2008).

The arguments of the necessity of non-violent resistance to imperial power as key to a Christian message that is consistent with the biblical Jesus suggest a very different interpretation of contemporary events and the obligations of the faithful (Horsley 2003; Crossan 2007). This theological discussion has intersected with the question of how imperial tropes link to interpretations of Islam and especially Christianity. More specifically this discussion suggests that analogies with Rome can be read as suggesting that Christianity is about resisting the power of empire rather than providing arguments for the exercise of imperial power. If America is understood as the new Rome rather than the new Jerusalem, then the appropriate mode of activity for Christian churches is not support for imperial wars in the Middle East, but rather resistance to the use of violence in attempts to remake foreign peoples in the image of a narrow interpretation of America.

Attempts by small groups of practicing Christians, in such groups as Christian Peacemaker Teams, have attempted to bring a very different religious message to the Middle East in the last decade in particular. Whether visiting Baghdad to suggest to Iraqis that not all Americans or Christians supported the invasion of 2003, or witnesses escorting Palestinians facing Israeli violence in the West Bank, this is a very different interpretation of religious obligation. Solidarity work suggests an alternative geopolitics, one where non-violence is key to social relations and the possibilities for peaceful action are worked out in numerous settings (Koopman 2008). It is not one that fits with the tropes of righteous violence, nor the necessity of imposing order in a chaotic world. These themes suggest forcibly the importance of thinking about non-violence and geopolitics (Megoran 2008). The also suggest the crucial importance of religion in struggles against oppression in American history, and the role of Christianity in the campaigns against slavery in particular. Simon Schama (2008) traces some of these key themes in his discussion of America as a land of promise, indeed as the land in which the new Jerusalem, the shining city on the hill, might in fact be built. These themes were in part mobilized in the Obama election campaign in 2008 where the possibilities of the future were once again key tropes used to considerable effect by a master of the necessary rhetorical techniques. But it is a rhetorical promise to overcome the discord of division, a direct challenge to the neo-conservative ontological commitments to violence and struggle rather than common purpose.

If Jesus of Nazareth's preaching is understood as non-violent opposition to the violence of the hierarchical social arrangements that perpetuate imperial power, and if some notion of divine grace is appropriate for all people, rather than only for those who are "saved", then a very different politics emerges. Such disputations of cultural identity are based on a very different epistemology than that invoked by the neoconservative versions of history, virtue and danger. It requires a recognition of violence as imperial and the possibilities that American actions abroad may be part of the problem rather than the solution. These arguments also delink Christianity from its geopolitics of America and Israel as promised lands and directly challenge the colonizing logic of Christian Zionism with all the political effects this movement has in linking American policy to the state of Israel (Cohn-Sherbok 2006).

In rethinking the Christian legacy in these terms emphasis is placed on the stories of deliverance, of exodus from slavery and of redemption. The story of the United States is frequently told in precisely these modes; the African American churches understood the promise of freedom in biblical tropes where the end of their time of bondage is analogized with the stories of Exodus. It is precisely in the contrast between deliverance from religious persecution in Europe, and the construction of America as a new Rome that many of the tensions in American identity play out. These are not new themes; they run back through the competing narratives of America from the early days through to contemporary times, but in recent years theology has returned explicitly to discuss these themes. While one does not need to follow the finer points of theological disputation, in particular

Horsley's (2003) reading of the Gospel according to Mark, which has Jesus involved in a social renewal of Israel in the face of the complicity of its rulers with the Roman occupation, is in stark contrast to the theology of end times and its rationalizations of American geopolitical violence.

In looking to America as Rome rather than as Jerusalem, theological disputations of the verities of Pax Americana also run in parallel with John Gray's (2008) condemnation of the neoconservatives' appropriation of religious tropes to reinvent rationales for such imperial actions as the invasion of Iraq in 2003. The reduction of complex geopolitical questions to a few "grand ideas" is precisely what has caused so many difficulties in the Middle East in particular in recent years (Kaplan 2008). Both theologians and political critics point to the dangers of appropriating religion to attempt to control history; the lessons of the last decade are powerful empirical validation of these misguided attempts.

Peaceful Futures?

To challenge the neoconservative view of a dangerous world of clashing forces, of the necessity of violence to perpetuate an extremely unequal global polity, it is clearly necessary to challenge the mappings of danger to the Pax Americana that the Project for New American Century advocated in the 1990s and brought to fruition in the years following 9/11. The neoconservative presupposition of a dangerous violent place in need of American violence clearly fits closely with the themes of most of the technothriller genre. The verities of its geopolitical knowledges portray numerous lurking threats to a basically benign American world order. When coupled to the supposedly superior revealed truth of biblical origin, with God on your side, geopolitical knowledge becomes even more compelling as a mode of interpretation. But the technothriller is about God being on the side of Americans, not on the side of the poor, the oppressed, those struggling with the difficulties of various occupations or injustices.

Challenging these interpretations in fiction is not difficult. Recent books and movies as varied as *Children of Men* (2006) and *Rendition* (2007) contain refutations of the simple moralities told by those who would support the existing political order. Spoofing the American obsession with security, as the Zombie genre, and in particular the movie *Fido* (2006) does with such humorous elegance (see Molloy 2009) or *The Day the Earth Stood Still* (2008) does in more mainstream contemporary criticism, makes it clear that the neoconservative framing of security is not hegemonic. Nonetheless it is remarkably persistent, precisely because as Debrix (2008) argues it reproduces moral tales that feed fear and suggest the audience is virtuous and vulnerable, positioning them as in need of protection from all manner of threatening forces, mostly out there, but always in danger of penetrating in here too. These themes are longstanding parts of the American cultural experience (Schama 2008). The irony of all this is, as Schama notes, there are other narratives in that experience that write other stories of virtue, ones that do not assume a given common sense approach that specifies vigilante violence as virtuous, or that it assumes it will ensure either salvation or peace.

But fictional alternatives are not enough as a mode of contesting the legitimations of violence. Challenging the presuppositions of the neoconservative framing of insecurity requires a larger engagement with the ways in which politics is contextualized. This requires a rethinking of ethics and in particular the mappings of spaces as vulnerable and external others as threatening; the cartographic imagination of contemporary international relations thinking needs to be tackled explicitly (Burke 2007); the logics of intervention and humanitarian cooperation do not escape the difficulties of geography either no matter how universal claims are invoked in rhetorical guise (Williams 2008). Other stories of virtue also require thinking much more about non-violence and the possibilities of international cooperation as key to security (Megoran 2008). But as this chapter has suggested critical geopolitics still also has much more work to do challenging contemporary fictional representations of danger. The technothriller articulations of virtuous violence are but a small part of the task, a task now made both more difficult and more important by the genre's current imbrications with violent Americanist prophesies.

References

Agnew, J. (1983), 'An Excess of 'National Exceptionalism': Towards a New Political Geography of American Foreign Policy', *Political Geography Quarterly* 2:2, 151–66.

—. (2006), 'Religion and geopolitics', *Geopolitics*, 11:2, 183–191.

Boyer, P. (1992), *When Time Shall Be No More: Prophecy Belief in Modern America* (Cambridge, MA: Harvard University Press).

Braudy, L. (2003), *From Chivalry to Terrorism: War and the Changing Nature of Masculinity* (New York: Knopf).

Burke, A. (2007), *Beyond Security, Ethics and Violence: War Against the Other* (London: Routledge).

Campbell, D. (1998), *Writing Security United States Foreign Policy and the Politics of Identity*, 2nd Edition (Minneapolis: University of Minnesota Press).

Clancy, T. (2003), *The Teeth of the Tiger* (New York: Putnam).

Cohn-Sherbok, D. (2006), *The Politics of Apocalypse: The History and Influence of Christian Zionism* (Oxford: One World).

Connolly, W. (2008), *Capitalism and Christianity, American Style* (Durham: Duke University Press).

Crossan, J.D. (2007), *God and Empire: Jesus against Rome, Then and Now* (New York: Harper Collins).

Dalby, S. (2003), 'Calling 911: Geopolitics, Security and America's New War', *Geopolitics* 8:3, 61–86.

—. (2006), 'Geopolitics, Grand Strategy and the Bush Doctrine: The Strategic Dimensions of U.S. Hegemony under George W. Bush', in David, C. and Grondin, D. (eds) *Hegemony or Empire? The Redefinition of American Power under George W. Bush*, (Aldershot: Ashgate), 33–49.

—. (2007), 'Regions, Strategies and Empire in the Global War on Terror', *Geopolitics* 12:4, 586–606.

—. (2008a), 'Imperialism, Domination, Culture: The Continued Relevance of Critical Geopolitics', *Geopolitics* 13:3, 413–36.

—. (2008b), 'Warrior Geopolitics: Gladiator, Black Hawk Down and the Kingdom of Heaven', *Political Geography* 27:4, 439–55.

—. (2009), 'Geopolitics, The Revolution in Military Affairs and The Bush Doctrine', *International Politics* 46:2/3, 234–52.

Davis, D. (2006), 'Future War Telling: National Security and Popular Film', in Martin and Petro (eds) *Rethinking Global Security*, 13–44.

Debrix, F. (2007), 'Tabloid Imperialism: American Geopolitical Anxieties and the War on Terror', *Geography Compass* 1:4, 932–945.

—. (2008), *Tabloid Terror: War, Culture, and Geopolitics* (New York: Routledge).

Der Derian, J. (2001), *Virtuous War: Mapping the Military-Industrial-Media-Entertainment-Network* (Boulder: Westview).

Diken, B. and Laustsen, C.B. (2004), '7-11, 9/11 and Postpolitics', *Alternatives* 29:1, 89–113.

Dittmer, J. (2005), 'Captain America's Empire: Reflections on Identity, Popular Culture and Post 9/11 Geopolitics', *Annals of the Association of American Geographers* 95:3, 626–43.

—. (2007), 'Intervention: Religious Geopolitics', *Political Geography* 26:7, 737–39.

Dittmer, J. and Dodds, K. (2008), 'Popular Geopolitics Past and Future: Fandom, Identities and Audiences', *Geopolitics* 13:3, 437–57.

Faludi, S. (2007), *The Terror Dream: Fear and Fantasy in Post 9/11 America* (New York: Metropolitan).

Fitzgerald, F. (2008), 'Who Would Jesus Vote For?' *The New Yorker* June 30, <http://www.newyorker.com/reporting/2008/06/30/080630fa_fact_fitzgerald>, accessed 16 April 2009.

Forsyth, F. (2007), *The Afghan* (New York: Signet).

Gillem, M. (2007), *America Town: Building the Outposts of Empire* (Minnesota: University of Minnesota Press).

Gogwilt, C. (2000), *The Fiction of Geopolitics: Afterimages of Culture from Wilkie Collins to Alfred Hitchcock* (Stanford: Stanford University Press).

Gray, J. (2008), *Black Mass: How Religion led the World into Crisis* (Toronto: Anchor).

Gregory, D. (2004), *The Colonial Present: Afghanistan, Palestine, Iraq* (Oxford: Blackwell).

Halsell, G. (1986), *Prophecy and Politics: Militant Evangelists on the Road to Nuclear War* (Toronto: NC Press).

Hannah, M. (2006), 'Torture and the Ticking Bomb: The War on Terrorism as a Geographical Imagination of Power/Knowledge', *Annals of the Association of American Geographers* 96:3, 622–40.

Hofstadter, R. (1967), *The Paranoid Style in American Politics and Other Essays* (New York: Vintage).

Horsley, R. (2003), *Jesus and Empire: The New Kingdom of God and the New World Disorder* (Minneapolis: Fortress Press).

Ingram, A. and Dodds, K. (eds) (2009), *Spaces of Security and Insecurity* (Aldershot: Ashgate).

Kagan, R. and Kristol, W. (eds) (2000), *Present Dangers: Crisis and Opportunity in American Foreign and Defense Policy* (San Francisco: Encounter Books).

Kaplan, F. (2008), *Daydream Believers: How a Few Grand Ideas Wrecked American Power* (Hoboken, NJ: John Wiley).

Knelman, F. (1985), *Reagan, God and the Bomb: From Myth to Policy* (Toronto: McLelland and Stewart).

Koopman, S. (2008), 'Alter-Geopolitics: Another Geopolitics is Possible.' Paper presented to the Critical Geopolitics 2008 conference at Durham University.

Kuus, M. (2007), *Geopolitics Reframed: Security and Identity in Europe's Eastern Enlargement* (New York: Palgrave Macmillan).

Lieven, A. (2004), *America Right or Wrong: An Anatomy of American Nationalism* (New York: Oxford University Press).

Lindsay, H. (1970), *The Late, Great Planet Earth* (Grand Rapids, MI: Zondervan).

Linker, D. (2007), *The Theocons: Secular America Under Siege* (New York: Anchor).

Malinen, M. (2006), *Tom Clancy and Orientalism: Arabs and Muslims in the Contemporary Technothriller Novel*, University of Jyväskylä Unpublished Thesis.

Martin, A. and Petro, P. (eds) (2006), *Rethinking Global Security: Media, Culture and the War on Terror* (Camden: Rutgers University Press).

Megoran, N. (2008), 'Militarism, Realism, Just War, or Nonviolence? Critical Geopolitics and the Problem of Normativity', *Geopolitics* 13:3, 473–97.

Molloy, P. (2009), 'Zombie Democracy' in F. Debrix and M. Lacy (eds) *The Geopolitics of American Insecurity: Terror, Power and Foreign Policy* (London: Routledge).

Murphy, C. (2007), *The New Rome?: The Fall of an Empire and the Fate of America* (New York: Houghton Mifflin).

Northcott, M. (2004), *An Angel Directs the Storm: Apocalyptic Religion and American Empire* (London: I.B. Tauris).

Pain, R. and Smith, S. (eds) (2008), *Fear: Critical Geopolitics and Everyday Life* (Aldershot: Ashgate).

Power, M. and Crampton, A. (eds) (2007), *Cinema and Popular Geo-Politics* (London: Routledge).

Project for the New American Century (2000), *Rebuilding America's Defenses: Strategy, Forces and Resources For a New Century. A Report of The Project for the New American Century* (Washington, DC: The Project for the New American Century).

Rosenberg, J. (2006), *The Ezekiel Option* (Wheaton, IL: Tyndale).

—. (2008), *Dead Heat* (Wheaton, IL: Tyndale).

Samet, E. (2005), 'Leaving No Warriors Behind: The Ancient Roots of a Modern Sensibility', *Armed Forces and Society* 31:4, 623–49.

Schama, S. (2008), *The American Future: A History* (London: Bodley Head).

Sharp, J. (2000), *Condensing the Cold War: The Reader's Digest and American Identity* (Minneapolis: University of Minnesota Press).

Sturm, T. (2006), 'Prophetic eyes: the theatricality of Mark Hitchcock's premillennial geopolitics', *Geopolitics* 11:2, 231–55.

Wallace, I. (2006), 'Territory, Typology, Theology: Geopolitics and the Christian Scriptures', *Geopolitics*, 11:2, 209–30.

Williams, J. (2008), 'Space, Scale, and Just War: Meeting the Challenge of Humanitarian Intervention and Transnational Terrorism', *Review of International Studies* 34:4, 581–600.

Williams, M. (2003), 'Words, Images, Enemies: Securitization and International Politics', *International Studies Quarterly* 47:4, 511–31.

—. (2007), *Culture and Security: Symbolic Power and the Politics of International Security* (London: Routledge).

Chapter 5
The 'New World Order' and American Exceptionalism

Michael Barkun

Millenarians – those ardent believers in imminent end-time events – are often mistakenly thought to be date-sensitive, anxious to know the precise moment of the Second Coming. There have, of course, been dramatic cases of date calculation, such as the followers of William Miller in the 1840s. Miller, a lay preacher in upstate New York, wrestled with biblical chronologies until he determined that Christ would return sometime between March 21, 1843, and March 21, 1844. When that year passed uneventfully, he accepted a follower's calculation that the real date was October 22, 1844. That, too, came and went, and the tens of thousands in the movement dispersed in what came to be called the 'Great Disappointment.' But such attempts to find a precise millennial date have been the exception, not the rule. Despite concerns that the passage from 1999 to the year 2000 would set off all manner of religious fanaticisms, no such upheaval occurred.

Instead, millenarians tend to be event-sensitive. They look for portents, what they sometimes term 'signs of the times,' happenings that signal the nearness of the final days of history when the scenes described in such scriptural texts as the Book of Revelation will be played out. Such events have come to be sorted out in certain stereotypical categories, so that end-time believers can fairly readily separate irrelevant happenings from those that have eschatological significance. Inevitably, the categories have explicit or implicit geographical significance as well, since the events must play out in some spatial location. These categories are generally accepted by all millenarians who subscribe to the dispensational premillennial system prevalent among American fundamentalists and evangelicals. This system lays out a sequence of end-time events in which the saved will be 'raptured' off the Earth rather than having to endure the seven-year period of Tribulation when the Antichrist will rule. That period will end with the battle of Armageddon and Christ's return, along with those who have been raptured, at which point the Last Judgment will take place and a New Heaven and New Earth will replace the old.

Recognizing the 'Signs of the Times'

For those who believe they might be raptured, knowing how close the moment is becomes a matter of high consequence. A website called 'Rapture Ready' provides a running index of rapture imminence, based on 45 indicators (Rapture Ready). As of mid-August 2008, the index stood at 163, fairly high, since the site defines anything above 160 as "fasten your seat belts." The record, however, was 182 in September 2001, shortly after the 9/11 attacks.

Of the 45 indicators, fully 13 (roughly 28 percent) have an explicit or implicit geographical focus. About half deal with the Middle East, mostly with conflict involving Israel. A few deal with a revived Roman Empire (millenarian shorthand for the European Union). The rest concern global issues, such as alliances, nuclear war, and terrorism. Nine indicators involve natural disasters—floods, earthquakes, and the like. That leaves roughly half of the indicators—23—which concern decay in domestic society. Some are indicators of economic stress, such as unemployment, inflation, and crashes. The rest purport to measure the society's path to religious and moral downfall. These include everything from belief in the occult and Satanism, to apostasy and ecumenism, as well as drug use and a rising crime rate. Many of the indicators are keyed to biblical citations as warrants for their legitimacy.

What is most striking about the indicators is their capacity to adapt to changing political conditions. There are enough internationally oriented indicators to respond robustly to periods of international tension, yet when such tension is absent, the numerous domestic indicators can take up the slack. Given the range of problems the indicators collectively encompass, it is hard to imagine a period when the rapture index would fall below 100, a sign of "slow prophetic activity," although in fact it did drop to a record low of 57 in December 1993, as the Cold War was ending. The reason is not difficult to seek. The index is a composite of 45 portents that range from crime and Satanism, to floods and earthquakes, to Israel and global turmoil, to Antichrist and Beast Government.

The end-time scenario of dispensational premillennialists assigns key events to the Middle East (see Chapter 6 in this volume). The climactic moments occur in and around Israel, where a new Temple must be built to fulfill prophecy and where the battle of Armageddon is to take place. Absent conflict in that region, no amount of tumult and moral decay elsewhere can fully compensate, although, as we shall see, millennialists have done their best to try. Consequently, those anticipating imminent end-time events have been frustrated during periods where international conflict seemed unlikely, particularly conflict in the Middle East. Since the dispensationalist script assigned Russia an important role in battles with and on the soil of Israel, the sudden collapse of the U.S.S.R. suggested that the end times, far from being imminent, lay somewhere in the dim, distant future.

The 'New World Order'

Thus the end of the Cold War in the early 1990s, while it was welcomed by many, was not necessarily welcomed by millennialists, for whom the trumpets of war are often a welcome sound. In this time of seeming millennial doldrums, the concept of the 'new world order' suddenly became attractive. Yet it was not a new concept, nor was it one that was particularly linked to conservative Protestants.

The phrase 'new world order' ran, as it were, on two tracks. It appeared both in the rhetoric of the extreme political right and in that of mainstream foreign policy discourse. On the extreme right, its first appearance seems to have been in Gary Allen's 1971 book, *None Dare Call It Conspiracy*. Allen argued that a cabal of the super-rich had been working for decades to implement a plan to socialize America as the first stage in a larger design to establish "a dictatorial world government." "The *Insider's* code word for the world superstate," Allen told his readers, "is 'new world order'..." (Allen 1971, 121). Although Allen's publisher claimed that five million copies were in print by the following year, the phrase did not immediately take hold. It did not appear with frequency in right-wing circles until the early 1990s.

The sudden popularity of the phrase among extremists was due in no small measure to its eventual appearance in mainstream politics, for it became a political catch phrase only after its use by President George H.W. Bush. What Bush meant by 'new world order,' however, was not a conspiracy but rather a post-Cold War global security arrangement. Speaking to a joint session of Congress just before the First Gulf War, he was referring to the possibility of cooperation rather than conflict with Soviet Premier Mikhail Gorbachev. Nonetheless, his popularization of the phrase inadvertently confirmed the fears of right-wing conspiracists, given their suspicions about Bush – his privileged Eastern upbringing, his university membership in Skull and Bones, his leadership of the CIA, all characteristics that appeared to validate his membership in some conspiratorial inner circle. To extremists, his use of the phrase in public speeches was a brazen act that signaled to them that the cabal was on the verge of making an open grab for power.

Bush's use of 'new world order' also coincided with the rise of the militia movement on the far right. These paramilitary organizations, deeply suspicious of the federal government, provided an organizational setting for the dissemination and reinforcement of increasingly bizarre conspiracy theories. The New World Order (now always capitalized) was said to operate through armadas of black helicopters; gun owners and dissidents were to be incarcerated in concentration camps operated by FEMA; and 'New World Order' was said to be the real translation of 'Novus Ordo Seclorum' on the back of the one-dollar bill (Barkun 1996). The phrase in fact means 'A new order of the ages.'

New World Order, which was now both a commonplace of mainstream political speech and an important element in the ideology of anti-government militants, was about to enter the religious realm. However, its migration into millenarian rhetoric was not without difficulties. In the first place, the concept had been created by secular conspiracy theorists as a comprehensive explanation for political decision-making. Religion played no part in its origins, and it had no inherent religious significance. Second, many versions of New World Order conspiracy theory were heavily anti-Semitic, placing international Jewish bankers at the center of the cabal that allegedly sought world power. As a result New World Order ideas were not only out of the political mainstream by virtue of their conspiracism; they were stigmatized because of their frequent association with anti-Semitism. In order for the New World Order to play a significant role in the end-time thinking of conservative Protestants, some way would have to be found around both of these obstacles.

The Role of Antichrist

New World Order's entry point into millennial thought was through the concept of the Antichrist. The Antichrist appears fleetingly in the New Testament, in brief references in the First and Second Epistles of John. These cryptic mentions, suggesting a deceiver or false prophet, were embellished by subsequent biblical expositors, who raised the Antichrist to a person of literally demonic proportions. Those who speculated about this mythic figure saw him as Satan's representative on Earth in the last days. This was especially true of John Nelson Darby (1800–82), who developed dispensational premillennialism. Darby argued that the Rapture would be followed by seven years of the Tribulation, at the midpoint of which the Antichrist would assume dictatorial control of the world preceding the battle of Armageddon. Darby's follower, C.I. Scofield, published his influential *Reference Bible* in 1909, which integrated the concept of Antichrist into a cohesive interpretation of scriptural texts for fundamentalist readers (Fuller 1995, 125).

The vast speculative literature that arose about Antichrist initially focused on who he might be, and there seemed no shortage of candidates. At one time or another figures as various as the Pope and Mussolini have been suggested (Fuller 1995, 37, 161). However, by the post-World War II period, another approach began to appear, in which the emphasis was less on the *person* of Antichrist and more on the Antichrist *system*. That meant linking Antichrist with institutions and forms of control. The Antichrist system was sometimes identified with the United Nations, although more often with the European Common Market (later, the EU), on the theory that it would be the Roman Empire, reborn. Sometimes, Antichrist was identified with a giant computer, and fanciful millenarian folklore claimed that an EU computer somewhere in Brussels would eventually control the world (Boyer 1992, 283; Lalonde and Lalonde 1994, 197). The proliferation of credit and debit cards, bar codes, and computers provided ample material for this new Antichrist literature.

Once the focus was on Antichrist as a system of control rather than on a single individual, two further developments became possible. First, conspiracy theories of all kinds became in theory compatible with Antichrist speculation, for global control meant eroding and then destroying the sovereignty of existing states. Hence New World Order theories became candidates for adoption by millenarians. Second, the system of control perspective allowed millenarians to concentrate on either international or domestic developments. Any evidence that state power was becoming diluted through the growth of international organizations, such as the

European Union, was evidence in favor of the Antichrist's rise. However, any evidence on the national level that power was becoming concentrated, that elites were gaining at the expense of the citizenry, or that new technologies might be used to exert control was also deemed important. Therefore, if traditional events important for end-time determination, such as conflict in the Middle East, were temporarily in abeyance, there were now ample alternatives available.

From this perspective, the obstacles that stood in the way of New World Order theories now seemed less daunting. The fact that they were out of the political mainstream and that many of their practitioners were anti-Semites now appeared manageable. If, after all, the activities allegedly explained by the theories were being attributed to the diabolical Antichrist, why be disturbed by such niceties?

The New World Order Enters the Prophecy Literature: Pat Robertson and John Hagee

One of the principle channels through which the New World Order became legitimized as a category of Christian end-time discourse was Pat Robertson's book, *The New World Order* (1991). Robertson was arguably the best known of the televangelists associated with the New Christian Right. He was then nearing the height of his fame. He had unsuccessfully sought the Republican presidential nomination in 1988. The following year he had turned the day-to-day operations of his political action organization, the Christian Coalition, over to Ralph Reed, an organizational genius, and the Coalition then entered a period of extraordinary growth through the mid-1990s. *The New World Order* sold in the hundreds of thousands and appeared in paperback racks in airports and bookstores throughout the country. Robertson's end-time theology has been inconsistent. He has vacillated between a premillennial expectation of catastrophes in the last days and a postmillennial demand for Christians to "take dominion" over the world until Christ comes (Weber 1983, 216, 232; Boyer 1992, 138, 303–04). Regardless of these inconsistencies, however, the scenario he presented in *The New World Order* fit perfectly into the vision of dispensationalists, with its picture of the onrushing forces of Satan.

Much of Robertson's book was a pastiche of conspiracy materials that had long circulated on the right. He stressed the evil, revolutionary designs of the Illuminati and the machinations of international bankers, most with Jewish surnames. He implied but did not state outright that a common plan existed linking global finance, Ivy League institutions, the Council on Foreign Relations, and New Age practitioners in order to simultaneously destroy national sovereignties and the Christian religion. The way would then be open for a world state dominated by an Antichrist Luciferian religion. The conspirators are presented as members of a cultural elite whose values and background have nothing in common with those of the American public. They also tend to be institutions associated with internationalism, pursuing or favoring links between America and other nations.

It is as if a giant plan is unfolding, everything perfectly on cue. Europe sets the date for its union. Communism collapses. A hugely popular war is fought in the Middle East. The United Nations is rescued from scorn by an easily swayed public. A new world order is announced. Christianity has been battered in the public arena, and New Age religions are in place in the schools and corporations, and among the elite. Then a financial collapse accelerates the move toward a world money system. (Robertson 1991, 176–77)

Only Christian America, Robertson says, stands in the way of Satan's plan. "(I)f America goes down, all hope is lost to the rest of the world" (Robertson 1991, 256).

There is here, as in many millenarian texts, a struggle between determinism and choice. On the one hand, as Robertson himself concedes, biblical prophecy requires, at least in a dispensationalist reading, that the Antichrist must rule for three and a half years. The Tribulation, with all its war and suffering, must take place. On the other hand, he wishes to hold its onset back as long as possible, in part to protect Israel and in part to allow politically mobilized Christians time to influence public policy in the United States. He suggests that his Christian Coalition may be the way to pry government loose from the hold of the "Establishment."

The New World Order's recurrent emphasis on strengthening America suggests the influence of American exceptionalism. From the late nineteenth-century on, the millennialism that dominated evangelical Protestantism had no role for America. Its geographical focus was, as indicated earlier, on the Middle East. Supporting roles were given to Russia and China as allies of Arab forces, but biblical expositors in mainstream conservative Protestantism could find no texts or place names that suggested the United States. Robertson, building on the concept of an Antichrist system, viewed the United States as the last bastion of defense against Satanic power.

When *The New World Order* appeared, it passed unnoticed as far as serious reviewing organs were concerned, despite Robertson's rising political significance. However, Robertson's book eventually burst on the consciousness of American intellectuals, most of whom had never heard of it, as a result of two long articles in *The New York Review of Books* in 1995 (Lind 1995; Lind and Heilbrun 1995). The authors, Michael Lind and Jacob Heilbrun, argued that much of the conspiracy material on which Robertson relied had been drawn from the work of two anti-Semitic writers, Nesta Webster and Eustace Mullins. Webster (1876–1960), a British author whose conspiracy writing in the interwar period, argued both for the continued existence of the Illuminati and for their involvement with "the Jewish power." Mullins (b. 1923), a protégé of Ezra Pound's, claimed that the Jews controlled the Illuminati. Although Robertson insisted that he intended no anti-Semitism and was a staunch friend of Israel, he never explained how Webster's and Mullins's material came to be used in the book (Robertson 1995).

John Hagee is a more complex case. Hagee is pastor of the 17,000-member Cornerstone Church in San Antonio, Texas, and, unlike Robertson, was little known outside evangelical and Christian Zionist circles until recently. He came to public notice during the campaign for the 2008 Republican presidential nomination after his endorsement of John McCain led opponents of McCain to retrieve anti-Catholic sermons Hagee had given. Hagee subsequently apologized to the Catholic community in a letter to William Donohue, the president of the Catholic League for Civil and Religious Rights. Among other recantations, Hagee promised to stop referring to Roman Catholicism as the "great whore." Nonetheless, McCain eventually repudiated the endorsement when it became known that Hagee had once delivered a sermon asserting that Hitler was part of God's plan to facilitate the creation of a Jewish state, and thus fulfill prophecy (Eilperin and Kindy 2008).

At times when Hagee might well have invoked the New World Order, he has not done so. For example, he issued a tract on the millenarian significance of the 9/11 attacks. He wrote of how, watching news reports of the planes crashing into the Pentagon, "I recognized that the Third World War had begun and that it would escalate from this day until the Battle of Armageddon" (Hagee 2001, 4–5). Much of the rest of the book, *Attack on America*, restates the familiar dispensationalist sequence of end-time events. There is a brief discussion of the characteristics of the Antichrist's tyranny, with due attention given to a worldwide economy made cashless by computerization; and a world government in which nation-states and patriotism will cease to exist (Hagee 2001, 191–94). A perfect opening for invoking the New World Order, one would think, but it goes unmentioned.

Yet that is not because Hagee is unaware of the phrase or unwilling to use it. A few years later, in a 2008 video sermon entitled "Novus Ordo Seclorum (New World Order) – The Final Warning," he integrated it seamlessly into his millenarian framework (Hagee 2008). "There's a New World Order coming," he predicted. He saw the war in Iraq as the catalyst for an Arab coalition which would, under Russian leadership, move against Israel, as Ezekiel had foretold. That led him to claim that 2008 might see the rise of Antichrist, "Satan's Messiah," from somewhere in Europe. As the title of the sermon suggests, he noted the motto on the reverse of the one-dollar bill as evidence that Masonic influence had infiltrated the U.S. government, and by implication, was preparing the way for the Satanist Illuminati operating through manipulation of the levers of financial power.

It was standard New World Order rhetoric, strikingly similar to the ideas Pat Robertson had advanced seventeen years earlier, except now John Hagee had the advantage of the lurid imagery video could produce, that might do some justice to the apocalyptic vision of Revelation. What is also worth noting about Hagee's sermon is how effortlessly it moves between international and domestic concerns. The apocalypse will play out on an international stage, as dispensationalists have always expected, on the territory of Israel, with invaders from Europe and Asia. But there is enough New World Order and Antichrist material describing end-time dictatorship to make clear that the state of America is also a significant concern.

Hagee's appropriation of the New World Order is also worth noting because of his pivotal position as a Christian Zionist. Notwithstanding his bizarre sermon about Hitler's role in God's plan, there is perhaps no other evangelical pastor for whom the defense of Israel is so central a part of his ministry. Hagee is the Founder and National Chairman of Christians United for Israel, the principal pro-Israel lobbying organization for evangelicals. Its Executive Board includes such figures as Gary Bauer and Jonathan P. Falwell. Robertson claimed similar pro-Israel views, but coupled his use of the New World Order with anti-Semitic supporting materials. Hagee, on the other hand, employed the concept in an entirely unselfconscious way, suggesting that he was totally unaware of the anti-Semitic baggage it carried.

The New World Order and the Geography of Control

New World Order conspiracy theories, in both secular and religious versions, claim that the plotters wish to create a world state in which individual nations and boundaries have disappeared. While some international relations theorists and foreign policy analysts have argued that the nation-state is gradually disintegrating, only visionary advocates of world government have actually advocated the dismantling of the state system itself. The claim made by predictors of the New World Order fall into a different category. They assert that as a matter of fact the conspirators, in the secular version, or the Antichrist's minions, in the religious version, will exert such intense and insidious power that governments will collapse. A combination of agents working from within and newly empowered international organizations will destroy a system which has existed since at least the Treaties of Westphalia in 1648.

The belief that the New World Order will destroy individual states makes New World Order scenarios particularly appealing to those motivated by nationalist appeals. The New World Order must be fought, the argument goes, because unless it is defeated, America (and all other nations) will cease to exist, absorbed into a homogeneous world state. Here, again, one finds the millenarian tension between determinism and choice. For, presumably, the rise of Antichrist is a necessity of the end-time plan, just as will be his inevitable defeat. But the rallying cry to arm against the forces of Antichrist in order to protect American sovereignty, paradoxically, is made by the same clergy who lay out the invariant millennial timetable. Antichrist and the Satanic Illuminati forces that do his bidding must be stopped. The apparent logical contradiction is nonetheless functional, for instead of passivity, it counsels mobilization and action against the forces of evil. The time gained offers additional opportunities to evangelize those who have not yet been saved. And it exerts a particular appeal to those on the political right open to calls for higher defense spending, a more aggressive foreign policy, and greater emphasis on internal security.

If the 'macro' element of New World Order control is the elimination of the state, surely the 'micro' element is the utilization of electronics to monitor individual behavior. In this respect, religionists were in advance of their secular counterparts, no doubt because of their desire to identify real world counterparts of the "mark of the beast" required in order to buy or sell during the Antichrist's reign (Revelation 13:16–17). Millennialists became caught up in the possibilities technology offered for realizing this prophecy, beginning with Mary Stewart Relfe's suggestion in 1982 that bar codes used by supermarkets and other retail outlets to scan products for price information could provide just such a mark (Boyer 1992, 288). However, the bar code vogue was short-lived among end-time believers. It quickly went out of fashion as it was overtaken by a new and more compelling technology. By 1991, at the same time the New World Order was making its appearance in millenarian circles, implanted microchips emerged as the Antichrist's control of choice (Fuller 1995, 180–81). That seemed not only more up-to-date but more insidious and consequently more in keeping with what might be expected of Satan's representative on Earth.

There were numerous advantages to these speculations, as far as chiliasts were concerned. First, they made millenarianism appear au courant, for in addition to drawing in the day's political news as millennialists always had, they could now digest the latest in information technology for its prophetic significance. Second, such speculation effectively collapsed the division between international and domestic affairs. The purpose of the beast's mark was to create a world economy (yet another way of destroying national sovereignty), but the means for doing so supposedly could be seen at the individual level, in the mundane transactions of daily life. Even if international affairs appeared temporarily quiescent, therefore, developments with world-shaking importance might be occurring right under one's nose. Finally, the emphasis on the technology of control and the bridging of the international and the domestic made clear (if that, indeed, was even required) the anti-globalization bias of contemporary millennialism. Globalization was not merely some automatic and impersonal process; it was the work of Antichrist. It was said to provide the tangible evidence that the New World Order was on the march.

This anti-globalization bias has given contemporary millennialism the appearance of a political ecumenism. As was noted earlier, the concept of the New World Order was appropriated from the political far right in the early 1990s. Although there are many on the right who also hold anti-globalization views as well, antipathy towards globalization is also intense on the left. There seems to be no evidence that millenarians' distrust of globalization has anything to do with a desire to curry favor with the political left. The similarities of viewpoint seem entirely inadvertent. Yet it once again demonstrates the willingness of millenarians to adopt language and viewpoints irrespective of their political associations.

The Continuously Receding End Times

Because most millenarians are not date-sensitive, they are not trapped by calculations as the Millerites were in the nineteenth century. On the other hand, those who adhere to the dispensationalist system – and they constitute the great majority of evangelicals and fundamentalists – have a clear idea of the events that most clearly portend the end of days, namely, those connected with war in and around the Land of Israel. They can not only point to biblical passages that mention armed conflicts in the Holy Land and adjacent areas; they see such clashes as catalysts that can make the confirmation of other prophecies possible. For this reason, the Six-Day War (1967) was of tremendous significance to Christian millenarians, not so much for the fact that it took place as for its consequences. It gave Israel borders that more nearly approximated those of the former Davidic kingdom and it reunified Jerusalem. The latter was singularly important, since it made possible the rebuilding of the Temple (although that would, of course, require the displacement of Muslim holy sites). Although ultra-Orthodox Jews might treasure the idea of a rebuilt Temple as well, it is essential for Christian millennialists because of John Nelson Darby's concept of "rightly dividing the word of truth." That is, Darby distinguished between those prophecies that had to be fulfilled before the end times proper (Rapture, Tribulation, and so forth) and those prophecies that applied to the end times themselves. He asserted that all prophecies that applied to the Jews had to be fulfilled before events relevant to Christian eschatology could take place.

Yet events in the Middle East have not unfolded in the linear fashion that the 1967 war suggested to millenarians. Their euphoric expectation dimmed in light of the Oslo process and the withdrawal from Gaza, just as it heightened again during the two Gulf wars. Intermixed with these events were long stretches where the direction of events in the region was uncertain. The leadership role given to Russia in the final battles was suddenly made problematic by the dissolution of the Soviet Union. Millennialists have felt chained to the region both by biblical texts and by their allegiance to Darby's interpretive system. Thus although not hostages to any specific date, they may be said to be hostages to events in a region where politics is notoriously difficult to predict and where sudden changes appear to be the norm. Just as the end times appear to be in sight, they seem to recede as yet another political obstacle prevents fulfillment.

Yet millenarians are nothing if not flexible. As much of the research on cognitive dissonance has shown (Dawson 1999), those who hold strong beliefs often make great efforts to preserve those beliefs, even in the face of an uncooperative reality. And where most millenarians are concerned, the issue is not prophetic failure or disconfirmation; rather, it is delay. It is true that writers and preachers often speak of theirs as 'the last generation,' but they usually build in some escape clause so that they do not become trapped in the kind of chronological cul-de-sac that destroyed the Millerite movement in the 1840s. Contemporary millennialism has preserved its elasticity, its ability to adjust to changed circumstances, and it has done so precisely because it is built on elastic texts.

Although the dispensationalist system is relatively restrictive, the biblical texts that lie behind it are often couched in general or obscure terms that invite multiple interpretations. This is particularly true of the Book of Revelation, the text most identified with end-time prophecies. Its symbols are powerful but at the same time sufficiently vague to be available to interpreters with a wide range of agendas.

For example, the late nineteenth-century Populist, Carl Browne, saw in the beast with seven heads, ten horns, and ten crowns (Revelations 13:1) a representation of a national bank, pieces of financial legislation, "the ten great trusts," and the plutocrats who controlled them (Barkun 1988). If so far-fetched a meaning might be extracted in the 1890s, it is little wonder that millenarians a century later might find similar flexibility.

The problem now is to account for history's detours and decelerations at a time of high expectation. The New World Order is perfectly suited to perform this function. Resting as it does on passages in the Epistles of John as cryptic as those in Revelation, the New World Order has come to represent the onrushing ambitions of Antichrist. Even when events in the Middle East appear to be in the doldrums or, worse yet, at times when Arab-Israeli peace impends, the New World Order provides evidence that the apocalyptic timetable remains on track. It does so, as we have seen, by subtly shifting the geographic focus.

When attention is paid to the New World Order, examples may be taken from the domestic sphere: from governance, information technology, alleged activities of secret societies, and the like. Since Antichrist is presumed to be seeking dominance whether or not war threatens the Holy Land, anything that might suggest his rise augurs the nearness of the Tribulation. Indeed, if millenarians believe there is such evidence, they become convinced that war in the Middle East may be only months away. For how could it not be, given the brazenness of Antichrist's encroachments?

Inverting America's Millenarian Role

Incorporating the New World Order into millennialism did nothing less than invert America's millennial role. That role derived from a religious history even older than the Republic. It is true, of course, that the end-time script of dispensationalism, to which most Protestant millenarians adhere, has no specific task for the United States. While the Arab states, Russia, and even China figure in eschatological events (Boyer 1992, 152–80), biblical expositors found no text that referred to America. Yet, as we have seen, millennialists were able to assign a powerful secondary role to the U.S. as the supporter and defender of the State of Israel, for only a strong Israel could act in a manner that would ensure that the 'prophetic clock' would continue to tick. In Darby's system, all prophecies concerning the Jews had to be fulfilled before those that concerned Christians, such as the Second Coming, could take place. Consequently, even though America's protective relationship toward Israel might not be scripturally grounded, it appeared to be essential for prophetic fulfillment.

Yet the dispensationalist approach took place within a broader cultural context. From the days of the Puritans, it had become an article of faith that America was not like other nations, that it had a redemptive role in history as God's 'new Israel.' As Robert Bellah pointed out in his seminal essay, this belief in a redemptive vocation became a central element in the "civil religion" that underlay American political institutions (Bellah 1967). This belief was never more vividly expressed than by Herman Melville in his 1850 novel, *White-Jacket or, The World in a Man-of-War*, a passage so remarkable that it deserves to be quoted at length:

> Escaped from the house of bondage, Israel of old did not follow after the ways of the Egyptians. To her was given an express dispensation; to her were given new things under the sun. And we Americans are the peculiar, chosen people – the Israel of our time; we bear the ark of the liberties of the world. Seventy years ago we escaped from thrall; and, besides our first birth-right – embracing one continent of earth – God has given to us, for a future inheritance, the broad domains of the political pagans, that shall yet come and lie down under the shadow of our ark, without bloody hands being lifted. God has predestinated, mankind expects, great things from our race; and great things we feel in our souls. The rest of the nations must soon be in our rear. We are the pioneers of the world; the advance-guard; sent on through the wilderness of untried things…Long enough have we been skeptics with regard to ourselves, and doubted whether, indeed, the political Messiah has come. But he has come in us, if we would but give utterance to his promptings. And let us always remember that with ourselves, almost for the first time in the history of earth, national selfishness is unbounded philanthropy; for we can not do a good to America but we give alms to the world. (Melville reprinted 2002, 150–51)

This extraordinary text does more than merely link America with ancient Israel. It also equates the American constitutional legacy with the contents of the Ark of the Covenant, the tablets bearing God's very writing, and claims that this legacy is so powerful that it will guarantee America a global empire, bloodlessly attained. The "political pagans" will joyfully lie at our feet in a democratic millennium, for the nation constitutes a potential collective messiah ("he has come in us"), because the selfish acts of America in reality lift up the rest of the world. It is Manifest Destiny as a religion. Melville was scarcely the only person to think in these terms. Shortly before, in 1846, William Gilpin had written of "The *untransacted* destiny of the American people…to subdue the continent" not for its own selfish gain but in order "to unite the world in one social family – to dissolve the spell of tyranny and exalt charity – to absolve the curse that weighs down humanity, and to shed blessings round the world!" (Smith 1950, 40) Like Melville's, Gilpin's is a millenarian vision for which America is the essential instrument.

Ideas like these, exalting an American sense of mission, had been deeply implanted in American culture long before dispensationalist premillennialism appeared on the religious horizon. The problem for contemporary millenarians has been, as Timothy Weber notes, that "In theory, at least, this millennial vision of America and premillennial eschatology are incompatible" (Weber 1983, 237). It may be said that the absence of scriptural references to an American end-time role has been softened for believers by the residual cultural power of earlier beliefs in national mission, so powerfully have these infused the cultural ethos.

The appearance in millenarian circles of New World Order ideas, however, represents a significant shift. It is true that in the past millenarians have predicted divine punishment for declining moral standards and spoken darkly about the coming rule of Antichrist (Boyer 1992, 230–90). Speculation about Antichrist is a long-running theme in conservative American religion, so much so that one of its leading chroniclers, Robert Fuller, describes "naming the Antichrist" as an obsession, "a persistent preoccupation with an irrational idea or feeling," (Fuller 1995, 191). However, the New World Order originally had nothing to do with Antichrist. It was a secular conspiracy theory that claimed to describe and explain all political evil. Antichrist was merely the religious 'hook' that brought it into the realm of millenarian speculation. But once that took place, the millenarian scenario was transformed. The importation of New World Order theory signaled a sea-change. Where previous gloom-and-doom scenarios concerning the United States emerged organically out of religious ideas, the New World Order constituted a comprehensive theory of political evil that was now engrafted onto religious conceptions. In so doing, it turned the millennial role of America upside down.

According to the New World Order construct, America has become a stage on which the forces of evil perform virtually uncontested. They control major institutions, both public and private: corporations, universities, media, and government agencies. Their operatives have infiltrated the corridors of power. They are in constant communication with counterparts in other regions of the world. They are said to have at their disposal an array of military and paramilitary forces, including UN peacekeepers, black helicopters, and concentration camps ready to receive dissenters. They await only a signal to make the final open grab for total power. While this scenario is certainly compatible with pre-existing ideas about Antichrist, it is laid out with a degree of detail seldom if ever seen in the religious literature. It often comes with an apparatus of names, dates, and places that provides a pseudo-scholarly appearance of validity.

The picture of America that emerges is one of a nation so permeated by the forces of evil that any sense of national virtue or of redemptive vocation has been lost. What remains is a country seemingly on the brink of destruction, that can be saved only by divine intervention. The salvationist role once claimed for America so eloquently by Herman Melville and William Gilpin can nowhere be seen. Instead, the importation of the New World Order into millennialism presents a picture of a country morally desolate, on the verge of conquest by the forces of Satan. Those millenarians who have found the New World Order construct congenial have not consciously turned their backs on older concepts of an American mission. Instead, they saw NWO theory merely as a convenient way of both expressing fears of the Antichrist and of shifting attention to domestic distress at times of torpor in the Middle East.

Whatever the expediential reasons, this inadvertent admission of right-wing conspiracism has introduced a powerful strain of pessimism. Although millenarians necessarily remain confident that God and Christ will triumph in the end, New World Order ideas greatly magnify their concept of Satanic power, make the victory of virtue more difficult, and portray America as a land in the grip of evil.

References

Allen, G. (1971), *None Dare Call It Conspiracy* (Rossmoor, CA: Concord Press).
Barkun, M. (1988), "Coxey's Army' as a Millennial Movement', *Religion* 18:4, 363–89.
—. (1996), 'Religion, Militias and Oklahoma City: The Mind of Conspiratorialists', *Terrorism and Political Violence* 8:1, 50–64.
Bellah, R. (1967), 'Civil Religion in America', *Daedalus* 96:1, 1–21.
Boyer, P. (1992), *When Time Shall Be No More: Prophecy Belief in Modern American Culture* (Cambridge: Harvard University Press).
Dawson, L. (1999), 'When Prophecy Fails and Faith Persists: A Theoretical Overview', *Nova Religio* 3:1, 60–82.
Eilprin, J. and Kindy K. (2008), 'McCain Rejects Pastor's Backing Over Remarks', *Washington Post* (May 23), A01.
Fuller, R. (1995), *Naming the Antichrist: The History of an American Obsession* (New York: Oxford University Press).
Hagee, J. (2001), *Attack on America: New York, Jerusalem, and the Role of Terrorism in the Last Days* (Nashville, TN: Thomas Nelson).
—. 'Novus Ordo Seclorum: The Final Warning' (added July 13, 2008) <http://www.youtube.com/watch?v=6_LXmq1ir7A>, accessed 16 June 2009.
Lalonde, P. and Lalonde, P. (1994), *The Mark of the Beast* (Eugene, OR: Harvest House)
Lind, M. (1995), 'Rev. Robertson's Grand International Conspiracy Theory', *The New York Review of Books* 42 (February 2), 21–25.
Lind, M. and Heilbrun, J. (1995), 'On Pat Robertson', *The New York Review of Books* 42 (April 20), 71–76.
Melville, H. (2002), *White-Jacket or, The World in a Man-of-War* (New York: Modern Library).
Rapture Ready. (2008), <http://www.raptureready.com/rap2.html>, accessed 16 April 2009.
Robertson, P. (1991), *The New World Order* (Dallas: Word Publishing).
—. (1995), 'Statement', *The New York Times* (March 4), 10.
Smith, H. (1950), *Virgin Land: The American West as Symbol and Myth* (New York: Vintage Books).
Weber, T. (1983), *Living in the Shadow of the Second Coming* (Grand Rapids, MI: Academie Press).

Chapter 6
Imagining Apocalyptic Geopolitics: American Evangelical Citationality of Evil Others

Tristan Sturm

> My interest in geography has aided me significantly in my study of prophecy. Bible prophecy involves many different nations and places. Knowing the historical locations of these places and their modern counterparts is indispensable in understanding the outline of end time events and places.
>
> (Mark Hitchcock, March 13th, 2006 interview with the author)

To the "Republican base", as Karl Rove referred to evangelical Christians, President Bush's axis of evil has a connotation of a particular geopolitical shape, what one popular dispensational premillennialist theologian, Mark Hitchcock (2002, 27), has called "God's 'Top-Ten Most Wanted List.'" Historically, President Reagan and members of his government were known to make apocalyptic and religiously motivated statements. It is not known whether Reagan's foreign policy decisions were consciously shaped by his interpretation of prophecy. His massive military spending and his hyper-moral religious language of 'evil empires' was, however, consistent with prophetic belief (see Halsell 1986).[1] Regardless of a political leader's true belief, these phrases and representations shape a highly moralized sense of space for those listening with real implications for foreign policy in an era when born-again evangelicals play a significant role in right-wing American politics.

The most worrying aspect of all this is, of course, is that dispensationalists are not the fatalists they propound to be. Rather they have the agency to take action, effectively moving from the space of biblical and predestined prophecy to that of self-fulfilling reality. Communications scholar, Barry Brummett (1991, 89) writes of dispensational apocalypticism that "because humans cannot alter history, then apocalyptic [discourse] can therefore have nothing historically grounded or relevant to say." Although dispensationalists believe they do not change history, it is a fact that people change history, dispensationalists change history, and dispensational politicians change history through various cultural, economic, and political networks of power. For example, since the 1967 Six-Day War,

---

[1] But during the 1984 presidential campaign he made it clear that his foreign policy was "never [a]... plan according to Armageddon" (Boyer 1992, 142).

dispensationalists have taken the lead in reaching out to Israel and lobbying for pro-Israeli causes because they believe that Israel will make substantial territorial gains based on God's promise to Abraham for all the land west of the Euphrates to the "river of Egypt" (Gen. 15:18). Many also believe that Solomon's Temple will need to be built on the site where the Dome of the Rock now sits.

This chapter is concerned with these geopolitical power relations as they are aligned with American foreign policy mappings of the world and how the changing geopolitical themes present in the dispensationalist writings of Cold War evangelical authors Hal Lindsey, Tim LaHaye, and John Walvoord are reterritorialized by post-Cold War evangelical authors Joel Rosenberg, Mark Hitchcock, and Charles Dyer.[2] All of these authors play the part of knowledgeable geographer, as the epigraph above suggests, and intellectual of statecraft. Using images, popular dispensationalist books, and interviews, this chapter offers a comparative cultural history of the protean geopolitical boundaries imagined by dispensationalists and their historical relation to American foreign policy from 1970 to 2008. Paul Boyer's (1992) major historical work on prophetic thought in America was published too early to thoroughly examine dispensationalism's most recent imagination as to where evil and impiety reside. Boyer (1992, 180) writes of Russia's declining importance after the Cold War: "in view of the prophecy writers' resourcefulness and long experience at adapting to changing realities... this restructuring will probably be successfully accomplished."

What is most interesting are the strategies employed to legitimate these protean geographies as the true, literal, and inerrant words of God in the face of changing geopolitical realities. This is achieved specifically through the use of geography for strategies of radical Othering that draw clearly defined lines between evil and good, and also through the citationality of themes in preceding books that 'prove' the reality of end times. As work on cognitive dissonance has made clear, those with a firm set of beliefs work hard to sustain these beliefs in light of a changing world (Festinger 1956; see Dawson 1999; Howard 2006). Harding (2000) argues that biblical prophecy needs to be constantly updated because prophecy expounders read history backwards, interpreting the present in line with the future. Furthermore, Barkun (2003) argues that it is the elastic or obscure texts of Revelation and Ezekiel that allow for a variety of interpretations. Complimentary to these explanations, the main argument of this chapter is that the use and abuse of geography as a signifier for end time fervor allows prophecy writers to continuously update their predictions as though each renaming of a region or

---

2 These authors are not, it must be said, the only voices in the evangelical community; many others hold much disdain for such interpretations of scripture (see Northcott 2004; Sizer 2004). Indeed they are of a radical right fringe but the resonance of their voices in all spheres of life should not be underestimated. That said, this chapter should not be read as an affront attack on evangelicals, I have made the normative claim elsewhere that there is much within the evangelical community to envision a more pluralistic and less territorially austere geopolitics (Sturm 2006; see also Connolly 2008).

movement of a border is evidence that the end is imminent. By investigating these geopolitical contradictions and religio-governmental resonances it is possible to see, functionally at least, the importance of the signification of these mythical geographies.

In the following section, it is argued that binary geographies of good-Christian and evil-Communist or Muslim help legitimate geopolitical constructions of the world that qualify Others on the basis of state or regional borders. The identification of these moral geographies is part of an internal dispensationalist geopolitical hegemony concerning the identification of places that continuously uses the same citations and patterns of identification, from compass location to linguistic similarity. Dispensationalists also strive to mirror American foreign policy within the limits of their biblical taxonomy of place names by cherry-picking sources from history that corroborate their claims. Three empirical sections follow the theoretical section below. The first briefly traces the evolution of conflating places and armies of Armageddon from medieval Christianity to Cold War evangelicalism of the 1970s. The second empirical section traces the post-Cold War apocalyptic geopolitics and the third section describes the post-Iraqi threat geopolitical imagination of contemporary evangelical writing of the apocalypse. It is concluded that dispensationalists use geographical citationality of evil Others as they relate to American political and cultural identifications of evil to prove the validity of their claims.

The Protean Geopolitics of Strategies of Naming Others

Spatial terminology is ever present in religious studies and historical and political analysis, although often as metaphors or descriptions (Foucault 1980). Geographies are dynamic entities with the power to change and interact with the social or historical processes that are being located and explained. Geographic metaphors are overarching 'meta'-products of our minds that are overlaid onto a world and used as inactive templates from which to define the world whether through politics or from the necessity of stabilized boundaries for empirical and statistical research (Strong 1984; see Chapter 7, this volume concerning governmentality). Such practices can make seemingly complicated issues understandable by virtue of transposing one set of densely packed histories on to another situation (Beer and de Landdtsheer 2004). They make familiar the unfamiliar, and thus provide a model for identifying evildoers and victims. They are used as abstract labels for political blaming and calls to action by being underwritten by the loaded meanings of the terms. Geographical imaginations, however, are ever-changing and the boundaries that 'contain' these imaginations, ideological or not, are not simply permeable or contingent, but very much part of a dynamic and historically traceable, in the case of this chapter, prophetic ontology (Agnew 1998). Contrary to realist academics, politicians, and others that concern themselves with policy formation, the world is not simply a static map of nation-states: it is a dynamic system undergoing continuous ideological change as it is redefined through a process of hegemonic and networked ebb and flow of power. In the apt words of Agnew and Corbridge (1995, 13–14), "to understand the world, therefore, requires that we understand its changing geography."

The changing prophetic geopolitics of the world serve as an example of how spatial constructions (in this case strongly tied to the territorial state) have also defined regional conglomerations of the Other, often made racial or given an evil historical future in relation to the other side of the binary: themselves. Commensurably, foreign policy and colonial discourse utilize metaphoric binary constructions such as civilized and barbaric, progressive and backward, capitalist and communist, and for the scope of this chapter, Christian and Muslim, that find their commonsense understandings in cultural myths often made plain by tabloid media sources. Shapiro (1997) has argued that the experience of American political subjectivity depends on the practice of radical Othering.

These religio-political binary metaphors raze the complex terrains of cultural geographies, in this case the Middle East and Russia, rendering them culturally and morally flat as Western society is led to believe that culture comes in grand geo-packages of us and them. For example, the Muslims of contemporary prophecy are represented as a biblically known threat with specific biblical names; they represent for many American evangelicals what Sophie Bessis (2003, 72) calls "unfathomable otherness" and Derek Gregory (2004, 161) calls a "hyper-villain" through which a category of a 'good-us' can be shored-up. Debrix (2008) argues that Muslims in the present extreme war time populism, have become the central topic of American trade paperback psyche as exemplified in Hansen (2002) and Ledeen (2002). The threat is not simply from without, it is also within America. Through immigration, Muslims have become the leading threat to America's so-called Judeo-Christian democratic tradition and the ostensible values therein (see Huntington 2004). In this way, prophecy has adopted, informed, and inflamed these binary qualifications between American capitalism and Soviet communism during the Cold War and American Christianity and Middle Eastern Islam in the post-Cold War period, rendering both sides into divisive geopolitical halves.

Agnew (2003) explains that it is through a "commitment to this or that ideology (apocalyptic Christianity, microeconomics, etc.) that does not require knowledge of real places", that such prophetic abstractions about the world can be made. In other words, poor geographic knowledge of places—in this case reliance on brief biblical passages written between 4000 and 1900 years ago and a largely tabloid media source—aids the process of making evil spaces of the Other. Circumnavigating the complexities of the local gives way to a generalizing and transcendent "Jesus-trick" process, as Sparke (2005, 308) terms it, which enables prophecy expounders to predict the future with no regard to its dehumanizing consequences. Representations of the 'Muslim world,' 'Roman Empire,' and Russia all make possible the dystopic categories and known justifications for prophesied wars that span entire continents of people based on claims that they are all the same: Just enemies (e.g., Land 2002; see Flint and Falah 2004; Megoran 2008).

To make the world intelligible from the perspective of biblical toponyms, the process of naming in religious terms is important; often state and regional borders assume the shape of biblical ones and vice versa. "In the beginning nothing exists because nothing was named", Gunnar Olsson (2004: 205) writes emulating Genesis, "all that exist are two coordinates, above and below." Naming and renaming from above make familiar the unfamiliar by bringing the world into nominal presence by closing the gap between the word and the place it represents (Gregory 1994). This space had to be written in biblical terms before it could be interpreted by prophecy expounders. Furthermore, place names, when inscribed into geographic scenarios and representations, can legitimize geographical identities for both the namer and the named (Cohen and Kliot 1992; Kadmon 2004). The act of naming is not only constituting the place and people named, but also constitutes the subjects who name. In this case, the 'evil' Muslims of the Middle East constitute the goodness of those describing them as such. Thus, this epistemological strategy of textualizing space in dualistic biblical terms can make the world knowable, as though the Bible was a Western, English, orientalist text. It is this act of geographical naming and renaming throughout prophetic history and through the act of distancing that is of interest.

Agnew (1998, 4) suggests that geopoliticians use history to "justify their decisions as if history simply repeated itself without the necessity of geographically informed judgments about the character of different world situations." Dispensational theologians identify spaces of evil based on historical presuppositions, and do so in relation to the showy enemy of the United States. Dispensationalists justify history via spatial (re)presentations of formal, popular, and practical geopolitics through a particular reading of the Bible (see the Introduction, this volume). Dispensationalist Hal Lindsey's (2002, 230–231) claim to a historically "ever-present" and "Everlasting Hatred" between Muslims and Judeo-Christian civilization is an example of Agnew's claim to a historical template from which to generalize and graft people as antipodal opposites, in this case on a map of hatred on the one hand and Bible-derived innocence on the other. This always-written and everlasting hatred mirrors an Orwellian double-speak strategy: Oceania has always been at war with Eurasia.[3] In terms of Islam, dispensationalists render spatial through maps and descriptions of the world the notion that God has always been at war with Muslims, often referred to biblically as 'Ishmael'.

To dispensationalists, writes George Marsden (1984, 96), "all that is significant about history can be found in the Bible. There, the pattern of history of all the coming ages was written in advance... Modern history is of interest only as it

---

3 Winston remembered that, "Oceania, four years ago, had been at war with Eastasia and at peace with Eurasia" however Julia could only recall that "we'd always been at war with Eurasia" (Orwell 1961 [Orig. 1949], 127).

produces some facts that document the cultural decline predicted in the Bible." However, the geographical crystallizations, most notably Israel, do have modern historic significance, although only in as far as they illuminate present geographical imminence. Marsden (1984, 99) furthers his argument that history becomes, then, "basically a matter of collecting and classifying facts that fit the biblically derived pattern." Prophetic geography (these 'facts') act as a mechanism and transaction used to reason an explanation, cherry-picking facts to make the case for what is believed with no independent basis, only a collection of disparate citations. Common among all of the texts examined in this chapter is the use of the frame of science in the form of footnotes and citations of academics, journalists, and politicians. Dispensationalists borrow this academic frame to legitimate their geopolitics, moving from their faith foundations to deductive observations.

Compounding this spatial privileging is a common pattern of citations or citationality. 'Citationality', a term borrowed from Butler (1990), is understood here as a complex web of overlapping references that give veracity to prophetic events, places, and representations but change from their original intended meanings and uses as they are employed in different cultural, historical, and (geo)political contexts. Citationality builds not just on elastic interpretations of the Bible, but also on successions of previous prophecy writers to the point that a citation of a colleague holds the same level of authority as the Bible. To provide veracity to their geopolitical claims, the authors analyzed in this chapter, Hal Lindsey, Tim LaHaye, John F. Walvoord, Mark Hitchcock, and Joel Rosenberg, all cite Herodotus, Josephus, William Gesenius, Arno Gaebelein, J. Dwight Pentecost, and even Voltaire's *The Philosophical Dictionary*, which was ironically written as a response to fanatical religion and millennialism but is nonetheless citationally hijacked by evangelicals to give veracity to their arguments (see Orieux 1981). However, it is not simply an intra-evangelical citationality; these writers also piggyback on what Debrix calls "tabloid geopolitics": tabloid-like op-eds, blogs, and popular fiction and non-fiction by zealous and Zionist media analysts, (amateur) Sovietologists, and self-described experts on Islam who are caught in the vertiginous world, and find stable ground in the prophetic possibility of a new war with Russia.

From Russia to The Middle East

Elite-level prophetic geographical imaginations in the United States have, since the end of the Cold War, undergone a significant conversion, one that sheds light on the political contingencies of prophecy previously thought to be predestined by Christian dispensationalists. Dispensationalist prophecy restructured its geographical imaginations as to where evil lies from the Cold War, when the United States' enemy was believed to be the Communist Soviet Union, to more recent tropes that have relocated the 'evildoers' to the Muslim Middle East. Prophecy is presently focused on what is called the "Muslim Alliance", otherwise referred to as the "King of the South". Furthermore, the "King of the North"—Rosh, Magog, Gomer, Meshech, Tubal, and Beth-togarmah, all of which are led by Gog (Ezekiel 38 and 39; Daniel 11: 40–45; Joel 2:20)—which were supposed to represent Russia and the Soviet Union are now restricted to Muslim states (see Table 6.1). This is, of course, not a new concept: Muhammad figured as the Antichrist more than a millennium before the Bolshevik Revolution.

It can be said that Gog and Magog were usually places of mystery set in opposition to what was known (save their biblically inspired evil character). Though the specific locations have changed frequently throughout history—the

Table 6.1   Changing dispensational toponyms, 1970–2008

Biblical Place	Cold War Association	Post-Cold War Association	Post-Iraqi Threat	Geopolitical Power Bloc
Rosh	Russia	Russia	Russia	
Magog (and Gog)	USSR	(Muslim) Central Asia	Russia and Iran coalition	
Meshech	Moscow, USSR	Turkey	Moscow, Russia	
Tubal	Tobolsk, USSR	Turkey, Iran	Tobolsk, Russia	King of the North
Gomer	Eastern Europe, East Germany	Turkey	Turkey, Gaul, Spain	
Beth-Togarmah	southern Russia	Turkey	Turkey, Armenia, Central Asia	
Persia	Iran	Iran	Iran	
Cush	black Africa	Sudan, Ethiopia	Sudan, Ethiopia, Eritrea	
Put	Libya, Algeria, Tunisia, Morocco	Libya	Libya, Algeria, Tunisia	Kings of the South
'Many Peoples with you'	Egypt	Iraq, Syria, Jordan, Egypt	Arabia, Spain, Syria, Lebanon, Jordan	
Babylon	Rome	Babylon, Iraq	Babylon, Iraq	King of the West

Huns, the Mongols, the Arabians, the Magyars, and the Turks—they were usually drawn east of Western Europe (see Werbeke, Verhelst, and Welkenhuysen 1988).[4] To various medieval cartographers, Magog and Gog were the tribes of and beyond the Caucasus Mountains, often Tartars and Saracens to the South (Boyer 1992, 153; see Scafi 2006).[5] However, by the thirteenth century and the ensuing Crusades, Gog represented the Byzantine and later the Ottoman Empire for most of continental Europe and England well onto the late-seventeenth-century (Cosgrove 2003, 56). Indeed the "appearance of Turks in Europe was seen as the unchained people of Gog" (Kumar 1995, 202). Christians at this time saw Jews as those who were the 'outsiders within' and the Muslims as the 'outsiders outside'; the latter's presence defined the limits of a Christian Europe (Buchanan and Moore 2003, 7). In the late nineteenth and early twentieth centuries, however, Arabs and Bedouins were not perceived as evil by American Christians, rather they were thought of as atavistic, living representations of the Israelites of Jesus' time (Shamir 2003, 51–52).

The Islamic world never won the unanimous identification as Magog by the United States that the Soviet Union did during the Cold War. Russia's association with Rosh, Magog, Gomer, Beth-togarmah, Tubal, and Meshech, all of whom will be ruled under the leadership of Gog, has its beginnings in Wilhelm Gesenius' (1979 [1824]) musings on modern toponyms of the Bible. This conflation was popularized in the United States with the help of John Nelson Darby, the founder of dispensationalism, and others' missions in the U.S. (see Introduction, this volume). After the Russian Revolution and the onset of the Cold War, the Soviet Union would remain in the spotlight for forty-five years as dispensationalism's evil atheist enemy through a citational web reaching back to Gesenius' associations.

Understanding the historical context from which Gesenius, a professor of theology at the University of Halle in Prussia, identified 'Rosh' as Russia, 'Tubal' as the Siberian capital 'Tobolsk' and 'Meshech' as 'Moscow' in his 1828 lexicon, is key in showing how the territorial politics of the day would influence biblical geography. After the Treaty of Tilsit (1807), which reduced Prussia's territory, Russia was deemed as the evil entity referred to in the Bible by many Prussians (Boyer 1992; Wilson 1977). For Darby in England, the Crimean War further exacerbated Russia's evil character. This is not to say that these were the sole determinative reasons for identifying 'Rosh' with Russia; certainly linguistic similarity was also a factor, which is evident in dispensationalists' use of

---

4   Ethnic groups are often confused as 'nations' and ultimately states, a convenient transformation to the modern nation-state system to retrofit biblical references of tribes, clans and other relatively small ethonyms to territorial registers that make sense today.

5   As Scafi (2006) notes, however, the relationship between good and evil, and history and eternity, were not always divided territorially. Replacement theology (or supersessionism) posited that Christ's crucifixion represented the victory of good over evil which was manifest in the New Testament. Therefore the covenant with the Church usurping the apocalyptic authority of the Old Testament covenants with the Jews.

'Russia' as opposed to the U.S.S.R. during the Cold War.⁶ Therefore, Russia as a historical template and association with the many names in Ezekiel 38 and 39 was cast one hundred years before the Cold War; however, Cold War geopolitics, geoeconomics, and fervent American anti-communism helped to further color in the prophetic spaces left blank in these times.⁷ Indeed, anti-communism helped solidify the evangelical movement (Powers 1995). Herodotus' writing is also cited as evidence of Russia and Persia's evil nature. Although dispensationalists are often led to believe that the East/West division has biblical beginnings between Isaac and Ishmael, representations separating the East from the West preceded any biblical renditions (see Hartog 1988).

In the twentieth and twenty-first centuries, Magog, Rosh, Gomer, and all the nations variously identified with these toponyms, are defined by contemporary dispensationalists in relation to the state of Israel. The late John Walvoord (1967), a popular End Times writer and former president of the Dallas Theological Seminary, wrote in regard to the significance of Israel:

> Of the many peculiar phenomena which characterize the present generation, few events can claim equal significance as far as Bible prophecy is concerned with that of the return of Israel to their land. It constitutes a preparation for the end of the age, the setting of the coming of the Lord for His church, and the fulfillment of Israel's prophetic destiny (26).

Dispensationalists embraced Zionism in the nineteenth century, and their prediction of the establishment of the Jewish state was borne out in 1948. In terms of the political pressure by dispensational evangelicals in Britain to re-establish Israel, both Anthony Ashley Cooper, who was leader of the Evangelical Party and the Seventh Earl of Shaftsbury, and F. Laurence Oliphant, member of Parliament and author of *The Land of Gilead* (1880), urged Britain to create an Israeli state in Palestine where persecuted Jews from Russia and Eastern Europe could settle (Cohen-Sherbok 2006). While not arguing that such religious pressure had any effect, Tuchman's (1956, 336) analysis of the colonial motives behind the Balfour Declaration suggests that Christian reasoning practices were used by Britain to "dignify" the acquisition of "the Holy Land with a good conscience." With the Six-Day War of 1967 and growing Arab hostility against Jews throughout the latter twentieth century, dispensationalists have increasingly focused on the role of Israel

---

    6 'Rosh' in many Bibles (The King James Version, the Revised Standard Version, the New American Bible, the New Living Translation, and the New International Version) is an adjective meaning 'chief.' However, Hitchcock, Lindsey, Dyer, Walvoord, and Rosenberg see 'Rosh' as a proper noun (as written in the Jerusalem Bible, the New English Bible, and the New American Standard Bible). Hitchcock (2002, 39) denies that the conflation of 'Russia' with 'Rosh' is derived from linguistic similarity (see Sturm 2006).
    7 During World War II, the Germans represented the evil Other (see Marsden 1980, 141–53).

within the prophetic end times scenario.[8] It is in part this Arab unwillingness to accept Israel, a view that identifies Arab nations as enemies of Israel and therefore the enemies of God, that dispensationalists are able to find their declaration of biblical 'evil' for the Arabs and Muslims of the Middle East. However, without a doubt 9/11 has been predominantly the polarizing catalyst in recent years for American dispensationalists who already held enmity for Arabs and Muslims (e.g., Hitchcock 2002; Lindsey 2002).

Using evidence gathered by Gesenius, John Cumming, and the fifth-century BCE Greek geographer Herodotus, Hal Lindsey outlined "four spheres of political power" in *The Late Great Planet Earth* (1970, 69–70), the best-selling book of the 1970s according to the *New York Times*. The first of Lindsey's power blocs was the 'King of the West' which were the nations that made up the European Economic Community or 'Roman Empire' (Dan. 9:26); the second was the 'King of the North' which was the Soviet Union (Dan. 11:40); the third was the 'Kings of the East' which were China and possibly Japan and India (Rev. 16:12); and the fourth was the 'King of the South', which was Africa and Iran, and possibly other Muslim nations (Dan. 11:40). Lindsey's modern prophetic map has further divided the world by attributing Japheth, Ham, and Shem's sons (Gen. 10:2), common among T and O maps, to nation-states within the larger blocks but interestingly he also added the region that has been deemed the 'Kings of the East.' These Kings divide the world into a simplistic compass quadrant system of race.

Lindsey's "power blocs" and strategic maps are still used to outline the post-Cold War prophetic scenario by Dyer, LaHaye, Hitchcock and Rosenberg; it is only the nations that make up these power blocs and therefore the boundaries of the blocs that have changed. For Lindsey it was less a change in place names, but rather a change in emphasis from Russia to the Muslim Middle East. Prophetic geographies are generally not discarded; they are simply reverbalized or reterritorialized to positions relative to current events. Dispensational theologians elevate that which fits and revise that which does not—Russia was too crucial in prophetic history to be dropped immediately following the Cold War. Lindsey's (2002, 230–231) shifting emphasis from Russia to the Muslim Middle East is obvious in his recent books on the subject, where he identifies the King of the South as "the Muslim sphere of power" rather than his previously predicted "African Alliance."

The determinism of Lindsey's (1970, 59) claim that "Russia is a Gog" is strengthened by his quoting Israeli General Moshe Dayan that "the next war will not be with Arabs, but with Russians." Similarly LaHaye, in 1974, even went so far as to give reasons why Russia and not the Middle East was a harbinger sign of the End Times: "the present Middle East crisis is not…predicted for the end time [because] Egypt, a prominent ally of Russia today, is not listed in the group" (73).

---

8 William Koenig, for example, correlates (selective) American catastrophes to moments in history when American support for Israel waned or when America pressured Israel toward a resolution over the Palestinian occupied territories, arguing that God was punishing America for not supporting Israel (Clark 2007, 252–55).

This is all placed in stark contrast to both Lindsey's and LaHaye's later focus on the "coming Islamic invasion." For example, Lindsey wrote in 2002 that "the last war will begin with a coordinated attack against Israel by the Iranian led Muslim forces joined by Russia" (235). The invasion is no longer led by Russia; instead it is *joined* by Russia and *led* by "Muslim forces."

The change from the Soviet Union to the Middle East was a gradual one beginning with the 1973 oil crisis and later with the Iranian Revolution. In Lindsey's major follow-up to *The Late Great Planet Earth* in 1980, he writes that the "backward and underdeveloped Arab nations...will gain more power to control the world situation" with the "power of oil" (57, 68). However, consistent with his earlier musings, he adds that "the power of Russia" will still be the major force in the invasion of Israel. This change was no doubt influenced by Walvoord's 1974 *New York Times* best seller, *Armageddon: Oil and the Middle East Crisis,* which outlines the rising power of the Middle East and Muslims in particular. Walvoord, former president of the Dallas Theological Seminary where Lindsey obtained a Th.M., foresaw what he called "Russia's downfall" (1974: 125). Walvoord observed that a "Dramatic realignment of political and economic power on the international scene is already in the making.... The power of Arab oil and European agriculture and industry may lead to a cartel that will eventually eclipse the power of both Russia and the United States in the Middle East" (1974, 20). However, Lindsey's clear obsession with Russia's prophetic role has continued to influence many contemporary prophecy writers. Lindsey's *Planet Earth 2000 A.D.* (1994) would concede that Islam and not the Russia would be Israel's central adversary mid-way through the Tribulation period, but he sustained that the fall of Communism was Gorbachev playing possum, quietly preparing for the great invasion of Israel. Prophecy books increasingly emphasized Islam as evil in these times. Boyer (1992), in this respect, confirms that premillennial dispensationalists will be among the last people in the United States to admit the Cold War is over.

Apocalyptic Geopolitics in the Post-Cold War Period

The Soviet Empire met its demise through secular means and not by the hoped and predicted biblical Armageddon. Many prophecy scholars thought it was an ephemeral decline while others denied the evidence and advised their readers to "not believe everything they read" (quoted in Weber 2004, 204). Most, however, shifted their analysis to new geographical spheres mainly the Islamic Middle East, after the fall of the Berlin Wall and the 1990 Iraqi invasion of Kuwait. Unlike Lindsey whose prophecy was centered on Russia, Hitchcock's (1994: 69) prophetic geography is about: "Muslims, Muslims, and More Muslims."

Hitchcock, a doctoral graduate of the Dallas Theological Seminary, popular pastor of the Faith Bible Church in Edmond, Oklahoma, and author of over twenty books, couches his knowledge of prophecy in both Lindsey and Walvoord (see Sturm 2006). In an interview on March 13, 2006, he explains that he became "interested in prophecy in the early 1970s when *Late Great Planet Earth* by Hal Lindsey was published ... . [Among others] who have had a major impact on me include, Dr. John F. Walvoord." In 2007 Hitchcock posthumously revised Walvoord's 1974 title, changing the title from *Armageddon: Oil and the Middle East Crisis* to *Armageddon, Oil and Terror* which both refocuses the book on terrorism as the newest obsession of the United States and reinforces the ostensibly historically ever-present Middle Eastern centrality of prophecy. Despite his teachings, Hitchcock admits that prophecy has undergone a geographical shift since the end of the Cold War. In *After the Empire* (1994, XI), he writes, "the purpose of this book, therefore, is to take a fresh look at Ezekiel 38 and 39 and to let the true God speak."

Hitchcock (1994, 9) retraces all of the biblical names of Ezekiel 38 and 39 and restructures Lindsey's entire geopolitical order. In this respect he writes, "all of the names mentioned in Ezekiel 38–39 are presently within the geographical boundaries of Islamic nations.... The Soviet Union has fallen, and Islam seems to be moving in to fill the vacuum." In 1993, Hitchcock wrote in contrast to the earlier work of Lindsey that the Communism of the Soviet Union was a "constraint on fundamentalist, radical Islam" and therefore a constraint on prophecy fulfillment (4).

Although Hitchcock continues to include Russia as part of 'Rosh', his books are linked intertextually to ideological currents and political events that have vilified Islam. In his book, *The Coming Islamic Invasion of Israel* (2002, 27, 59) he outlines what he calls "God's 'Top-Ten Most Wanted List'" and compares it to President George W. Bush"s "axis of evil".[9] The reason for this list is obvious in relation to post-Cold War American affairs when he writes, "all of these nations today [Russia excluded] have one thing in common: Islam" (2002, 32–33). However, Turkey (Beth-togarmah, Gomer, Meshech, and Tubal) seems to be overrepresented in relation to where "fundamentalist" Muslims exist. Hitchcock (ibid., 47; see Table 6.1 above) admits that "Turkey today is not as hard-core Muslim as some of the other nations in Ezekiel's list" but he reminds his readers, having visited Turkey three times as of March 13, 2006, that Turkey is still "an Islamic nation."

In tracing the genealogy of Magog, Hitchcock (1994, 23–26) asks, "Who are the peoples who inhabit the ancient territory of these barbaric savages today?" and answers, "Kazakhstan, Kirghizia, Uzbekistan, Turkmenistan, Tajikistan, and the Ukraine. All of these former republics, which are now independent nations, are Muslim nations except the Ukraine." Ethiopia, to Hitchcock, is not symbolic of "Black Africa" as Gesenius and Lindsey had put forth, but it is rather Sudan which qualifies as the symbolic center. As Hitchcock endlessly argues, Sudan is "one of only three Muslim nations in the world with a militant Islamic government" (ibid., 79). 'Put', to Hitchcock, is Libya: "it takes no great imagination to conceive that Libya

---

9   North Korea does not make a cameo appearance, however, in Hitchcock's books. Quite simply North Korea does not fit his geopolitical focus on the dangers of the Islamic world.

would join with the other nations" as "Arabic is the official language of Libya, and the official religion is Islam" (ibid., 85). Although China was to be the leader of the 'Kings of the East' to Lindsey, Hitchcock does not mention it explicitly. To him, the 'Kings of the East' are first and foremost nations with large Muslim populations: "India, Afghanistan, Pakistan, and all the nations of the Orient, or Far East" (ibid., 149). To Lindsey, the Euphrates River had marked the division between the East and the West. Therefore all of the nations east of the Euphrates (with the exception of Iran) were part of the Kings of the East (Lindsey 1970, 81). This is consistent with the prevailing thought that the kings of the east were Chinese or the Mongol invaders on the Steppes of Central Asia as fulfillment of Revelation 16:12 which writes of the Euphrates River as drying up to make way for the 'kings of the east' (McGinn 199, 150). What is clear is that unlike many of the Realist geopoliticians who struggled with adapting to the transnational phenomenon of the post-Cold War whom focused on regional alliances, like Samuel Huntington's "Clash of Civilizations" (1993) religious or cultural groupings, dispensationalists like to stick with the state-centrism of classical Realism, later grouping them into biblically inspired regional coalitions or racial lines.

The biblical territory 'Gomer', during the Cold War was thought to be East Germany specifically and the regions of Eastern Europe swallowed by the Soviet Union after Word War II more generally. For Lindsey, Gomer's location made good American sense in the historical and geopolitical context of the Cold War. However, more recent prophecy scholars have revisited the origins of these toponyms in light of the changing political landscape. As early as five years after the Cold War, Hitchcock (1994; 2002, 48) relocated Gomer to 'Muslim Turkey.' LaHaye (1972, 68), one of the proponents of this new geographical positioning of Gomer as Turkey, in 1972 had sided with Lindsey citing him that indeed Gomer "is in the area of modern Poland, Czechoslovakia, and East Germany."

Babylon, which was traditionally to be understood as the embodiment of apostasy and Satan's religion, has also changed. Babylon has represented to Christian millennialists generally places and events from liberalized New York City to the surrender of French troops in Canada at the Battle of the Plains of Abraham in 1759 (Boyer 1992, 118, 72). For Lindsey (1970, 133) and others, however, Babylon was Rome: "there is no question about where the city [of Babylon] will be—it was Rome." However, with the wars in Iraq and Saddam Hussein's efforts to reconstruct the city of Babylon, a preferred literal reading of the Bible was made possible. Dyer (2003 [1991]), Provost and Dean of Education at America's most theologically conservative Bible institute, Moody, is perhaps best known for this reterritorialization of Babylon (on Dyer, see Chapter 9, this volume). Dyer writes that Babylon is "the geographical 'X'" (ibid., 21) and "we might conclude that the major end time evil empire would be centered in the Middle East in the country of Iraq, in the city of Babylon" (ibid., 61). Perhaps the most blatant example of knowing the 'evil Other' is found on the cover epigraph on Charles Dyer's major work. It reads: "Inside the mind of the Iraqi dictator: How new satellite images reveal what's really behind this Mideast menace." Dyer's (ibid., 97–104) use of twelve satellite photos of Babylon combines the truth of maps with the ability to know the real reason Saddam Hussein is rebuilding Babylon.

The concept that Babylon was literally the Babylon in Iraq would later be shared by Hitchcock (2003) and others, such as *Left Behind* series authors LaHaye and Jenkins. *Left Behind*, a series that has sold more than 60 million copies in ten years, locates the Antichrist Nicolae Carpathia's Global Community headquarters in the New Babylon in Iraq.[10] Nicolae Carpathia is most likely based on Nicolae Ceausescu, the Communist President of Romania until 1989 who also referred to himself as a Carpathian, which is suggestive of the sustained role that Communism would play in the end times for the authors.[11] However, Lindsey's *Late Great Planet Earth* (1970) did not mention Iraq and his book *The 1980s* (1980, 68) predicted that it would become part of Syria. Given that with the regime change in Iraq the chances of Babylon being rebuilt are slim, Dyer (2003) argues in this case that "if Hussein disappears tomorrow from the international scene, someone else will pick up the battle cry. And there will be one more martyr to the cause of Arab unity" (ibid., 126).

The Post-Iraqi Threat and the Return to Russia

Apocalyptic geopolitics have recently undergone yet another shift in emphasis, this time from the key role of Iraq in the end times scenario to that of Iran, whose sphere of influence has greatly expanded since American-led Iraqi destabilization. Joel Rosenberg (2006), whose books have ranked number one on Amazon.com and in the top ten in the *Wall Street Journal* and the *New York Times* hardcover fiction bestseller lists, continues Dyer and Hitchcock's Babylonian story line in the face of evidence to the contrary (see Chapter 4, this volume). Rosenberg (ibid., 174) writes that the "Scriptures are clear… Iraq will be rich and powerful", and Babylon will be a "great city", and a center of "extravagant luxury." This might seem improbable, Rosenberg agrees, but we need to look past the forlorn state of Iraq by seeing "Iraq through the third lens", that is, the scriptural lens (ibid., 173).

As evidence, Rosenberg's citational strategy recenters onto himself by using his own fiction books as the prophetic authority for which he bases his "uncanny" ability to prophesize events and the "the geopolitical landscape" (2006, 7): "as I

---

10 An extensive 2006 study by the *Baylor Institute for Studies of Religion* found that one in five people in America have read a *Left Behind* book (2006, 19). This in no way means, however, that readers were dispensationalist or let such fiction influence their everyday lives (see Frykholm 2004). Dittmer (2008) argues that there is a tremendous amount of diversity in the views of *Left Behind* readers.

11 It is also possible that 'Nicolaitans', or early rival churches of Ephesus and Pergamos, referred to by John in Revelation 6 and 15, are also the reference or a clever double entendre.

described in *The Last Days*, *The Ezekiel Option*, and most recently *The Copper Scroll*, Iraq is about to see a political and economic renaissance unparalleled in the history of the world" (173). In all of these novels Rosenberg describes a scenario in which Russia, Iran, and other Middle Eastern followers of Magog are annihilated, with the exception of Iraq. Left out of the first battle of the Tribulation which destroys the Russian and Middle Eastern armies, Iraq will be the world's last major oil supplier. Such oil wealth provides the pre-text for rebuilding Babylon. What we see here is not only a citationality of Dyer's assertion that Babylon will rise in Iraq, but also a self-citationality and cross-promotion of Rosenberg's own novels as evidence for an unlikely proposition, but one that has become central to the prophetic party-line: that the Babylon in Iraq will be the seat of power of the Antichrist and the economic center of the world. Like Russia, this prophetic commitment has become a literalist's indispensable reality in their geopolitical world view.

Rosenberg's employment of Dyer and others is used as proof of Iraq's rise to geoeconomic prominence using a complex web of overlapping citations that purport to confirm the narrative. Indeed Rosenberg uses self-citation to justify that he himself is a sage, spoken to by God to reveal events before they happen. He claims to have imagined September 11, 2001 days before it happened as God was speaking to him, "somehow telegraphing future events" (ibid., 9). He gives a short list of events he predicted in his novels that came true, from Yasser Arafat dying to Iranian nuclear aspirations, and revels in being called a "Modern Nostradamus" by *U.S. News & World Report* (Bedard 2003). The fictional geopolitics provides Rosenberg an opportunity to put forth leaping predictions without implication or ridicule for being wrong, after all it is only fiction. However if he is right, he is crowned a sage by his evangelical peers

Rosenberg also claims he predicted Iran would rise to prominence in the Middle East. Rosenberg is not alone in this new found Iranian antagonism; Mark Hitchcock predicted Iran's role in Armageddon as early as 1993 in his book, *The Silver Kingdom: Iran in History and Prophecy*. In a telephone interview on 22 March 2006, Hitchcock mentioned that he had a new book in press (released in late 2006) entitled, *Iran: The Coming Crisis: Radical Islam, Oil and Nuclear Threat*, that relates Iran's nuclear program and the recent tensions between Iran and the United States to prophecy. He went on to suggest that when he published his first book, *The Silver Kingdom* (1993), he was "ahead of his time" in terms of getting the prophecy right, even if it is 15 years in the making.[12] Implicitly he was revealing that he could be wrong about prophecy, and thereby revealed the political contingencies his work is based on (that of the current tensions between

---

12 Hal Lindsey, for example, predicted the end would come by 1988 (1970, 54). On January 2nd 2007, Pat Robertson prophesized that a terrorist attack of nuclear proportions would hit the United States in the next year (Associated Press, Jan. 3). Such predictions are common but such predictions are useful in stirring fear in the moment and also easily forgotten or ignored perhaps for fear of public condemnation.

Iran and the United States). But more than this, like Rosenberg, he was revealing that his authority was justified as not just an interpreter of scripture but also a vicar for God.[13]

Excluded from this convenient citationality is the war in Afghanistan, the complicity of Pakistan, and the Saudi Wahhabist role in 9/11 as a means of destabilizing the region. Quite simply, none of these places figure in the Bible and therefore require that they be ignored. Rather, radical Islamic militarism is repackaged and placed squarely in the borders of Iran. 9/11 and Afghanistan are floating signifiers that are connected to Iran by avoiding explicit associations and instead describing the issues on the same pages as the geopolitical discussion with Iran. Similar geographical dissonance and rhetoric by George W. Bush, Richard Cheney, and Colin Powell led to the absurd notion that Al Qaeda was synonymous with Iraq.

Most interesting in this recent prophetic reterritorialization is the resurgent speculation that Russia is again orchestrating an attack on Israel. Russia's recent attempts to restore and expand its influence in Central Asia and the Caucasus has granted prophecy writers the legitimacy to arc back to their closely held prophecy that Russia is Magog, led by a communist leader Gog from Moscow (the presumed biblical Meshech) (see Table 6.1). Islam will continue to figure centrally in the post-Iraqi threat. However, Gog, thought to be Putin or ultranationalist Vladimir Zhirinovsky, will continue the "Islamic Initiative" but the central reigns of power have been taken back by Russia. Rosenberg (2006, 167) explains:

> It is clear that God intends this War of Gog and Magog as a divine judgment of Israel's mortal enemies. It is also clear from the text that the countries certain to be affected by this judgment include those that have in recent times warred against the very existence of God (Moscow-driven Communism and atheism) and those that have warred against the One True God, the God of the Bible, the God of Abraham, Isaac, Jacob, and Jesus (Mecca-driven Islam). ... The wrath of God will fall on the Kremlin and the Red Army. And the world will witness the end of radical Islam as we know it.

The crux here resonates with Wallace's (2006) point that much of this eschatology has more to do with America than it does with Christianity. 'Communism', 'Kremlin', 'Red Army', and 'radical Islam', fitting the American enemies in their day, but the anxiety concerning issues of the Cold War has carried on past its reference. Rosenberg (2006, 104) in this regard takes issue with a quote from Richard Cheney, "none of us believe that Russia is fated to become an enemy."

---

13 The framing of Iran as biblically evil is revealing of an Americanistic underwriting. Jews were freed from Babylonian enslavement by the Persian Emperor, Cyrus, who freed the Jews and sent them back to Judea to rebuild their Temple (Isa. 44: 28). This is why Cyrus is said to be the "messiah" (meaning anointed one) of the Jewish people. "Thus says the Lord to Cyrus, His anointed one, whose right hand He has grasped" (Isa. 45:1).

Rosenberg (2006) responds, "but that...line troubled me, for when one looks at Russia through not only the political and economic lenses but also through the third lens of scripture, one sees that Russia is in fact, destined to become an enemy of the West and particularly of Israel, in part because of its alliance with Iran."

We find out in the above quote that Iran is placed in the crucial role of evil Russian sidekick because (and for no other reason) in a list of nations to join Magog Persia is listed first. Rosenberg attempts to shore-up the connection between Iran and Russia by citing LaHaye's book, *The Coming Peace in the Middle East* (1984), which argues that peace in the Middle East will only come by its destruction in Armageddon, seemingly reverbalizing Tacitus' oft-cited criticism of Roman doctrine: "to ravage, to slaughter, to usurp under false titles, they call empire; and where they make a desert, they call it peace." But LaHaye never explicitly puts Russia and Iran together as a conspiring coeval force. In fact the closest LaHaye gets is that "Russia will... mount up its forces, stir up Arab hatreds, and attack Israel" (as cited in Rosenberg 2006, 9). LaHaye makes no mention of the biblical Persia being part of Russia's grand plan. In a sermon in Jerusalem at the *King of Kings Community* (2008) Pentecostal congregation, the Pastor Wayne Hilsden interviewed Rosenberg on stage. Inquiring if Rosenberg felt the end times were near, Rosenberg answered, "this may not happen in our lifetime, let alone soon", and continued ignoring Soviet support for Iran at the end of the Iran-Iraq War, "But this is the first time Russia has made an alliance with Iran in 2500 years, so it looks that way." On the mere grounds of order in a list of names, then, Rosenberg clamors the congregation to his observation that the end is surely just around the corner.

Conclusion

In a press conference on mid-October 2007, George W. Bush, in an attempt to conjure up world support for a preemptive war with Iran said, "if you're interested in avoiding World War III, it seems like you ought to be interested in preventing [Iran] from having the knowledge necessary to make a nuclear weapon." What might have been an off hand comment concerning 'World War III', a moment evangelicals call Armageddon, is none the less the vernacular Bush chose to use to describe the future. Ritter (2007) calls the 'World War III' comment an 'imprinting' or citational summary of the discourse about Iran circulating the Oval Office, far from a proverbial slip-up. That James Dobson was invited to the White House to discuss President Bush's Iran policy, or lack thereof, can make sense as a serious engagement with apocalyptic discourse or an attempt to assuage the Republican base (see Kuo 2006). Reagan too made a point to cite Hal Lindsey's *Late Great Planet Earth* (1970) many times over, although not explicitly, but borrowing from this common geopolitical tradition in the evangelical community (O'Leary 1994). Stephen O'Leary (ibid., 177) argues that Lindsey's geopolitics and foreign policy proposals "[were] part of Reagan's campaign." Bush's references to American planned military intervention in the Middle East as modern 'Crusades', his reference to an axis of evil, or the many other apocalyptic statements he made (see Afterword, this volume), are all examples of citational imprinting about the geopolitical state and shape of global politics and America's position toward it that take on particular meaning for American evangelicals and the prophecy writers who glean this discourse through their scriptural lenses. Such elite geopolitical statements give the prophetic claims veracity as to God's intended geopolitical shape of the world.

Geopolitics has been constructed in different ways over history as the world it has purported to reflect or anticipate has changed (Agnew 1998). These apocalyptic geopolitics have changed in prophecy, continuously refashioning a new enemy. Apocalyptic geopolitics is formed by wider socio-political events abroad and at home. The place where someone is from will have an impact on how the Bible is interpreted as these socio-political events are filtered through a local nationalist and cultural prism (Agnew 1987). American ethnocentrism purports a kind of Realist Manichaeism that beyond the American horizon evil forces are gathering to harm the righteous.

Clues to the names of the end times places and nations are not simply provided for by American mappings of foreign policy, but also a citationality of Herodotus, Voltaire, and a relatively recent panoply of dispensational authors who identified biblical tribes and regions. Using order, direction, and historical places read literally and free from what Augustine called a 'spiritual' or allegorical reading of the Bible, biblical tribes are assigned to modern city and state toponyms. Like the biblically justified and citationally sanctioned thought that the universe was centered on earth rather than a heliocentric solar system, centering the world on Israel and having Russia or the Middle East spin around it is little different. This citational strategy is combined with a cultural obsession with the location of evil, easily judges the character of place based on the monikers 'communist' and 'Islamic', which in turn stokes the American war machine and a sentiment of hate toward others. Of course Edward Said's great empirical thrust in *Orientalism* (1979) was that orientalism was perpetuated from a few overly cited and read sources, which mirrors the citationality of prophecy writers. The Cold War prophetic scenario certainly reinforced the idea that Communists were a biblical evil, as have the newly drawn boundaries circumscribing Muslims in the Middle East. Most dispensational books on matters of prophetic signs focus on the conflation of prophetic toponyms in the Bible with current nation-states and their roles for the end times.

The chapter has shown that despite a literal and certain interpretation of biblical geopolitics, such signs seem to change freely and without recourse. It has also shown that the protean American geopolitical world view and othering practices are funneled through the dispensationalist prism cast of the Bible, refracting images onto and off the page to reconcile citational expectations of previous apocalyptic geopoliticians. For dispensationalists, it is concluded, prophecy is not only about making the world in the image of their biblically inspired maps, but it is also about retracing the biblical taxonomy of toponyms to make global politics the geopolitics of God.

References

Agnew, J. (1987), *Place and Politics: The Geographical Mediation Between State and Society* (Boston: Allen Unwin).
—. (1998), *Geopolitics: Re-visioning World Politics* (London: Routledge).
—. (2003), 'Commentary III: Learning from the War on Iraq,' *The Arab World Geographer*, 6:1, 58–60.
Agnew, J. and Corbridge, S. (1995), *Mastering Space: Hegemony, Territory and International Political Economy* (London: Routledge).
Barkun, M. (2003), *Culture of Conspiracy: Apocalyptic Visions in Contemporary America* (Berkeley: University of California Press).
Baylor Institute for Studies of Religion (2006), 'American Piety in the 21st Century: New Insights to the Depth and Complexity of Religion in the U.S.',<http://www.baylor.edu/content/services/document.php/33304.pdf>, accessed 26 January 07.
Bedard, P. (2003), 'Washington Whispers: Modern Nostradamus,' *U.S. & World Report*, Nov. 3.
Beer, F.A. and de Landdtsheer, C. (eds) (2004), *Metaphorical World Politics: Rhetorics of Democracy, War and Globalization* (East Lansing: Michigan State University Press).
Bessis, S. (2003), *Western Supremacy: Triumph of an Idea?* trans., P. Camiller (New York: Zed Books).
Boyer, P. (1992), *When Time Shall Be No More: Prophecy Belief In Modern American Culture* (Cambridge: Harvard University Press).
Brummett, B. (1991), *Contemporary Apocalyptic Rhetoric* (New York: Praeger).
Buchanan, A. and Moore, M. (2003), 'Introduction: The Making and Unmaking of Boundaries,' in Buchanan, A. and Moore, M. (eds) *States, Nations, Borders: The Ethics of Making Boundaries* (Cambridge: Cambridge University Press), 1–18.
Butler, J. (1990), *Gender Trouble* (New York: Routledge).
Clark, V. (2007), *Allies for Armageddon: The Rise of Christian Zionism* (New Haven: Yale University Press).
Cohen, S.B., and Kliot, N. (1992), 'Place Names in Israel's Ideological Struggle Over the Administered Territories,' *Annals of the Association of American Geographers*, 82: 4, 653–80.
Cohen-Sherbok, D. (2006), *The Politics of the Apocalypse: The History and Influence of Christian Zionism* (Oxford: Oneworld).
Connolly, W.E. (2008), *Capitalism and Christianity, American Style* (Durham, NC: Duke University Press).

Cosgrove, D. (2003), *Apollo's Eye: A Cartographic Genealogy of the Earth in the Western Imagination* (Baltimore, MD: The Johns Hopkins University Press).

Dawson, L. (1999), 'When Prophecy Fails and Faith Persists: A Theoretical Overview,' *Nova Religio*, 3:1, 60–82.

Debrix, F. (2008), *Tabloid Terror: War, Culture, and Geopolitics*. (New York: Routledge).

Dittmer, J. (2008), 'The Geographical Pivot of (the End of) History: Evangelical Geopolitical Imaginations and Audience Interpretation of *Left Behind*,' *Political Geography*, 27:3, 280–300.

Dyer, C. (2003) [1991], *The Rise of Babylon: Is Iraq at the Center of the Final Drama?* (Chicago: Moody).

Festinger, L. (1956), *When Prophecy Fails* (London: Harper and Row).

Flint, C. and Falah, G.-W. (2004), 'How the United States Justified its War on Terrorism: Prime Morality and the Construction of a "Just War,"' *Third World Quarterly* 25, 1379–99.

Foucault, M. (1980), 'Questions on geography,' in C. Gordon (ed.), *Power/Knowledge: Selected Interviews and Other Writings, 1972–1977* (New York: Pantheon) 63–77.

Frykholm, A.J. (2004), 'What Social and Political Messages Appear in the *Left Behind* Books: A Literary Discussion of the Millenarian Fiction,' in Forbes, B.D. and Kilde, J.H. (eds) *Rapture, Revelation, and the End Times: Exploring the* Left Behind *Series* (New York: Palgrave) 167–96.

Gesenius, H.W.F. (1979), [1824], *Gesenius' Hebrew-Chaldee Lexicon of the Old Testament* (Grand Rapids, MI: Baker Books).

Gregory, D. (1994), *Geographical Imaginations* (Oxford: Blackwell).

—. (2004), *The Colonial Present: Afghanistan, Palestine, Iraq* (Oxford: Blackwell).

Halsell, G. (1986), *Prophecy and Politics: Militant Evangelicals on the Road to Nuclear War* (Chicago: Lawrence Hill).

Hansen, V.D. (2002), *An Autumn of War: What America Learned from September 11 and the War on Terrorism* (New York: Anchor Books).

Harding, S. (2000), *The Book of Jerry Falwell: Fundamentalist Language and Politics* (Princeton: Princeton University Press).

Hartog, F. (1988), *The Mirror of Herodotus: The Representation of the Other in the Writing of History* (Berkeley: University of California Press).

Hitchcock, M. (1993), *The Silver Kingdom: Iran in History and Prophecy* (Oklahoma City: Hearthstone).

—. (1994), *After the Empire: Bible Prophecy in Light of the Fall of the Soviet Union* (Wheaton, IL: Tyndale).

—. (2002), *The coming Islamic invasion of Israel* (Sisters, OR: Multnomah).

—. (2003), *The Second Coming of Babylon* (Sisters, OR: Multnomah).

—. (2006), *Iran: The Coming Crisis: Radical Islam, Oil and Nuclear Threat* (Sisters, OR: Multnomah).

Howard, R. (2006), 'Sustainability and Narrative Plasticity in Online Apocalyptic Discourse After September 11, 2001,' *Journal of Media and Religion* 5:1, 25–47.

Huntington, S. (1993), 'Clash of Civilization?' *Foreign Affairs*, 72:3, 22–49.

—. (2004), *Who Are We? The Challenges to America's National Identity* (New York: Simon and Schuster).

Kadmon, N. (2004), 'Toponymy and Geopolitics: The Political Use and Misuse of Geographical Names,' *The Cartographic Journal*, 41:2, 85–87.

King of Kings Community, Jerusalem (2008), 'On Stage Interview Between Wayne Hilsden and Mark Rosenberg', 18 November.

Kumar, K. (1995), 'Apocalypse, Millennium and Utopia Today', in Bull, M. (eds) *Apocalypse Theory and the Ends of the World* (Oxford: Blackwell), 200–24.

Kuo, D. (2006), *Tempting Faith: An Inside Story of Political Seduction* (New York: Free Press).

LaHaye, T. (1972), *The Beginning of the End* (Wheaton, IL: Tyndale).

—. (1984), *The Coming Peace in the Middle East* (Grand Rapids: Zondervan).

LaHaye, T.F. and Jenkins, J.B. (1996), *Left Behind: A Novel of Earth's Last Days*. (Wheaton, IL: Tyndale House).

Lindsey, H. (1970), *The Late Great Planet Earth* (Grand Rapids, MI: Zondervan Books).

—. (1980), *The 1980's: Countdown to Armageddon* (King of Prussia, PA: Westgate).

—. (1994), *Planet Earth 2000 A.D.* (Palos Verdes: Western Front).

—. (1998), *Planet Earth: The Final Chapter* (Beverly Hills, CA: Western Front).

—. (2002), *The Everlasting Hatred: The Roots of Jihad* (Murrieta, CA: Oracle).

Marsden, G.M. (1980), *Fundamentalism and American Culture* (Oxford. Oxford University Press).

—. (1984), 'Evangelicals, History and Modernity,' in Marsden, G.M. (ed.), *Evangelicalism and Modern America* (Grand Rapids, MI: WM. B. Eerdmans), 94–102.

McGinn, B. (1998), *Visions of the End: Apocalyptic Traditions in the Middle Ages* (New York: Columbia University Press).

Northcott, M. (2004), *An Angel Directs the Storm: Apocalyptic Religion and American Empire* (London: I.B. Tauris).

Land, R. (2002), 'The So-Called 'Land letter,' *The Ethics and Religious Liberty Commission of the Southern Baptist Convention*. <http://erlc.com/article/the-so-called-land-letter>, accessed 1 December 2008.

Ledeen, M. (2002), *The War Against Terror Masters: Why it Happened, Where We Are Now, How We'll Win* (New York: St. Martin's Press).

Megoran, N. (2008), 'Militarism, Realism, Just War, or Nonviolence? Critical Geopolitics and the Problem of Normativity,' *Geopolitics* 13:3, 473–97.

O'Leary, S.D. (1994), *Arguing the Apocalypse: A Theory of Millennial Rhetoric* (Oxford: Oxford University Press).

Olsson, G. (2004), 'Placing the Holy,' in Mels, T. (ed.) *Reanimating Places: A Geography of Rhythms* (Burlington, VT: Ashgate), 201–214.
Orwell, G. (1961) [1949], *Nineteen Eighty-Four* (New York: The New American Library).
Orieux, J. (1981), *Voltaire* (Garden City: Doubleday).
Powers, R.G. (1995), *Not Without Honor: The History of American Anti-Communism* (New York: Free Press).
Ritter, S. (2007), 'On the Eve of Destruction,' *Truthdig*, Oct. 22,
Rosenberg, J.C. (2006), *Epicenter: Why the Current Rumblings in the Middle East Will Change Your Future* (Carol Stream, IL: Tyndale).
Said, E. (1979), *Orientalism* (New York: Vintage).
Scafi, A. (2006), *Mapping Paradise: A History of Heaven on Earth* (Chicago: University of Chicago Press).
Shamir, M. (2003), '"Our Jerusalem": Americans in the Holy Land and Protestant Narratives of National Entitlement,' *American Quarterly* 55:1, 29–60.
Shapiro, M.J. (1997), *Violent Cartographies: Mapping Cultures of War* (Minneapolis: University of Minnesota Press).
Sizer, S. (2004), *Christian Zionism: Road-Map to Armageddon?* (Leicester: InterVarsity Press).
Sparke, M. (2005), *In the Space of Theory: Postfoundational Geographies of the Nation-State* (Minneapolis: University of Minnesota Press).
Strong, T.B. (1984), 'Language and Nihilism: Nietzsche's Critique of Epistemology,' in Shapiro, M.J. (ed.) *Politics and Language* (New York: New York University Press), 239–63.
Sturm, T. (2006), 'Prophetic Eyes: The Theatricality of Mark Hitchcock's Premillennial Geopolitics,' *Geopolitics* 11:2, 231–55.
Tuchman, B. (1983), *Bible and Sword* (London: Macmillan).
Wallace, I. (2006), 'Territory, Typology, Theology: Geopolitics and the Christian Scriptures,' *Geopolitics* 11:2, 192–230.
Walvoord, J.F. (1967), *Israel in Prophecy* (Grand Rapids, MI: Zondervan).
Walvoord, J.F. (1974), *Armageddon: Oil and the Middle East Crisis* (Grand Rapids, MI: Zondervan).
Walvoord, J.F. and Hitchcock, M. (2007), *Armageddon, Oil and Terror: What the Bible Says About the Future Of America, the Middle East, and the End of Western Civilization* (Carol Stream, IL: Tyndale).
Weber, T.P. (2004), O*n the Road to Armageddon: How Evangelicals Became Israel's Best Friend* (Grand Rapids, MI: Baker Academic).
Werbeke, W., Verhelst, D., and Welkenhuysen, A. (eds) (1988), *The Use and Abuse of Eschatology in the Middle Ages* (Louvain: Louvain University Press).
Wilson, D. (1977), *Armageddon Now! The Premillennial Response to Russia and Israel Since 1917* (Grand Rapids, MI: Baker Academic).

PART III
Missionary Geopolitics

Chapter 7

The Problematic Synergy Between Evangelicals and the U.S. State in Sub-Saharan Africa

Hannes Gerhardt

> Worldwide, religious trends have the potential to reshape political assumptions in a way that has not been seen since the rise of modern nationalism. While we can imagine any number of possible futures, a worst-case scenario would include a wave of religious conflicts reminiscent of the Middle Ages, a new age of Christian crusades and Muslim jihads ... In responding to that prospect, we need, at a minimum, to ensure that our political leaders and diplomats pay as much attention to religions and to sectarian frontiers as they ever have to the distribution of oil fields.
>
> (Philip Jenkins 2002, 13)

Evangelical forms of Christianity are growing rapidly throughout the world, perhaps nowhere more dramatically than in Africa (Brouwer, Gifford, and Rose 1996; Jenkins 2002). In many ways this swift growth of evangelicalism portends to the emergent future of Christianity as a whole, as it moves towards greater passion, revivalism, and dogmatism while also becoming increasingly focused on non-Western societies. Evangelicalism is often characterized as a faith grounded in the conviction that the Bible is the inerrant word of God and that humans, being born in original sin, require a personal, spiritual rebirth by recognizing Jesus Christ as their savior. In addition to these fundamental values, however, another key aspect of the evangelical faith, and one that goes a long way to help explain its dramatic spread, is the perceived biblical command to engage in the Great Commission, that is, to spread the content of the Gospel to 'all nations of the world'.

Committed to a religion staunchly dedicated to proselytizing, American evangelicals have been adamantly embracing and fostering the global boom in their beliefs and practices. This has been particularly true for conservative evangelicals who, beyond maintaining an unwavering conviction that their religion represents the only true faith, view the spread of their beliefs as the march of Good in a forlorn and often literally evil world. These conservative evangelicals generally identify themselves with the religious wing of the Republican Party and they tend to support a proactive U.S. foreign policy grounded in military power. Based on a recent survey by the *Pew Forum on Religion and Public Life* (2008), it can be inferred that about 40 percent of all white evangelicals can be linked to such conservative beliefs, which amounts to about 10 percent of the entire U.S. population. It is these conservative evangelicals and the institutions that they support that are the primary force in exporting a U.S.-flavored evangelical faith to the rest of the world (Brouwer, Gifford, and Rose 1996).

The issue to be pursued in this chapter concerns the positioning of the U.S. state in regard to this evangelical movement with a particular focus on sub-Saharan Africa, where fervent evangelical missionary work is increasingly coming into conflict with the aims of another absolutist religion: fundamentalist Islam. Of particular interest here is the fact that U.S. evangelicals have been allowed to become a key component of a U.S.-supported governmentality that is currently taking root in Africa. The basis of this development is a perceived synergy between U.S. global interests and conservative evangelical aims. Whereas the U.S. is seeking a partner that can help provide solutions to the gaping governance vacuum in Africa, conservative evangelicals are using U.S. support in an attempt to completely remake the African soul through a continent-wide evangelical revival.

Yet despite any synergies that may exist between the U.S. state and conservative evangelicals, I argue that there are also inherent and significant ideological differences that inform the geopolitical projects of these juxtaposed global actors. The core distinction here lies in the fact that conservative evangelicals generally do not share the liberal normativity that has historically animated the U.S. state's global vision for the world. More specifically, the U.S. state's commitment to pluralism and its relegation of religion to the private sphere fundamentally conflicts with the moral absolutism of biblical inerrancy cherished by evangelicals (see Howe 2008). This difference is further exemplified in the clashing (geo)political positions inherent in the contrasting normative world views held by these opposing actors. Where such a clash is particularly evident, I contend, is in the way that the U.S. state and conservative evangelical actors relate to what Carl Schmitt (1996) identifies as the essence of the political, that is, the decision on the identification of the enemy.

Roughly following observations made by Schmitt (1988; 1991; 1996; 2003), the U.S. can be understood as a nation devoted to a liberal-universalism that aims to achieve global hegemony by dissolving the friend-enemy distinction. In contrast, conservative evangelicals can be seen to be actively taking up the friend-enemy distinction in their confrontation with Islam. As the U.S. state increasingly shifts development and humanitarian governance operations to non-governmental and often evangelical organizations, however, the U.S. risks becoming identified as a main actor in an unfolding global clash between Christianity and Islam. Indeed, in its sanctioning of evangelical discourse and practice in Africa, there is a risk that the U.S. state will become an unwitting actor in a perceived clash of civilizations that is based not on necessarily incompatible religious differences but on the fact that the concept of such a confrontation has been widely adopted as a powerful heuristic by both some evangelicals and some Muslims on the ground.

Global Governmentality and the Political

To fully grasp the synergies and clashes between the U.S. state and conservative evangelicals it is helpful to consider these two actors in terms of a broader, contested global governmentality. The conceptualization of governmentality (Foucault 1991) was essentially devised to explore the subtle and constructive forms of power involved in shaping subjectivities through rationalizations centered on the 'art of government.' Governmentality is thus about the rationality involved in the control, regulation, and direction of the self and the population. These rationalizations, in turn, must be recognized as constructions that are formed beyond the intentionality of any one actor; they emerge out of a mix of forces within society; the state and the evangelical movement are thus but two nodes in this mix.

The concept of governmentality can also effectively be applied to the global level in various ways (see for example Larner and Walters 2004). One of the earliest and most influential examples is Hardt and Negri's (2000) work *Empire*, which posits a globalized capitalism that transcends the nation-state and yet infuses social life everywhere. According to Hardt and Negri, Empire's ultimate function is the extraction of surplus, which is seen to occur ubiquitously at the level of life lived. In other words, a global governmentalization has taken shape that works to service the demands of capitalism. Yet what is interesting and useful about Hardt and Negri's work is that beyond this rhizomatic saturation of power on the micro-level, they also recognize an integrated nexus of key political-economic organizations where power coagulates. Outside of nation-states this nexus includes transnational corporations as well as international governmental and non-governmental organizations. Taken together, Hardt and Negri claim, these various institutions comprise a "new form of sovereignty... that governs the world" (Hardt and Negri 2000, xi).

Bringing in the concept of sovereignty at this point rests somewhat uneasily within Hardt and Negri's otherwise much more diffuse concept of power. Nonetheless, I agree with their impulse to retain the concept of sovereignty as a necessary component of governmentality – namely as a site of decision making where the norm and the exception still must be determined, backed by the authority, capacity, and willingness to use violence and ultimatley to kill (see, Agamben 1998). It is with this recognition that the thought of Carl Schmitt, whose work has experienced a resurgence of interests in the last few years (undoubtedly related to the crass reappearance of sovereign power on the world stage, but also the appearance of new translations into English) becomes particularly pertinent. Schmitt's thought provides a useful lens for considering how a U.S.-sponsored global governmentality, which involves the rationales and practices of organizing worldwide populations, intersects with U.S. geopolitical aims and strategies that are squarely focused on the interests of the sovereign state and its power in the world.

According to Schmitt (1988; 1991; 2003), who commented extensively on U.S. global power, the U.S. strategy for world hegemony is based on the execution of a totalizing legal vision of the world that strives to overcome difference via a universalized and all encompassing liberalism. For Schmitt, a world organized in such a way is grounded on an attempt to expel completely the realm of the political, that is, the friend-enemy distinction. In this sense, the United States seeks world dominance on the basis of a governmentality that is purely economic and legal, embodied today in the neo-liberal program of international governmental institutions and the humanitarian legal agenda of the U.N. It is precisely in such a technocratic world order that, according to Schmitt (2003), it is no longer possible to recognize a just enemy; only the criminalization of the enemy is possible.

There is, however, a problem in this development that Schmitt identified and that has since been ardently taken up by Chantal Mouffe (1988; 1991; 1993a; 1993b; 1993c; 1993d). Mouffe's argument, building on Schmitt, is simply that the friend-enemy distinction cannot, in fact, ever be truly dissolved. Following Schmitt (2003; 2007), attempting to dissolve the essence of the political, by for example criminalizing the enemy, leads instead to a global civil war characterized by what he termed the *partisan*, the rebel who fights clandestinely, unattached and unhindered by sovereign state institutions and territories. Indeed, as followers of Schmitt like to point out, we are seeing precisely such a world coming to fruition in the West's clash with Islamism, characterized by an open-ended global war on terror and a blossoming transnational, underground terror network united in their drive to undermine Western power (see, for example, Kochi 2006; Mouffe 2005; Müller 2006; Shapiro 2005).

For Mouffe, then, a key challenge for the West is the pursuance of radically democratic institutions that are capable of dealing with friend-enemy realities as a way to diffuse their potential violence. Thus, following this line of thinking, Islamism, understood as the political project to socially enact a fundamentalist interpretation of Islam, should not be criminalized; it should rather be recognized as a formal challenger to Western thought and practice. In other words, the West must come to terms with political Islam's ideology and demands. Democratic institutions in which these demands can be expressed and possibly enacted must, in turn, be created. The hope here is that political legitimacy will diffuse the radical elements within particular antagonistic groups, thereby tying these groups to the ground rules of a given political community. From this perspective, for example, Hamas should be allowed to govern Gaza democratically with the hope that this experience will bring them into the fold of the international political community, similar to the trajectory that the Fatah movement underwent.

Yet what is often ignored or underplayed by Schmitt's followers is that sovereignty is not a power that resides isolated and context-free within the institution of the state. In this sense, it does not make sense to speak of liberalism as an uncontested sovereign project – both the governmentality of liberalism and the sovereign power embracing this governmentality are challenged and internally fractured. Following from this, it must be acknowledged that the normative baggage of liberalism's universalism, grounded in an absolutist faith in rationality, is not the only obstacle to minimizing the dangers of the friend-enemy grouping that constitutes the realm of the political. Attention must also be paid to an array of illiberal and undemocratic forces within global governmentality as well as within the domain of sovereignty, that is, the political. One of these forces, which looms large in the world today, is religion, particularly in its politicized and absolutist form.

In his work, Schmitt (2003) focused on the dangers of the religiously informed politics that held sway in the Middle Ages during what he termed the *respublica Christiana*, linking this political order to both the incessant calls for crusades against the Muslim infidels and the Thirty Years War in which Europe essentially devoured itself. Yet today, many conservative evangelicals again cling to the age-old friend-enemy dichotomy that pits Christians against Muslims (see Chapter 6, this volume). The contemporary conservative evangelicals' vision, however, also differs from the ideal of a *respublica Christiana*. In the case of conservative evangelicals the final and ultimate authority is seen not to rest in the Pope but in the eschatologically weighted word of the Bible. The execution of what is seen as God's will is here, furthermore, heightened to the level of the political, that is, to the level where a friend-enemy grouping is established. Yet as in the past, due to its religious and absolutist underpinnings, the enemy (here the Muslim) is transformed into the foe. The difference is that the foe must be fully vanquished; it is not possible to push the enemy back to a specific territory upon which civil relations may once again commence after hostilities have ended. The battle can never end until Jesus' return to earth. This is a battle, furthermore, that many evangelicals believe the U.S. has a providential responsibility to take part in.

The Synergy Between the U.S. State and Conservative Evangelicals

When considering the intersection between conservative evangelicals and the U.S. state we can conceive of a number of possibilities. First, the state apparatus itself can be occupied by evangelical interests, the most extreme manifestation being a theocracy. Second, evangelicals can organize interest groups that succeed in pressuring the state to make particular concessions. Third, it is possible to conceive of a type of synergy, in which state interests establish a symbiotic relationship with evangelicals to further particular ends, although they do not necessarily share the exact same goals. In considering the foreign policy approach of the United States with regard to Africa all three of these types of intersection can be discerned, although I contend that it is the third form of intersection that is the most prevalent. In this sense I loosely adhere to the idea offered by William Connolly (2005) of a 'resonance machine' in which the power of the U.S. state and that of the evangelical movement each feed off of each other without as such being caused or determined by the other.

In what follows I will unpack this resonance, or synergy, by considering what it is that the U.S. state and the evangelical movement hope to achieve in Africa, and the challenges and opportunities that they face in terms of their mutual interaction in shaping an emergent African governmentality. In the last section I will explicate the limitations and dangers that exist for the U.S. state in its support of such a governmentality.

The U.S. State and the Governmentalization of African Space

Ever since its independence, largely between the 1950s–1980s, sub-Saharan Africa has posed a challenge for the Western world. Over time, this challenge has been presented as one of delayed, then lagging, and ultimately of failing 'development'. The main culprit that came to be identified by Western governments and institutions to explain this failure was the African state. As a result development discourse and practice became increasingly controlled by international governmental organizations, such as the International Monetary Fund and the World Bank, which were used to reign in 'irresponsible' African states and their leaders by dictating development policy aims and agendas via strict loan conditionalities and project supervision. In this way a global neo-liberal governmentality was executed that opened the way to the flows of capital, technology, and ideas, which essentially plugged Africa into a U.S.-dominated global economic system (see, for example, Harvey 2007).

Yet, apart from the macro-level rules and measures imposed by this global system, neo-liberal governmentalization was generally intended to be a passive affair in which invisible hands, that is, markets, do all the work. Nevertheless, as the failure of Africa became increasingly stark throughout the 1990s, the *laissez-faire* approach encapsulated in neo-liberalism became increasingly scrutinized. In many ways Africa was beginning to look like the consequence of what Alain Joxe (2002) has succinctly referred to as an 'empire of disorder'. For Joxe, neo-liberal globalization is characterized by an unfocused U.S. domination that lacks a clear political agenda. According to Joxe, the world dis-order is based on a rationality whose imperative is the "constant management of a calculated massacre as an act at directly regulating, not politics . . . but demographics and the economy" (2002, 12). Such a state of affairs, incidentally, very much resembles Schmitt's original depiction of the United States as a power that is pursuing a liberal-universalist project that downplays the political in favor of the economic.

Yet the response to the worsening development situation in Africa has primarily been a strategic rethinking within a still fundamentally liberal outlook. In this rethinking, the African state, which was deemed to have failed in responding positively to Western guidance and pressure, was now to be circumvented altogether. Thus, as the 1990s progressed a new development paradigm emerged, known as the New Policy Agenda (Bebbington 2004; Bebbington and Riddel 1995; Hulme and Edwards 1997; Dicklich 1998), which marked a shift towards a greater emphasis on the role that non-state actors, particularly non-governmental organizations, were to play in the development regime imposed on Africa. Coinciding with this shift, a new perception developed that the only way forward, within what was still a fundamentally neo-liberal agenda, was a transformation of

African society itself. As Joseph Stiglitz, former vice president of the World Bank, stated: "Development represents a transformation of society, a movement from traditional relations . . . to more 'modern' ways" (quoted in Rojas 2007, 107). It is in such a challenge that we can discern the roots of a critical opening for religious, particularly evangelical, actors in shaping the governmentality of African space.

To begin, it should be noted that the above shift in thinking on development emerged at the same time that the discourse on compassionate conservatism was created and promoted by the first G.W. Bush Administration, later crystallizing in the administration's faith-based initiative. This initiative encouraged faith-based organizations to seek federal funding for programs addressing particular social issues. As Bush had already declared in his first run for the Presidency,

> In every instance where my administration sees a responsibility to help people, we will look first to faith-based institutions, charities and community groups that have shown their ability to save and change lives. We will rally the armies of compassion in our communities to fight a very different war against poverty and hopelessness. (quoted in Helms 2001, n.p.)

It is precisely such a sentiment, translated to the New Policy Agenda, that offers evangelical organizations an opportunity to seek government funding and support for their foreign initiatives (Lindsay 2007).[1] For example, in the last six years the proportion of federal foreign aid dollars going to faith based organizations in the world has doubled, from 10 to 20 percent (Stockman et al. 2006). Ninety-eight percent of the beneficiary organizations, furthermore, are Christian (Stockman et al. 2006) and a very large proportion of these are active in Africa, which witnessed a tripling of U.S. aid monies under the presidency of George W. Bush (Martinez and Fisher-Thompson 2007).

It is important here, however, to stay focused on the U.S.' aim in this turn to faith based groups in Africa. Despite the isolated presence of evangelical designs and desires within the U.S. state, I contend that the openness towards faith-based development is primarily a strategy aimed at stabilizing Africa in order to incorporate it more fully into a U.S.-dominated global economic system. In relation to this sentiment, it is interesting to consider the steadily growing Weberian discourse

---

1 It has been argued that the faith-based initiative has been mostly rhetorical with an aim primarily to capture the evangelical vote (Kuo 2007). Whereas such an argument may have some validity with regard to the openness that the U.S. government has shown towards evangelical organizations in the developing world, it cannot be denied that the flow of money to these organizations has indeed increased in the last few years. Beyond these financial contributions, however, it is my contention that it is the legitimization of faith-based organizations as somehow being exceptionally apt at addressing humanitarian and development issues that is of critical importance. In the case of Africa, this enables the U.S. to dissociate itself from responsibility, thereby giving free reign to the evangelical movement that is eager to step in.

that has been emerging in the last decade or two.[2] The oft well-funded research endeavors that fuel this discourse are generally focused on the transformative potential of Protestant evangelicalism in the developing world, particularly with regard to economically relevant attitudes such as entrepreneurialism and work ethics. The increasing number of publications and conferences that are appearing as a consequence of this research tend to present evangelicalism as a transformative belief system that fosters personal responsibility, belief in personal achievement, family values, and an ethical code that dissuades impoverished people—particularly poor men—from engaging in unproductive activities such as drinking alcohol and gambling (Bernstein 2002; Council on Foreign Relations 2006; Grier 1997; Sherman 1997; Martin 1990; 2002; Pew Forum on Religion and Public Life 2006).

Evangelicalism is thus portrayed as a movement of potential empowerment that can help poor people in the developing world succeed in an increasingly competitive and interconnected globalized world. The belief in such empowerment is certainly widespread within the evangelical movement in Africa. For instance, African evangelical believers frequently associate their faith with the promise of being blessed with a certain material lifestyle, often very Western in orientation (Brouwer, Gifford, and Rose 1996). As one in-depth ethnographic study in Tanzania showed, for example, even in an evangelical congregation where material pursuits were downplayed, the born again identity that was being offered was widely associated with cultural capital, upward mobility, and even with gaining a greater benefit from global capitalism (Stambach 2000), a trend that I have seen corroborated in my own encounters with evangelicals in Ghana and Rwanda.

In this way evangelicals and their churches must be recognized as critical instruments of micro-level discourses and practices that are working to fundamentally change African subjectivities. For the U.S. state this evangelical influence is generally viewed as a positive development, that is, as a means to transform African societies into capitalism-minded and U.S.-friendly populations. As one embassy official in Rwanda told me,

> French and Catholic influence in Africa is waning, Africans are increasingly turning to Protestantism (evangelicalism) and to the United States. Here in Rwanda this is explicitly clear, it won't be long and Rwanda will have switched from being a Francophone country to being an Anglophone one, with the U.S. as its primary ally. (Anonymous 2005)

What is happening, in other words, is an acknowledgement on the part of the U.S. state, either explicitly or implicitly, that faith-based development and missionary activities are serving as a form of soft power for the United States.

---

2 Particularly relevant here has been the work of Peter Berger and the Institute on Culture, Religion and World Affairs at Boston University, which he heads.

Soft power, of course, has become the new buzz word in foreign policy circles after eight years of unilateralism, which has left the United States an increasingly isolated nation with high unfavorability ratings throughout the world. More and more emphasis in academia, think tanks, and even within government has turned towards the perceived need for developing an alternative to sheer military might, with a focus instead on economic and especially cultural capital to win hearts and minds (Edwards 2007; Khanna 2008; Nye 2004a; 2004b; Rice 2008). Yet, in a context in which the liberal paradigm calls for diminished state engagement, as for example via the New Policy Agenda, it is clear that it is increasingly only non-state actors that are in a position to take on this challenge. The cultural power of faith-based organizations is especially potent here. Not only do U.S. evangelical groups provide services associated with the U.S., they also offer identities that embrace Western economic values and attitudes as well as the Western power that defends these.

Building on this religiously grounded cultural capital, furthermore, the U.S state holds the hope of expanding its geopolitical alliances in Africa. Indeed there has been evidence to support such a hope. Polling conducted by the Pew Forum on Religion and Public Life (2006), for example, showed that Pentecostals in Nigeria, as well as in other parts of the world, were adamant supporters of the U.S. 'War on Terror'. This finding was particularly true in countries where evangelical populations live together with large significant Muslim populations, as in Nigeria. In some instances, such as in evangelically oriented Uganda, this popular support for U.S. geopolitics has also been transformed into state support, as Uganda became one of the few African members of the 'coalition of the willing' supporting the U.S. invasion of Iraq. This is precisely the type of geopolitical support, largely manifested in diplomatic allegiances that pay high dividends in the United Nations, that the U.S. would like to garner and expand.

Evangelical Aims in Africa

Turning now to evangelical motivations and aims in Africa, it is worthwhile to first consider the extent to which evangelicals have become present and influential in Africa. Indeed, it appears that one of the most momentous Christianizing efforts in human history is passing largely under the radar screen of academia. Although the nineteenth century is often presented as the heyday of global missions, the fact is that worldwide missionary activity, in terms of missionaries deployed, has increased fivefold since the end of that century (Johnstone 1993). A key characteristic of this surge has been the increased role and power of U.S. evangelical Protestantism. Today more than 90 percent of Protestant foreign missionaries are evangelical, with Africa being a preferred target. In 1990 Africa received the most evangelical missionaries of any continent (Hearn 2002).

Paul Gifford, one of the best known academics on Christianity in Africa has stated: "Christian missions are arguably the biggest single industry in Africa" (quoted in Hearn 2002, 40). This is both evident in terms of the presence of people and the supply of money. In 1990, there were nearly 18,000 Protestant missionaries in Africa, most of these American born evangelicals (Hearn 2002). Julie Hearn (2002), for example, points out that 95 percent of the Americans registered at the American embassy in Nairobi are missionaries. She also draws attention to the fact that the U.S. evangelical missionary project's income, at two billion dollars, is "equivalent to one fifth of aid transferred by NGOs worldwide" (Hearn 2002, 40). Increasingly it is the money rather than the people that matter most in terms of what U.S. evangelicals supply. This is due to a recent change in focus towards "co-operative resource sharing partnerships" (Siewert and Kenyon 1993, 55), that is, there is a greater focus on supporting indigenous evangelical initiatives, such as 'planting churches' as opposed to establishing U.S.-run projects. Such a strategic shift explains why the number of U.S. evangelical missionaries in the world has decreased slightly over the past ten years, whereas the funding has remained constant—not a surprising development considering that over 75 percent of evangelicals now reside in the developing world (Nussbaum 2006).

Nevertheless, due to the close connections to U.S. evangelical institutions and their funding, indigenous evangelical movements generally embrace a particularly American flavored form of Protestantism. As Brouwer, Gifford, and Rose (1996, 178) conclude in their research on evangelicalism in Africa:

> A particular kind of Christianity is presented under the label "biblical", and this claim is taken at face value. This form of Christianity is new and fundamentalist and American. Through its resources, personnel and technology, it may be exerting an influence every bit as great as the colonial Christianity of the last century.

Thus, the suggestion of 'independent' African churches should be taken with a grain of salt. Instead, what the authors suggest is that the evangelical revival unfolding in Africa is leading to a significant soft power associated with the United States, one which, in addition to being passionately religious is also materialistic and morally conservative.

To understand the impetus behind this rapid spread of evangelicalism we must look past the favorable conditions provided by U.S. policy mentioned earlier. Attention must focus, instead, on the evangelical faith itself, namely in the strong conviction that the African continent needs to be evangelized. This conviction, furthermore, is sedimented in the belief of a God-given command, the Great Commission, to convert all nations to the Christian faith (Matt 28:16–20; Acts 1:4–8; Luke 24:44–49; John 20:19–23). Importantly, in conservative forms of evangelicalism this command is also very frequently tied to an eschatological understanding of history as an unfolding divine plan that progresses via a number of dispensations in which humans fulfill various aspects of God's will. Although following the rules of conduct demanded by each dispensation is not the path to salvation as such (which is achieved only through the grace of God when accepting Jesus as one's personal savior) it is nonetheless viewed as a stewardship of faith in which obedience to God is transformed into practice. In the current dispensation, the age of the Church, the Great Commission becomes a key aspect of divinely ordained action. Indeed, the spreading of the Gospel to all nations on earth is seen by many evangelicals as a key prerequisite for the final dispensation, the millennial rule of Christ over earth (see Introduction, this volume).

It should here briefly be clarified what evangelicals mean when they reference the biblical command to spread the Christian faith to every nation of the world. For evangelicals the term 'nation' is understood as indicating a distinctive 'people group', or what is commonly called an ethnic group. To help identify the various people groups of the world, an extensive U.S. evangelical research endeavor, know as the Joshua Project, has been launched. The intention of this project is to create an online storehouse of geographically ordered information on the various people groups of the world, particularly those considered 'unreached' by the Gospel. In its own words, "The mission of Joshua Project is to help bring definition to the unfinished task of the Great Commission by identifying and highlighting the people groups of the world that have the least exposure to the Gospel and the least Christian presence in their midst" (Joshua Project 2008).[3] Sub-Saharan Africa figures prominently into this project as it is seen to contain close to 1000 unreached people groups. Although this is less than in Asia, Sub-Saharan Africa, which is not home to any of the major world religions, is viewed as a relatively conducive site for the spread of evangelical Christianity.

Following from this biblically mandated imperative, many conservative evangelicals have also developed a very particular geopolitical vision centered on the Great Commission. This is clearly illustrated in the geographical imaginary of the 10/40 window embraced by many evangelicals (see Chapter 8, this volume). The 10/40 window is a rectangular swath of land between 10 and 40 degrees north of the equator where the least evangelized populations of the world live. This geographical imaginary, and the subsequent maps produced to present it, serve as a tool to depict undeveloped, impoverished, and frequently also demonic spaces

---

3   The Joshua Project is one among several Christian organizations that is engaged in the construction of instrumental forms of geographical knowledge. Mapping is here a critical tool used in order to make sense of and ultimately intervene in distant populations. Global Mapping International (GMI 2008), for example, is a company specializing in mapping solutions aimed at servicing global evangelization, offering a wide range of choropleth maps that help make the case for the need of evangelical interventions, particularly in the 10/40 window. GMI also offers numerous highly detailed GIS maps of specific populations and their religious affiliations. In many ways the knowledge being created by these Christian organizations is reminiscent of the colonial forms of knowledge maintained by imperial powers over their territories and subjects. The difference, however, is that the governmental target of evangelicals is not so much controlling and directing these populations to service colonial and mostly material aims; the goal is rather the conversion of these populations to evangelical Christianity, although it should be noted that, as I discuss in this chapter, both evangelicals and their state sponsors do also perceive an overlap between religious faith and materially productive behavior.

that are seen as desperately in need of intervention. Due to its place in the 10/40 window, and the perception that it is a space particularly conducive to evangelical conversion, parts of Africa have long been viewed as the next natural space for evangelical expansion, albeit one that is being vehemently contested by Islam – an issue that I will expound further below.

Consider, for example, Map 7.1: a map generated using the databank and mapping tools offered online by the evangelical NGO, Global Mapping International (GMI).[4] The reliability of the data used in this map is certainly an issue of concern, but I include it here not so much for the information that it depicts but rather for the particular aspiration that it reveals. The map is a visual presentation of evangelical growth in Africa based on the percentage increase of evangelicals in each country. In many of the countries shown on this map, however, particularly in Muslim countries, the actual number of evangelicals is minuscule, meaning that even a very small number of conversions—which are easy to exaggerate

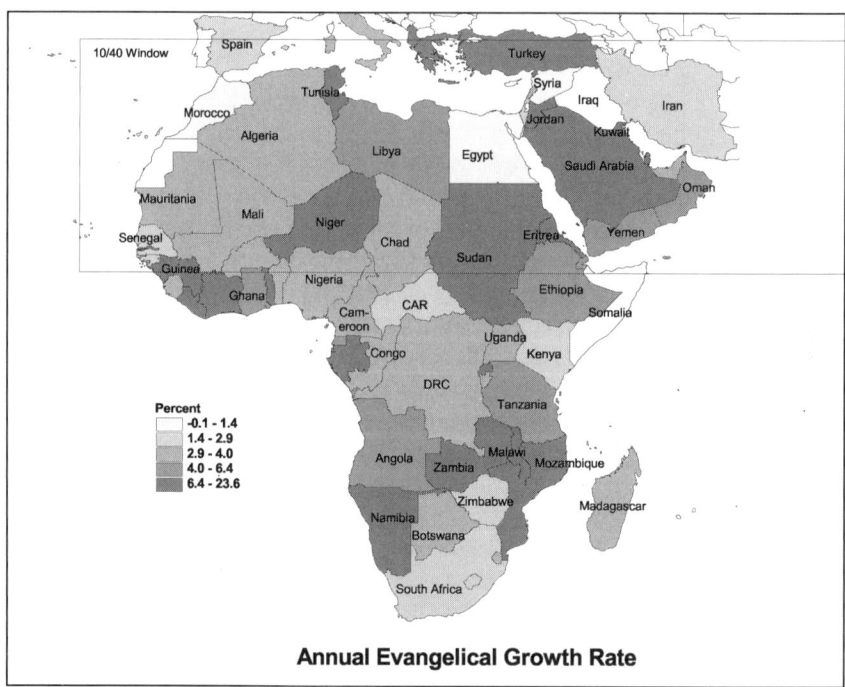

Map 7.1 Evangelicals' self-projected growth in Africa

Note: Self-generated map using hidden data from Global Mapping International's Mission Info Bank

---

4 The author is responsible for adding the 10/40 window to this otherwise automatically generated map.

on such a small scale—will appear as a significant growth in that country. Thus, mapping evangelical growth in this way makes it appear that Evangelicalism is rapidly spreading and gaining influence throughout the Islamic world, which, to a large extent, is more fancy on the part of the evangelical community than fact. Nonetheless, such maps offer a sense of hope and purpose for the evangelical community dedicated to fulfilling the Great Commission.

In Africa, then, the supreme goal of conversion that lies at the root of U.S. evangelical engagement takes on different forms depending on the particular country in question. In areas where Christianity is a minority religion or where it is essentially equal to Islam, as in much of the Sahel, western Africa, and Tanzania, there is extensive support for the 'planting' of indigenous evangelical churches. In these regions, furthermore, American evangelical churches and organizations tend to engage in development and aid services as a way to expose populations to the Gospel. Undergirding these efforts has been a concerted attempt to protect minority Christian populations from persecution, which is generally done by publicizing any maltreatment of foreign Christians and lobbying the U.S. government to prioritize the issue of religious freedom.

Conversely, in African countries already much under the sway of evangelicalism, such as the Ivory Coast, Kenya, Uganda, and Zambia, the focus of U.S. evangelical groups, beyond consolidating and expanding their base of believers, has been to deepen the influence of evangelical thinking on the society as a whole. Thus, U.S. conservative evangelicals who have been stymied in their attempts to create a more devout evangelical society in the United States, often see Africa as a promising and less complicated space, where it is believed that evangelicalism can still become the essential fabric of society (Rice 2004). In Uganda, for example, which is a country that receives much support and praise from U.S. evangelical churches and groups, the separation between church and state is diminished with much of the state apparatus firmly in the hands of devoted evangelicals; abortion is illegal; patriarchy is unquestioned, and AIDS prevention and population control are increasingly addressed through abstinence programs (Batard 2008; Epstein 2005; Rice 2004).

Yet regardless of their specific local interventions in Africa, evangelical actors in the United States have been broadly active in shaping the governmentality of African space by creating and supporting particular discourses and governmental technologies that they view as favorable to their underlying mission. For one, evangelical actors have been strong proponents of the emergent view that non-state actors, particularly faith-based organizations, are best positioned to engage in the provision of social, developmental, and humanitarian services in the developing world. For conservative evangelicals the state has always been viewed as a potential barrier to their ultimate task of converting individuals to Christianity. These evangelical actors were quick to realize that the New Policy Agenda outlined by the United States offered a critical opening for an evangelical revival in Africa and elsewhere (Hearn 2002). Through the New Policy Agenda missionary groups, many if not most of whom have now become registered as

NGOs, are being legitimized by U.S. government discourse while at the same time becoming eligible for large government sponsored development contracts (see Lindsay 2007; Stockman et al. 2006).

To give one example, Franklin Graham's humanitarian organization Samaritan's Purse, which is involved in over 10 countries in Africa and whose self stated mission is "serving the Church worldwide to promote the Gospel of the Lord Jesus Christ" (Samaritan's Purse 2008) received 31 million dollars in government funding between 2001 and 2006 (Canellos and Baron 2006). It should be noted that federal regulations stipulate that this money may only be used for non-religious expenditures such as building supplies and food. Nevertheless, projects that are partially funded with federal dollars may have an evangelizing component to them, for example the building of a church or the distribution of bibles, as long as these overtly religious activities are privately funded. In this way the government is able to claim the maintenance of a separation of church and state and evangelical organizations can reassure themselves that the government is not contaminating their primary effort of seeking Christian converts.

Another measure sought by conservative evangelicals to influence the governance of African space has been to sediment religious freedom as a foundational and primary human right. In pursuance of this goal evangelicals formulated and pushed for the passage of the International Religious Freedom Act, which was passed in 1998 (see Hertzke 2004). One of the results of this law was the establishment of the U.S. Commission of Religious Freedom. This commission is charged with maintaining a constant watch on religious persecution throughout the world and has developed into a powerful surveillance mechanism of international space—evaluating and categorizing global space on the grounds of religious freedom alone. The U.S. Commission of Religious Freedom has become a critical governmental technology that has enabled conservative evangelicals to push for U.S. intervention, both diplomatic and military, in countries where Christian populations are persecuted for their beliefs or their actions, such as proselytizing. For example, the U.S. Commission of Religious Freedom was a key player in exerting pressure on the G.W. Bush Administration to deal more forcefully with Sudan in its civil war with the animist and Christian south. In its first report the U.S. Commission of Religious Freedom branded Sudan as the worst persecutor of religious freedom in the world.

Opposing Conceptions of the (Geo)political in Africa

Having presented the synergies between the U.S. state and conservative evangelicals in Africa, I want now to turn to the deep running incompatibilities that nevertheless persist between these two global actors. These incompatibilities, I suggest, are rooted in competing normative differences that essentially put the secular U.S. state and conservative evangelicals at odds with each other in terms of long term visions for a future global governmentality. In his seminal work on governmentality Dean (1999) highlighted the normative content inherent in forms of governmentality, making the point that governmentality is intimately linked to underlying thoughts of an end goal toward which governance is oriented. Such an end goal is inevitably utopian in the sense that it entails a notion of an ethically grounded vision of a better world. It is precisely with regard to such a vision that the liberal democratic state and the evangelical movement tend to clash.

Although there clearly exists some overlap between the Christian faith and nationalist sentiment in the United States, for example in the sense of an American civil religion (Bellah 1967), the general impetus of the liberal state nonetheless has been to tame the absolutist tendencies of religious doctrine. In this sense, liberal democratic governmentalities are generally dedicated to a universalism in which contentious metaphysical and normative issues are pushed into the relativism of the private sphere, allowing for a diverse but peaceful coexistence in which public matters are decided through rational discussion and debate. The utopic world that is envisioned here is in many ways procedural, focused on conditions that allow the diffusing of conflict as opposed to a commitment to, and belief in, the achievement of a final and perfect society. Within the evangelical movement, on the other hand, there is a dedication to a homogenous society in which a metaphysical understanding of right and wrong becomes the guide for public action and where the end goal is the return of the Christ. In this belief there is a very concrete sense of an end to come that will bring about both salvation (and damnation) as well as a utopian, perfect world here on earth.[5]

With regard to this opposition in normative outlooks, it is my contention that liberalism has largely won the day in the United States and Western Europe – despite continuous challenges by the Christian Right – whereas in Africa it is an essentially anti-liberal evangelicalism, competing with an equally anti-liberal Islamism, that is becoming ever more prevalent. The argument that I want to make in the rest of this chapter is that this development, in which liberalism is losing out to absolutist belief systems, is largely due to a lackluster political commitment to

---

5   It is worth noting here John Gray's (2008) work, *Black Mass*, which essentially argues that secular liberalism shares a deeply embedded conviction with Christianity in a teleologically bound, better world to come. Furthermore, both Christian and secular liberal thought, Gray argues, are destined to promote dangerously regressive practices due to their absolutist beliefs in teleological 'progress'. In a sense Gray is in line with Dean's depiction of governmentality regimes as always possessing a normative utopic vision. Yet a key point I want to make is that these visions are very different and, indeed, in conflict. The ultimate question with regard to Gray's work, however, is whether liberalism will or will not succumb to Christianity's absolutism. Gray mentions the neoconservatives' obsession with spreading liberal democracy throughout the world as such an absolutist tendency, in which key obstacles to an envisioned normative end are identified and eradicated. An alternative to such an illiberal execution of liberalism, however, and one more true to the core and historically sedimented tenants of liberalism, would be the acceptance of difference in the world with an emphasis on the procedures that enable society to deal with such difference.

Africa on the part of the United States and the consequent openings that are being created for both evangelical and Islamist engagements and influence.

Positioning the Enemy

There are numerous concerns that can be raised about the evangelical revival in Africa. For one, the frequent emphasis on the achievement of wealth and health through faith is criticized as a veritable opiate of the people, in which a turn away from secular/scientific approaches to development only works to obscure the identification of structural problems that continue to go unaddressed (Brouwer, Gifford, and Rose 1996). Furthermore, the conservative ideals of evangelicalisms are seen to hinder the advancement of women or the acceptance of non-normative sexuality (Bates 2006). Finally, the restrictive evangelical view of family planning and sexuality has been shown to backfire in places like Uganda, once a model in the fight against AIDS in Africa, which is now witnessing serious setbacks after a policy turn that emphasizes abstinence and fidelity as opposed to the propagation of condom use (Epstein 2005; Goldberg 2007).

However, the starkest difference between the global project of the U.S. state and the evangelical movement is found in their opposing geopolitical visions, that is, their normatively informed, instrumental views of the world. This difference can be well grasped when making use of Schmitt's conception of the (geo)political as ultimately constituting the determination of the friend-enemy grouping, which is essentially grounded on competing commitments to a particular form of social organization. Following Mouffe, the key challenge to liberal democracies, and thus also western hegemony, is precisely the ability to take sides in such inevitable friend-enemy formations. On the global scale a clear example of such a friend-enemy grouping is the juxtaposition between the proponents of Western liberal democracy and those of an anti-Western Islamism. It is precisely with this potent political challenge in mind that the burgeoning American-flavored evangelical revival in Africa becomes especially relevant.

To begin, I want to suggest that there are two possible friend-enemy configurations that can take shape with regard to the West's relation to Islamism in Africa. The one is a juxtaposition between Islamism and a Western-backed democratic pluralism and the other is one between Islamism and a political, conservative Christianity. Taking up the first confrontation and building on Mouffe's thought, the best way for Western democratic states to deal with its juxtaposition to Islamism would be to confront it as a just enemy and to work towards its defeat, as opposed to casting it as criminal, which is ultimately a refusal to recognize the friend-enemy grouping. For the U.S., treating Islamism as a just enemy would require a concerted effort to establish a political project grounded on the ideals of pluralism and bottom-up democracy.

In Africa, such an effort would first and foremost entail establishing significant diplomatic and economic ties that are committed to addressing the social inequality and justice issues that provide the fodder for violent, religiously-based identity politics (Appleby 2000; Bandarage 2004; Cramer 2003; Ostby 2004). Second, support would have to be given to organizations within domestic civil society that represent the ideals of secular pluralism and which consciously offer a viable, tolerant alternative to the confrontational identities pushed by radical, absolutist faiths. Finally, and most challengingly, an effort would need to be made to support governmental institutions and regimes that confront Islamism in discourse and in practice. Such a confrontation would need to remain open and transparent, thereby constantly allowing for the possibility of negotiation and compromise.

As Mouffe (1991) has also pointed out, certain lines nevertheless need to be drawn in terms of what can be accepted within a political community, and force may be necessary in defending these lines. The challenge is thus to engage in the correct diplomacy, which would allow conflicting differences to be diffused within democratic institutions, while at the same time finding the correct judgment as to when this effort has truly failed. If it has failed then the correct action would be, following Schmitt, to engage in a war that is deemed necessary out of political expediency, rather than out of moral or religious passion.[6]

Yet, according to Mouffe, the shortcoming of liberalism as it stands today is precisely its inability to actually internalize the friend-enemy grouping. Such a position, to which the U.S. is still dedicated in its execution of a neo-liberal global governmentality, only exacerbates existing conflicts. Thus, with the insistence on the part of the U.S. to criminalize the enemy, the radical democratic approach addressed above remains a mere hypothetical. Rather, the political challenge of Islamism in Africa is largely ignored, being read instead as a global law and order issue that must be addressed through military means. Such a stance, in turn, creates a situation in which Islamism is forced into the position of the criminalized partisan, a clandestine resistance that perceives itself to be in a war of survival, in which there is only victory or death.

Further complicating this situation, however, is the fact that the political resistance to Islamism is now largely being left in the hands of a booming evangelical revival. The danger that this poses is the emergence of a friend-enemy grouping in which Islamic partisans see themselves not only engaged in a war that criminalizes them, but a war that is centered on passionately held religious

---

6 There is, following Schmitt, a conceivable second option for the Western state in dealing with a global friend-enemy confrontation, namely the creation of large spaces (*gross Räume*) that are bracketed off as being outside of the ethical juridical jurisdiction of the hegemonic West (see Mouffe 2005). In the case of radical Islam in Africa, for example, it would be possible to conceive of, alternately to Schmitt, medium spaces (*mittelgrosse Räume*), that is, isolated spaces that have embraced non-liberal forms of governance such as Sharia law, which nevertheless are recognized as spaces outside of a western-based international law. Such a formal and practical recognition would ideally serve as the foundation for an international peace between fundamentalist Islam and western secularism.

differences. In this way, the U.S. state, given its own absent political project, is increasingly being associated with the political project of conservative evangelicals, who increasingly have been making Islam their mortal enemy. Thus, instead of adopting a friend-enemy grouping that is based on a defense of its deepest held, ideological beliefs, the liberal democratic U.S. state is here being forced into a friend-enemy confrontation beyond its choosing, namely one that is grounded on the juxtaposed faiths of conservative evangelicalism and conservative Islam.

Such a development is absolutely counter to the United States' vision of a future, liberal world order, and it is also devastating for the ethical aim of reducing violent conflict and mass suffering in Africa. For, even though it is sometimes opined that conservative Muslims are more averse to secular liberalism than they are to evangelicalism, it is my contention, following Mouffe's lead, that it is only by establishing truly democratic institutions, that is, ones that embrace a certain relativism, that the potential violence of a confrontation with Islamism is best hedged. In other words, religiously conservative groups sometimes expound a shared commitment to a metaphysically based absolute. Yet such a commonality will always remain superficial since the fact that this absolute is interpreted differently in the most important matters means that there is always an underlying divergence that is fundamentally incompatible, thus establishing the conditions for an endless and ruthless war to the end.

The Evangelicals' Clash with Islam in Africa

In Africa, the potential of such an unhedged conflict, fueled by a clash between Christianity and Islam, is particularly acute. These two faiths have an extended history of competition in Sub-Saharan Africa, particularly in the east of the continent and in the Sahel, where early Christian missions associated with European colonization were confronted by an already-present Islam. Initially, Islam was often presented as an improvement to the African pagan religions, but it was quickly recognized as a serious competitor. With this realization Islam was portrayed as a false religion, and thus as particularly evil in its deception. Yet at the same time, in those regions where Islam dominated, Christian missionaries eventually learned that they needed to adapt to the Islamic hegemony in place or risk being expelled, which, in turn, led to a toning down of the rhetoric against Islam. In the Sahel, for example, the Sudan Interior Mission's early aggressive approach to Islam in the late nineteenth century gave way to a practical tolerance several decades later (Cooper 2006).

Such pragmatism can also be witnessed in an isolated number of contemporary cases. In the Kaduna Peace Declaration of Religious Leaders in Nigeria, for instance, vehemently antagonistic Christian and Islamic leaders have agreed to work towards a diffusing of religious confrontation in a state that has witnessed the worst African religious riots in recent memory (Crawley 2003).[7] Another example that I have explored elsewhere is the case of southern Sudan, where a particular ethics of care developed on the part of U.S. evangelicals for the local population (Gerhardt 2008a). This development, in turn, led to a more pragmatic and compromised approach to the Islamic regime in Khartoum, thereby allowing these otherwise hawkish evangelicals to push for a peace deal between the North and the South of the country. Such developments are important to take note of as they go to show that certain impulses, such as external pressures to accept pluralism, the facilitation of inter-faith dialogue, or even the fostering of an ethics of care, can lead to the taming of radically incompatible belief systems, even if such taming remains a never-ending process.

Yet with an evangelical revival flourishing within Africa today, anti-Islamic discourse is again on the rise. This spike in confrontation can be related to the intense skepticism with which Islam is currently viewed by U.S. evangelicals, who have significant sway in Africa. It should be noted, however, that the negative rhetoric against Islam goes well beyond a simple identification of Islam with terrorism following 9/11. The current antagonism is also linked to a growing interest in eschatology within the conservative evangelical movement, largely popularized through a burgeoning number of U.S. Christian media productions. Although largely emanating from the United States, however, the belief in the End Days and the implications for action that this belief entails is also widely popular in Africa (Brouwer, Gifford, and Rose 1996; Pew Forum on Religion and Public Life 2006).

Within such dispensationalist approaches to Christianity Islam is often interpreted as the work of Satan, intent on postponing or minimizing the salvaging effect of Jesus' return. In this sense Islam is viewed as a sham religion that blinds believers to the true path of salvation. Islam's negative effects are seen to be achieved through its extensive proselytizing activities and its vehement defense against Christian missionary work, thereby hindering the Great Commission and thus, in some views, the return of Christ[8]. There is also the more abstract rationale

---

7 It is believed that up to 2,500 people were killed in 2000 when Kaduna state introduced *sharia* law despite protests by local Christians (Crawley 2003).

8 It should be noted that there is some friction within these variously held views. Dispensationalism, for example, is generally associated with premillennialism, that is, a belief in an impending Apocalypse and subsequent Rapture in which the world's true Christians will be delivered directly to Heaven. The Great Commission, in turn, is associated with postmillennialism, which generally denies an impending apocalypse. Instead, the focus is on creating the right conditions for establishing the Kingdom of Christ here on Earth. Yet, despite these apparently crass differences, it is often difficult to make clear cut distinctions between these various eschatological positions. In my own research I have found that evangelicals often combine pre- and postmillennial beliefs. For a further discussion of this issue see Gallaher's chapter in this volume as well as the exchange between Sturm (2008) and Gerhardt (2008b) in *Environment and Planning D: Society and Space*.

that the Islamic societies of northern Africa need to be kept in check by southern Christian peoples in an attempt to ease the pressure on the Jewish state of Israel, which must be maintained and even expanded in order for the final dispensation of Christ's return to unfold.

In Africa, such combative views foster and support a resurgent attempt to challenge Islam. Such challenges take numerous forms, such as increased proselytizing activities and direct attacks on the Islamic faith via fiery sermons by evangelical preachers (see for example Cooper 2006). Many of these sermons are broadcast by radio stations to cover large spaces within the African continent to the ire of Muslim populations. There have also been reports of increased discrimination against Muslims within African countries where evangelicalism is a powerful force, for example in Uganda, Ivory Coast, Kenya, and Tanzania (see for example Chirot 2007; Filkins 2002; Rice 2004; Heilman and Kaiser 2002).

This is not to say that religion on its own poses the predominant threat to African stability. As is often the case, religious identities that 'other' opposing groups are not embraced in a vacuum. Such identities are internalized in the context of already existing conflicts—whether it is the struggle over limited local resources and jobs between autochthonous groups and immigrants as in Nigeria and the Democratic Republic of Congo, or issues such as access to political representation and economic development as in the Ivory Coast, Kenya, and Tanzania. It is religion, however, that is increasingly providing the absolutist identities that can be harnessed for political means in securing leverage in these struggles and which, once embraced, make it extremely difficult to contain the battles that ensue.

Conclusion: The Challenge of Africa for the U.S. State

Following from the arguments above, it would be incorrect to embrace the Huntingtonian conception of an ontologically given and inevitable "clash of civilizations" within Africa (Huntington 1996). Rather, following Edward Said (2001), it is a question about the creation of such a clash. The key issue then becomes, how can such a perception of a clash of civilizations be avoided? As mentioned earlier, there are cases in which conservative evangelicals and Muslims have accepted a shaky peace. Yet such a peace needs external support, hence the critical role of the U.S. state with regard to the friend-enemy grouping that forms the basis of the West's conflict with a resurgent Islamism. As argued above, however, the U.S. has generally shirked placing itself within such a friend-enemy grouping, preferring instead a neo-liberal governmentality that is grounded on an all-encompassing normative legalism as opposed to a clearly defined political project.

Nevertheless, U.S. foreign policy has on occasion engaged in a superficial pandering to already existing religiously informed political discourses. For example, the U.S. has regularly tapped into biblically resonating constructions of a battle, or even a crusade (Ford 2001), of the Good against the Evil. The reason for such rhetoric can be seen, as Schmitt (1996) suggested, as a way to bind the body politic of a challenged nation together. Yet, what the U.S. state must acknowledge is that this religiously loaded discourse, which it uses as an opportunistic ersatz for its own empty political project, is, in fact, a language that has been internalized by hundreds of thousands of evangelical Christians worldwide for whom an epic political-religious struggle is already underway. By tapping into their discourse the U.S. is compounding its foreign policy challenges in Africa by being increasingly associated with this evangelical cause. In this sense, beyond the challenge of meeting a partisan resistance to its totalizing legalistic hegemony, the U.S. increasingly risks being confronted with challenges that are based on intractable and uncompromising religious positions as well.

The U.S. state's association with the evangelical cause is further exacerbated by a new governmentalization of African space in which conservative evangelicals are quickly filling the void left by increasingly defunct African states. U.S.-sponsored missionaries and evangelical aid organizations are thus poised to become the public, cultural face of the United States. As a consequence, the ideologies of secular democracy and pluralism are losing out to an American-flavored conservative evangelicalism. Such a situation, in turn, is sowing the seeds for a global foreign policy context that is very different than that ideally envisioned by the U.S. state.

In Africa, for example, it is quite possible that confrontational religious identities will begin to trump ethnic identity as the critical rallying point for conflicts grounded in much broader and entrenched problems. Such a development not only threatens to make these problems significantly more intractable, but due to the global nature of these identities local conflicts threaten to jump scale to include actors at the regional and even global level, making the demand for U.S. involvement more likely. It is therefore not unthinkable that the U.S. state may soon find itself at the center of a number of unwanted political upheavals and confrontations in Africa. More specifically, 'friendly', evangelically-oriented regimes may suddenly find themselves dealing with Islamic insurgencies. Or, conversely, the increasingly bold sermons and confrontational posturing of minority evangelical populations may lead to an Islamic backlash that spirals out of control, leading to claims of genocide and demands for intervention.

Thus, as the U.S. continues to ignore Africa politically, and, perhaps later when it is compelled to intervene based on this earlier inaction, the chance of engaging Africa in a way that diffuses the violent potential of the friend-enemy confrontation diminishes. In addition to this, and perhaps most disturbing, as the U.S. gets sucked into a religious friend-enemy (i.e. friend-foe) conflagration, the normative content of secular Western society—grounded on individual rights, a respect for diversity, and true democracy—is being fundamentally lost. The degree and permanence of such a loss in Africa remains to be seen.

References

Agamben, G. (1998), *Homo Sacer: Sovereign Power and Bare Life* (Stanford: Stanford University Press).
Anonymous, (2005), Discussion on Development in Rwanda (personal interview, 22 June, 2005).
Appleby, S.R. (2000), *The Ambivalence of the Sacred* (Lanham: Rowman and Littlefield).
Bandarage, A. (2004), 'Beyond Globalization and Ethno-Religious Fundamentalism', *Development* 47:1, 35–41.
Batard, A. (2008), 'Uganda's God Business', *Le Monde Diplomatique* (English version), 10 January 2008.
Bates, S. (2006), *A Church at War: Anglicans and Homosexuality* (New York: I.B. Tauris).
Bebbington, A. (2004), 'NGOs and Uneven Development: Geographies of Development Intervention', *Progress in Human Geography* 28:6, 725–45.
Bebbington, A. and Riddell, R. (1995), 'The Direct Funding of Southern NGOs by Donors: New Agendas and Old Problems', *Journal of International Development* 7:6, 879–94.
Bellah, R.N. (1967), 'Civil Religion in America', *Journal of the American Academy of Arts and Sciences* 96:1, 1–21.
Bernstein, A. (2002), 'Globalization, Culture, and Development. Can South Africa be More than an Offshoot of the West' in Berger, P.L. and Huntington, S.P. (eds) *Many Globalizations: Cultural Diversity in the Contemporary World* (Oxford: Oxford University Press).
Brouwer, S., Gifford, P. and Rose, S.D. (1996), *Exporting the American Gospel: Global Christian Fundamentalism* (New York: Routledge).
Canellos, P.S. and Baron, K. (2006), 'A U.S. Boost to Graham's Quest for Converts', *The Boston Globe* [website], 8 October. <http://www.boston.com/news/nation/articles/2006/10/08/a_us_boost_to_grahams_quest_for_converts/>, accessed 31 July 2008.
Chirot, D. (2006), 'The Debacle in Côte d'Ivoire', *Journal of Democracy* 17:2, 63–77.
Connolly, W. (2005), 'The Evangelical-Capitalist Resonance Machine', *Political Theory* 33:6, 869–86.
Cooper, B. (2006), *Evangelical Christians in the Muslim Sahel* (Bloomington: Indiana University Press).
Council on Foreign Relations (2006), [Online] 'The Nexus of Religion and Foreign Policy: The Global Rise of Pentecostalism', Transcript of meeting held in Washington D.C., 10 October. <http://www.cfr.org/publication/11758/nexus_of_religion_and_foreign_policy.html>, accessed 30 July 2008.
Cramer, C. (2003), 'Does Inequality Cause Conflict?', *Journal of International Development* 15:4, 397–412.

Crawley, M. (2003), 'Two Men Create Bridge over Nigeria's Troubled Waters', *Christian Science Monitor* [website], 28 February. <http://www.csmonitor.com/2003/0228/p07s01–woaf.html>, accessed 31 July 2008.
Dean, M. (1999), *Governmentality. Power and Rule in Modern Society* (London: Sage).
Dicklich, S. (1998), *The Elusive Promise of NGOs in Africa: Lessons from Uganda* (Basingstoke: Macmillan).
Edwards, J. (2007), 'Reengaging with the World', *Foreign Affairs* 86:5, 34–35.
Epstein, H. (2005), 'God and the Fight Against AIDS', *The New York Review of Books* [website] 52:7. Available at: <http://www.nybooks.com/articles/17963>, accessed 31 July 2008.
Filkins, D. (2002), 'For Kenya's Muslims, Resentments Local and International' *New York Times*, 1 December. <http://query.nytimes.com/gst/fullpage.html?res=9D02E3DA1138F932A35751C1A9649C8B63>, accessed 31 July 2008.
Ford, P. (2001), 'Europe Cringes at Bush 'Crusade' Against Terrorists', *The Christian Science Monitor*, 19 September. <http://www.csmonitor.com/2001/0919/p12s2–woeu.html>, accessed 31 July 2008.
Foucault, M. (1991), 'Governmentality' in Burchell, G., Gordon, C. and Miller, P. (eds) *The Foucault Effect: Studies in Governmentality* (Chicago: University of Chicago Press).
Gerhardt, H. (2008a), 'Geopolitics, Ethics, and the Evangelicals' Commitment to Sudan', *Environment and Planning D: Society and Space* 26:5, 911–28.
—. (2008b), 'The Role of Ethical Conviction and Geography in Religiously Informed Geopolitics: A Response to Sturm', *Environment and Planning D: Society and Space* 26:5, 935–38.
GMI (2008), 'Global Mapping International', <http://www.gmi.org/> (home page), accessed 12 October 2008.
Goldberg, M. (2007), 'How Bush's AIDS Program is Failing Africans', *The American Prospect*, 10 July. <http://prospect.org/cs/articles?article=how_bushs_aids_program_is_failing_africans>, accessed 31 July 2008.
Gray, J. (2008), *Black Mass: Apocalyptic Religion and the Death of Utopia* (New York: Farrar, Straus and Giroux).
Grier, R. (1997), 'The Effect of Religion on Economic Development: A Cross National Study of 63 Former Colonies', *Kyklos*, 50:1, 47–62.
Hardt, M. and Negri, A. (2000), *Empire* (Cambridge: Harvard University Press).
Harvey, D. (2007), *A Brief History of Neoliberalism* (Oxford: Oxford University Press).
Hearn, J. (2002), 'The "Invisible" NGO: U.S. Evangelical Missions in Kenya', *Journal of Religion in Africa* 32:1, 32–40.
Heilman, B. and Kaiser, P.J. (2002), 'Religion, Identity and Politics in Tanzania', *Third World Quarterly* 23:4, 691–709.
Helms, J. (2001), 'U.S. Senator Jesse Helms (R-SC) at The American Enterprise Institute', *NomadNet*, <http://www.netnomad.com/helms3.html>, accessed 30 July 2008.

Hertzke, A.D. (2004), *Freeing God's Children. The Unlikely Alliance for Global Human Rights* (Lanham: Rowman and Littlefield).
Howe, N. (2008), 'Thou Shalt Not Misinterpret: Landscape as Legal Performance', *Annals of the Association of American Geographers* 98:2, 435–60.
Hulme, D. and Edwards, M. (eds) (1997), *NGO's, States and Donors: Too Close for Comfort?* (Basingstoke: Macmillan).
Huntington, S.P. (1996), *The Clash of Civilization and the Remaking of World Order* (New York: Simon & Schuster).
Jenkins, P. (2002), *The Next Christendom: The Coming of Global Christianity* (New York: Oxford University Press).
Johnstone, P. (1993), *Operation World*, 5th Edition (Carlisle: OM Publishing).
Joshua Project (2008), 'What is Joshua Project?', <http://www.joshuaproject.net/joshua-project.php>, accessed 31 July 2008.
Joxe, A. (2002), *Empire of Disorder* (New York: Semiotext[e]).
Khanna, P. (2008), *The Second World: Empires and Influence in the New Global Order* (New York: Random House).
Kochi, T. (2006), 'The Partisan: Carl Schmitt and Terrorism', *Law and Critique* 17:3, 267–95.
Kuo, T. (2007), *Tempting Faith: An Inside Story of Political Seduction* (New York: Free Press).
Larner, W. and Walters, W. (eds) (2004), *Global Governmentality: Governing International Spaces* (New York: Routledge).
Lindsay, D.M. (2007), *Faith in the Halls of Power: How Evangelicals Joined the American Elite* (Oxford: Oxford University Press).
Martin, D. (1990), *Tongues of Fire: The Explosion of Protestantism in Latin America* (Oxford: Blackwell).
—. (2002), *Pentecostalism: The World Their Parish* (Oxford: Blackwell).
Martinez, M. and Fisher-Thompson, J. (2007), 'U.S. Aid to Africa Triples During Bush Administration', *America.gov*, <http://www.america.gov/st/washfile-nglish/2007/February/200702231136161EJrehsiF0.311886.html>, accessed 30 July 2008.
Mouffe, C. (1988), 'Radical Democracy: Modern or Postmodern?', in A. Ross (ed.) *Universal Abandon? The Politics of Postmodernism* (Minneapolis: University of Minnesota Press).
—. (1991), 'Pluralism and Modern Democracy: Around Carl Schmitt', *New Formations* 14, 1–16.
—. (1993a), 'On the Articulation Between Liberalism and Democracy', in *The Return of the Political* (London: Verso).
—. (1993b), 'Rawls: Political Philosophy Without Politics' in *The Return of the Political* (London: Verso).
—. (1993c), 'Towards a Liberal Socialism' in *The Return of the Political* (London: Verso).
—. (1993d), 'Politics and the Limits of Liberalism' in *The Return of the Political* (London: Verso).

—. (2005), 'Schmitt's Vision of a Multipolar World Order', *The South Atlantic Quarterly* 104:2, 245–51.

Müller, J. (2006), 'An Irregularity that Cannot be Regulated', *Notizie di Politeia: Rivista di Etica e Scelte Pubbliche* 112:84, 65–78.

Nussbaum, P. (2006), 'Evangelical Christianity Shifting Outside West', *Philadelphia Enquirer*, 20 February. <http://www.religionnewsblog.com/13711/evangelical-christianity-shifting-outside-west>, accessed 31 July 2008.

Nye, J. (2004), 'The Decline of America's Soft Power: Why Washington Should Worry', *Foreign Affairs* 83:3, 16–20.

Ostby, G. (2004), *Do Horizontal Inequalities Matter for Civil Conflict?* (Oslo: International Peace Research Institute).

Pew Forum on Religion and Public Life (2006), *Spirit and Power: A 10–Country Survey of Pentecostals* (Washington D.C.: Pew Forum on Religion and Public Life).

—. (2008), *U.S. Religious Landscape Survey Religious Beliefs and Practices: Diverse and Politically Relevant* (Washington D.C.: Pew Forum on Religion and Public Life).

Rice, A. (2004), 'Evangelicals vs. Muslims in Africa: Enemy's Enemy', *The New Republic* 239:6, 18–21.

Rice, C. (2008), 'Rethinking the National Interest', *Foreign Affairs* 87:4, 2–27.

Rojas, C. (2004), 'Governing Through the Social: Representations of Poverty and Global Governmentality' in Larner, W. and Walters, W. (eds) *Global Governmentality: Governing International Spaces* (New York: Routledge).

Said, E. (2001), 'The Clash of Ignorance', *The Nation*, 4 October. <http://www.thenation.com/doc/20011022/said>, accessed 31 July 2008.

Samaritan's Purse (2008), 'About Us', <http://www.samaritanspurse.org/index.php/Who_We_Are/About_Us>, accessed 31 July 2008.

Schmitt, C. (1988), *Grossraum Gegen Universalism in Positionen und Begriffe im Kampf mit Weimar- Genf- Versailles 1923–1939* (Berlin: Duncker & Humblot).

—. (1991), *Völkerrechtliche Grossraumordnung mit Interventionsverbot für Raumfremde Mächte: Ein Bitrag zum Reichsbegriff im Völkerrecht* (Berlin: Duncker & Humblot).

—. (1996), *The Concept of the Political* (Chicago: University of Chicago Press).

—. (2003), *The Nomos of the Earth in the International Law of the Jus Publicum Europaeum* (New York: Telos Press).

—. (2007), *Theory of the Partisan* (New York: Telos Press).

Shapiro, K. (2005), 'Carl Schmitt's Nomos of the Earth: Liberalism, Global Sovereignty and the Rule of the Exception', paper presented at the *Annual Meeting of the International Studies Association*, Honolulu, Hawaii.

Siewert, J. and Kenyon, J. (eds) (1993), *Mission Handbook*, 15th Edition (Monrovia: MARC).

Sherman, A.L. (1997), *The Soul of Development: Biblical Christianity and Economic Transformation in Guatemala* (New York: Oxford University Press).

Stambach, A. (2000), 'Evangelicalism and Consumer Culture in Northern Tanzania', *Anthropological Quarterly* 73:4, 171–79.

Stockman, F., Kranish, M., Canellos P.S., and Baron, K. (2006), 'Bush Brings Faith to Foreign Aid: As Funding Rises, Christian Groups Deliver Help – With a Message', *The Boston Globe*, 8 October. <http://www.boston.com/news/nation/articles/2006/10/08/bush_brings_faith_to_foreign_aid/?page=2>, accessed 30 July 2008.

Sturm, T. (2008), 'The Christian Right, Eschatology, and Americanism: A Commentary on Gerhardt', *Environment and Planning D: Society and Space* 26:5, 929–34.

Chapter 8
Reaching the Unreached in the 10/40 Window: The Missionary Geoscience of Race, Difference and Distance

Ju Hui Judy Han

> [U]sing visual aids helps encourage prayer. Perhaps a large world map or colourful flags would help. If you know people who have traveled widely, consider borrowing souvenirs from them which help people see or feel a foreign culture. ... Perhaps you can think of a symbol which will help illustrate prayer. To depict that prayer helps break through spiritual darkness in the 10/40 Window, one church set up a large map with black paper over each of the 10/40 Window countries. Each time they prayed, they removed one of the black coverings to reveal the colourful country underneath.
>
> (Johnstone et al. 1996, 26)

> Mission never has been the province solely of those who go; it also belongs to those who send—by their prayers and by their gifts.
>
> (Moreau et al. 2004, 203)

> Orientations shape not only how we inhabit space, but how we apprehend this world of shared inhabitance, as well as 'who' or 'what' we direct our energy and attention toward.
>
> (Ahmed 2006, 13)

The first passage above is excerpted from *Praying through the Window III: The Unreached Peoples,* a prayer guide that recommends displaying a map of the '10/40 Window' to orient church members toward world evangelization. The 10/40 Window, as will be discussed in more detail later in this chapter, is a priority evangelical destination and cartographic demarcation between ten and forty degrees above the equator, an area deemed most urgently in need of evangelization. To set the mood for prayer, the book suggests watching videos highlighting a 'foreign culture' from the Window or borrowing travel souvenirs to construct a feeling of proximity and intimacy. The book also encourages evangelicals to befriend immigrants to familiarize themselves with foreign languages, food and cultural customs. However contrived these methods may seem, they are techniques intended to create a feeling of connection and attachment to faraway places in order to 'reach' them even if only symbolically. Instead of pinning flags on the map as in colonial conquest, church members are encouraged to bask in self-satisfaction after prayer by unveiling each country from darkness—forgetting for a moment that they themselves had staged the whole covering and uncovering. To paraphrase Sara Ahmed (2006), how do these 'orientating' techniques shape how evangelicals apprehend the world and toward whom and what do they direct their energy?

My own encounter with the practice of missionary orientation took place in 2003 during ethnographic research of an immigrant Korean American church in California. It was here where I first learned about the 10/40 Window from a colorful map hanging on the wall outside the pastor's office (Han 2005). The Korean-language map was produced by a South Korean mission agency specializing in proselytizing to Muslims and featured a bright title emblazoned in bold: 'Swift World Evangelization: To the Last Frontier!' The church Elder who showed it to me was visibly amused that he was explaining a map to a geographer. He dramatically swept his hand westward from the Korean peninsula on the east to Israel on the west, and with his finger tracing a red horizontal line drawn on the map across Asia, he declared, 'We will follow the Silk Road through the 10/40 Window and deliver the Gospel to the unreached people. Ultimately, when Jerusalem returns to Jesus, when Jews finally accept the Gospel, then we will know that the time has come for the Second Coming of Christ our Savior.'

The moment is reminiscent of the rhetorical question that opens John Pickles's *A History of Spaces*—what is geography, if not fundamentally about fingers, lines, and eyes (Pickles 2004, 3)? In my case, the impassioned church Elder's finger, literally indexical, called my attention to a rectangular window on the map that compels people to take action by praying, fundraising and by participating in short-term mission trips. The westward direction of Christian missions, the demarcation of the 10/40 Window, the re-imagination of the Silk Road as a new proselytizing route and the eschatological significance of Jerusalem and its 'return to Jesus'— these together constitute a contemporary world geo-coded with a missionary mandate. Today we see the emergence of a new evangelical ethics of alterity (Han 2009a), coordinated through purpose-driven encounters with modernity's strangers in need of rescue and love (Ahmed 2000; Warren 2002). Rather than forceful persuasion and outright conversion, current rhetoric stresses reaching out and creating connections, emphasizing humanitarian aid and cultural exchange as the preferred mode of missionary encounter. The 10/40 Window obscures its power-laden histories and practices behind the insistence on the virtues of generosity, sharing and 'caring at a distance' (Barnett and Land 2007).

Such evangelical orientations continuously shape how one apprehends 'this world of shared inhabitance' and toward whom or what to direct the missionizing energy (Ahmed 2006). Ahmed's phenomenological provocations are instructive for thinking about the 10/40 Window as a cartographic imperative that normalizes particular technologies of navigation and direction-setting (Ahmed 2006). The directions and orientations produced by the 10/40 Window—as well as the attendant discourses like the 'unreached people groups' and 'cultural distance' discussed later in this chapter—suggest that in addition to analyzing how geopolitical visions and geographical imaginaries amplify or reflect certain motivations and intentions, we have much to gain by also considering *how* and *what kinds* of relations and connections are elicited in the process. After all, the 10/40 Window is not just a technology of designation, but more significantly a normative one, orienting one's attention in aiming towards it, targeting it and pursuing it. If militarized mapping practices envision the world as a target and equate seeing with destroying (Chow 2006; Kaplan 2006), evangelicals are instructed to target the world through the 10/40 Window ostensibly in order to save it. The activities that accompany this vision tell us much about the political praxis of evangelical connection and global integration.

I make every effort in this chapter to avoid generalizing evangelicals as a monolithic entity or associating all evangelicals with fundamentalism or the Christian Right. Protestantism is certainly a culturally and theologically diverse movement and Protestant missions are by no means led exclusively by white British or American actors or guided by a single, coherent agenda for action. Some evangelicals would no doubt disavow the mission strategies discussed in this chapter. I contend, nonetheless, that the imperative to evangelize and proselytize— in whatever method, whether to family and co-workers or to strangers on overseas mission field—is in itself a fundamental part of what it means to be an evangelical Christian. Consider for instance *Christianity Today*'s classic self-definition of an evangelical Christian: a person who "has had a born again conversion, accepts Jesus as his or her personal Savior, believes that the Scriptures are the authority for all doctrine and *feels an urgent duty to spread the faith*" (1978).[1] This chapter concerns precisely this 'urgent duty to spread the faith' and how it is operationalized in a discussion of the project of world evangelization, not evangelicals, *per se*. First, I locate the political-theological contexts and milestone international gatherings where the rhetorical strategies of the 10/40 Window and unreached peoples were fomented. Second, I discuss how these strategies orient missionizing efforts and rationalize racial difference through a science of evangelism. Finally, focusing on evangelization as an attempt at worldwide connectivity and incorporation, I argue for a careful consideration of both the intention and the process of the reaching.

World Evangelization in Historical Context

It is by now a common contention that religious missions have played a complex and contradictory role within the workings of colonial domination and capitalist development (Comaroff and Comaroff 1991; Cooper 2005; Keane 2007). Civilizing and evangelizing often went hand in hand, with mission schools seen as 'beachheads of Christian civilization in pagan territory which had helped in vanquishing pagan culture' (Saayman 1991, quoted in Kritzinger 2003, 26). Colonial missionaries

---

1 My emphasis. Christian Smith (1998) adds to this definition, 'Evangelicalism today is composed of Baptists, Presbyterians, Methodists, Pentecostals, Charismatics, Independents, Lutherans, Anabaptists, Restorationists, Congregationalists, Holiness Christians, and Episcopalians' (13).

were compelled by the so-called 'White Man's Burden' to enlighten and uplift the dark 'heathen regions,' and to 'save' poor souls from eternal damnation (Kipling 1899). In a remarkable resemblance to the logic behind the 10/40 Window, the Vatican Mission Exposition in 1925 presented itself as a 'window on the world,' displaying curios from mission efforts to win "the thousand million souls still unconverted" (Considine 1925, 28).

Over the last two decades, the 10/40 Window has emerged as a popular evangelical term for a region suffering from the "greatest degrees of poverty, illiteracy, disease and suffering" and the "least exposure to Christianity" (Bush 1996a, np). Consider the following descriptions:

> The core of the spiritually and materially neediest people of our world live in a rectangular-shaped window. [It] is where humanity suffers more than any other region in the earth... due to the historic bondages and alliances made with Satan himself. (Bush 1995, np)

> The majority of those enslaved by Islam, Hinduism, and Buddhism live within The 10/40 Window. From its center in The 10/40 Window, Islam is reaching out energetically to all parts of the globe; in similar strategy, we must penetrate the heart of Islam with the liberating truth of the gospel. We must do all in our power to show Muslims that the highest prophet described in the Koran is not Mohammed, but Jesus Christ. (Bush 1996b, np)

A rectangle on a world map—invariably using the much-despised Mercator projection[2]—the 10/40 Window is commonly defined as a cartographic designation that stretches from ten to forty degrees latitude north of the Equator in the Eastern Hemisphere, covering nearly four billion people and 62 countries in North Africa, the Middle East, West to Central Asia and East Asia. The 10/40 Window is said to encompass "most of the world's areas of greatest physical and spiritual need, most of the world's least-reached peoples and most of the governments that oppose Christianity" (Johnstone et al. 2001, 755). Faith in anything other than Christianity is likened to spiritual impoverishment and outright enslavement, as demonstrated by the excerpts above and elsewhere (Aikman 1995; Bush 1996b, 1996c). Human suffering refers to conditions of both spiritual and material poverty and Christianity is equated with assurances of prosperity. The saying that 'the poor are lost, and the lost are poor' echoes this equation between prosperity and Christianity (Bush 1996, np). The only antidote

---

2 Gerhardus Mercator's map, originally created in 1569, became commonly used throughout the eighteenth century as an icon of Western superiority and ideological centrality with its distorted enlargement of Europe. Despite well-known criticisms, the Mercator remains to this day as the most ubiquitous and popular world map projection including virtually all missionary maps featuring the 10/40 Window. See Harley (1989), Wood (1992), Kaiser and Wood (2001), and Pickles (2004).

to poverty is the Christian Gospel. Spreading Christianity is conflated with sharing the wealth and both operate alongside a deeply apolitical conception of foreign aid as gifts, well-intentioned acts of state-level generosity.

The evangelistic approaches of the 10/40 Window privileges foreign missions over local missions and emphasizes eternal salvation in a world to come as opposed to healing from economic exploitation and political oppression in this world. As such, it occupies the conversionist end of the conversionist-liberationist continuum (Kritzinger 2003). The conversionist orientation is well illustrated by the following quote attributed to Dwight L. Moody (1837–1899), the most influential evangelist of the late nineteenth century:

> I look upon this world as a wrecked vessel. God has given me a lifeboat and said to me, 'Save all you can.' God will come in judgment and burn up this world... If you have any friends on this wreck unsaved, you had better lose no time in getting them off (Kritzinger 2003, 21).

The image of the world as a sinking ship does not simply refer to widespread misery and devastation. In theological terms, this reference to dispensational premillennialism also portrays a limited human capacity to bring about the Second Coming (see Introduction, this volume). Since world evangelization or any other human action is not about preventing the world from 'burning up'—apocalypse is destined to happen regardless of human action—what is telling in this passage is that Moody himself is seen as already having been saved. Further, the evangelist is assigned the difficult task of saving others by sharing with them the lifeboat he has been given. Considering the finite capacity of human endeavor and the presumably limited capacity of the one single lifeboat, Moody is aware that he cannot save everyone. A certain degree of failure is guaranteed. The ethical position he embraces is to 'do all you can,' with the humbling awareness that it will not be sufficient but is acceptable nonetheless. While Moody the missionary performs his role as the heroic rescuer in direct communication with God, the ill-fated others face certain death unless they are reached.

Moody effectively fused dispensational premillennialism with the belief in the superiority of the United States through an unabashed pursuit of wealth and prosperity (Brouwer et al. 1996). The rise of American evangelicalism coincided with the rise of the United States as a nation and an emerging world power (Marsden 1987). The early twentieth century saw numerous voluntary and mission societies in the U.S. thrive in their effort to bring the gospel to the needy, buoyed by a sense of destiny and 'guided by a vision of spiritual and moral progress' (Bornstein 2003, 18). The emergent American Fundamentalist movement in post-World War I era rallied in protest against the modernist embrace of science and liberalizing theological tendencies that were seen as unacceptable doctrinal compromises. In essence, it was galvanized as a theological-political opposition to Darwinian evolutionism and Communism. Concerned with the perceived decline in cultural mores, the movement reaffirmed a fundamental set of Christian beliefs including the infallibility and inerrancy of the Bible and the insistence on the virgin birth of Christ as historical truth.

The 'new evangelicalism' of the 1940s later embraced and modified elements of early and fundamentalist evangelicalism (Marsden 1991). Between 1945 and 1975, the heyday of American liberalism in economics, politics and religion, evangelicals breathed new life into the old gospel of wealth and resurrected a gospel of prosperity based not only in an emphasis on hard work, but also on the idea that Christians were entitled to material blessings on the sole basis of their faith (Brouwer et al. 1996). As I argue elsewhere, 'prosperity gospel,' a theology that holds that material prosperity is evidence of God's favor, is applied not only to individuals but also operates on a national or people group scale of prosperity (Han 2007). Such effort often relies on calculations of national wealth and poverty through GDP statistics, counting the number of adherents and degree of 'reachedness' in what some have derided as 'statistical Christianity' (Smith 1972).

The post-World War II resurgence of Pentecostal and charismatic movements and rise of global evangelical Christianity culminated in the 1974 International Congress on World Evangelization in Lausanne (often referred to simply as Lausanne), a historical landmark of lasting consequence. During the Cold War and thereafter, including the so-called Decade of Evangelism in the 1990s, influential books such as *Spiritual Warfare: Victory over the Powers of This Dark World* (Warner 1991) and *Confronting the Powers: How the New Testament Church Experienced the Power of Strategic-level Spiritual Warfare* (Wagner 1996) promoted explicitly geographical and geopolitical analysis with a renewed emphasis on strategy and expertise. The following excerpt exemplifies this trend.[3]

> I have come to believe that Satan does indeed assign a demon or corps of demons to every geopolitical unit in the world, and they are among the principalities and powers with whom we wrestle. ... [D]ealing with territorial spirits is major league warfare. If you do not know what you are doing, and few whom I am aware of have the necessary expertise, Satan will eat you for breakfast (Wagner 1989, 47).

Lausanne I and Billions of Unreached People

It is difficult to summarize in this chapter all the various and divergent contexts of world evangelization, spanning not only the Catholic-Protestant divide but also what has been characterized as the ecumenical-evangelical divide (Bosch 1991) or the liberationist-conversionist continuum (Kritzinger 2002). There is considerable and ongoing debate concerning the meaning of mission, which is no minor issue as it concerns the very understanding of salvation. For the purposes of this chapter,

---

3 The religio-political implications of 'spiritual warfare' movements are well-documented in Diamond (1999) and Brouwer, Gifford and Rose (1996).

I focus on how the projects of world evangelization were conceptualized and mobilized in response to, or in opposition against, parallel developments in the Roman Catholic Church as well as the ecumenical movement represented by the World Council of Churches (WCC).

Ecumenicals gathered at the 1968 Uppsala Assembly of the WCC had concurred with the controversial Vatican II decree in de-emphasizing foreign missions and stressing solidarity with the poor and oppressed as a central priority of Christian missions and proposed instead an increased participation in secular programs for urban renewal and civil rights movements. After a five-year study project titled 'Salvation Today,' WCC presented its findings at the Bangkok Conference of 1972–1973, clearly outlining salvation in four dimensions: economic justice, human dignity, solidarity and hope (Thomas 1995, 122). Conservative, conversionist evangelicals railed against this liberationist and this-worldly direction and objected to WCC's 'preoccupation' with 'political liberation, land distribution, better pay for factory workers, [and] the downfall of oppressive systems of government' (MacGavran 1972a quoted in Thomas 1995, 129). To this day, many conservative evangelicals regard the WCC as a bastion of theological deviations influenced by Marxism and socialism and criticize it for being overly concerned with dialogue with non-Christians (Diamond 1999, 212). Against the current of WCC's proposals, evangelical dissenters were in fact laying the foundation for the gospel of wealth and prosperity that would come to represent a significant part of world evangelization in recent decades.

Donald McGavran, a missionary from India and chief architect of the Church Growth Movement, issued the following challenge to the WCC, entitled 'Will Uppsala Betray the Two Billion?'

> By 'the two billion' I mean the 'that great number of [persons] at least two billion, who either have never heard of Jesus Christ or have no real chance to believe in Him as Lord and Savior.' These inconceivable multitudes live and die in a famine of the Word of God, more terrible by far than the sporadic physical famines which occur in unfortunate lands. The Church, to be relevant, ... must plan her activity, marshall her forces, carry on her campaign of mercy and liberation, and be faithful to her Lord with the two billion in mind. If the sufferings of a few million in Vietnam, South Africa, Jordan, Buchenwald, or the slums of Rio de Janeiro or Detroit rightly excite the indignation and compassion of the Church, how much more should the spiritual sufferings of two thousand million move her to bring multitudes of them out of darkness into God's wonderful light (McGavran 1972b quoted in Thomas 1995, 158–159).

Rejecting the ecumenical position of mission as this-worldly improvement of philanthropy, famine relief, medicine and dialogue, conversionist evangelicals like McGavran criticized the WCC's focus on 'good deeds' and concern for social justice as secondary to the essential task of mission, which they defined more narrowly as 'discipling the peoples of the earth' (McGavran 1986). A few years later in 1974, when likeminded evangelicals gathered at the landmark Lausanne conference, world evangelization was defined as church planting involving proselytization across 'cultural frontiers.' The Lausanne Covenant, signed by the conference participants, stated that neither 'social action evangelism' nor 'political liberation salvation' was sufficient, but conceded that Christian social responsibility should include a measure of social-political involvement. This concession reflected the reality of political upheavals and social changes taking place throughout the world, especially in the decolonizing Third World, something the evangelicals could no longer ignore.

It was here at the 1974 Lausanne conference that Ralph Winter—professor of anthropology and linguistics at the School of World Missions at Fuller Seminary in southern California and named one of the 25 Most Influential Evangelicals in America by *Time Magazine* in 2005 (Van Biema et al. 2005)—followed up on McGavran's opposition against the WCC and concern for the 'two billion,' and introduced the concept of billions of 'people groups' unreached by the Gospel. Winter also put great emphasis on militaristic strategy development and anthropological approach to evangelism (Rynkiewich 2007). Declaring that 'the shattering truth is that at least four out of five non-Christians in the world today are beyond the reach of *any* Christian's evangelism,' he admonished Christians for their lethargy in foreign missions (Winter 1975). Hired by McGavran to teach at the influential Fuller School of World Mission, Winter went on to establish the U.S. Center for World Missions in 1976, referring to it as the headquarters for missionary research and 'a Pentagon for mission agencies around the world' (Diamond 1999, 214). With hundreds of mission strategists working together to 'gather information, coordinate efforts and develop the most effective plans possible to reach the nations,' Winter was at the center of a new scientific wave of world evangelization (Diamond 1999, 214).

Lausanne II, GCOWE and the 10/40 Window

While Winter is credited for jump starting the modern missionary movement and introducing the concepts of 'unreached people groups', the term '10/40 Window' was coined by Luis Bush, an Argentine-American mission strategist who addressed the plenary session of the 1989 International Congress on World Evangelization in Manila, Philippines (Lausanne II 1989). The concept gained traction at another international evangelism conference later in the same year, the 1989 Global Consultation on World Evangelism (GCOWE) in Singapore. Attended by many of the same people who had previously participated in Lausanne I and II but also drawing significant numbers from evangelicals in Asia, GCOWE was organized by Thomas Wang, a Chinese-born American leader of the AD2000 Movement.[4] As

---

4 Wang recently led the Great Commission Center International in San Francisco, and gained some notoriety as a leading opponent against same-sex marriage in California. Identified as a member of the missionary Chinese Ministerial Committee in Mountain

a project committed to building, in Wang's own words, a 'network of networks' (quoted in Koop 1995), the AD2000 Movement set out with the goal as the 'harvest of the unreached people groups throughout the world' under the slogan, 'A Church for Every People and the Gospel for Every Person by the Year 2000' (Bush 1996a, np). As the year 2000 drew near, the organization extended its expiration date and renamed itself as AD2000 *and Beyond* Movement (my emphasis).

In 1995 the AD2000 and Beyond Movement organized the second GCOWE in Seoul, South Korea, drawing nearly 4,000 Christian leaders from 186 countries over nine days 'to develop strategies and plans to achieve Christian evangelization of the entire world' (Aikman 1995, np). It was at this gathering that the 10/40 Window was formally selected to be the primary focus for evangelism for the coming decade (Coote 2000). Ralph Winter called it 'the most strategic Christian gathering in history' (Bush 1996d, np).

The Science of 'People Groups'

Overtly racialized and blatantly imperialist language of 'heathens' and 'civilizing missions' may no longer be in vogue among most mainstream evangelicals today, but still prevalent are comparable ideas that downplay the flexible and historically contingent character of race and ethnicity. Reproducing variations on the so-called objectivism of race, several projects of world evangelization produce racial definitions and profiles in technocratic, pseudo-scientific terms. Nowhere is this more evident than in the proliferating discourse of people groups and unreached people groups. The Lausanne Strategy Working Group in 1982 defined a people group as follows:

> a significantly large grouping of individuals who perceive themselves to have a common affinity for one another because of their shared language, religion, ethnicity, residence, occupation, class or caste, situation, etc. or combinations of these… [It is] the largest group within which the Gospel can spread as a church planting movement without encountering barriers of understanding or acceptance. (Winter 1989, 18)

Similarly, the evangelical almanac *Operation World* identifies 'ethnolinguistic people group' as a category of a people group, claiming that 'a person's identity and primary loyalty' is determined by language and/or ethnicity (Johnstone et al. 2001, 757). The idea is that these discrete units of people groups share a common cultural affinity and an internally coherent identity within which there are no

---

View, California, Wang told the *San Jose Mercury News*: 'We will not stand by to watch biblical marriage be destroyed by a radical agenda' (quoted in Ostrom 2004). Wang also told the *San Francisco Chronicle* that 'legalizing same-sex marriage would eventually lead to polygamy, multipartner marriage and incest' (quoted in Torassa 2004).

barriers to understanding or acceptance. The term essentially assumes an ontology of ethnicity-language nexus as foundational to individual and collective identity formation. Such a doctrine of racial typology and classification has little support in contemporary social science (Banton 2000), as scholars of race and ethnicity have thoroughly debunked the modernist view of the social order and the imaginary of the world as consisting of discrete groups—each with discernible ethnic identity as biology, language as heritage and culture as descent. With 'ethnicity' replacing the more vexed idiom of race, these 'practices of naming and knowledge construction deny all autonomy to those so named and imagined, extending power, control, authority and domination over them' (Goldberg 2000, 155).

Nonetheless, the absurd and impossible task of defining and accounting for all discrete people groups in the world has kept numerous evangelization projects busy over the years. Theological effort to justify the demarcation of ethnolinguistic people group as a biblically based terminology begins with the pivotal passage in the Bible known as the Great Commission (Matthew 28:18), which says, 'Go therefore and make disciples of all nations.' Claiming that the Greek word *ethne* was translated as 'nations' in English despite the original meaning referring to a distinct 'people with a unifying ethnic identity,' the Great Commission is interpreted as discipleship of all 'ethnic groups,' not nations in the contemporary sense (Piper 2004, 161). While the etymology of the word 'ethnicity' does support this argument to some extent, the problem is that the idea of an ethnolinguistic people group is taken for granted as an anthropological and scientific fact. On the contrary, the term 'ethnicity' was first introduced by researchers in the early 1940s who sought to avoid the word 'race' that had become inextricably associated with anti-Semitism, the so-called 'Jewish race' and Nazi genocidal policies (Hiebert 2000). The ethnic people group is a depoliticized variation of a racial taxonomy project.

This people group idea accomplishes several important objectives for the project of world evangelization. First, it posits that enumerating and quantifying the world's population, as well as measuring evangelization, is a new and empirical approach to a centuries-old task. 'Evangelistics,' for instance, is explained as 'the science that studies the propagation of Christianity' (Barrett and Johnson 2001, 741) and used to validate the need, purpose and activity of missions. Second, recognizing differences in language, religion, ethnicity, residence, occupation, class or caste as possible barriers to proselytizing, the promoters of 'people group thinking' reduce difference to a matter of cultural barriers and cross-cultural communication, altogether eviscerating power from the discussion. Third, like the 10/40 Window mapping, the empiricist pseudo-scientific study of people groups rationalizes the target of the mission mandate. It demonstrates that what is measurable is ultimately controllable.

Levels of Reachedness

The idea of 'reachedness' refers not to an individual person's conversion, but instead applies to a collective people group. Note that the unit of analysis does not involve territorial nation-states but rather focuses on the occupants. As one website puts it: 'God sees faces, not places. His Son, Jesus, died for people . . . not countries!' (Adopt-a-People Clearinghouse 2009). The five stages of reachedness are outlined as follows (Camp 1993). The first is 'Reported' or when the people group is simply brought to the attention of a Christian research group which then tries to verify their unreached status. The second, 'Selected,' refers to the stage when a denomination or mission agency takes on the responsibility to reach the people group. Third is the 'Adopted' stage when churches or fellowship groups agree to support the people group with prayers and financial assistance. The fourth or 'Engaged' stage occurs when cross-cultural missionaries are on site in order to plant churches. And finally, the fifth is the 'Reached' stage, achieved when a strong church-planting movement of sufficient size and strength has been established and can evangelize the rest of the group with no (or very little) outside help. There exists some disagreement in the precise definition, but a people group is usually deemed 'reached' not when every single member or the majority are converted, but rather, when mission efforts have established a local, indigenous church with the capacity and resources to continue to evangelize the rest of the group (Johnstone et al. 2001).

Reaching the Unreached

How then to reach the unreached? Literally, the word 'reaching' suggests making overtures to contact. If the previous discourse of 'civilizing mission' focused on uplifting and transforming the targets of evangelism, the contemporary idea of reaching stresses encounter, communication and connectivity as non-dominant expressions of Christianity's global reach.

Unreachedness is Danger

The discourse of the 10/40 Window and the unreached peoples is strikingly reminiscent of the 'neoliberal geopolitics' of Thomas P.M. Barnett, a prominent military analyst and influential advisor to the Pentagon (Roberts, Secor & Sparke, 2003). Barnett's book, entitled *The Pentagon's New Map*, draws explicitly from Thomas Friedman's 'smooth spaces,' Samuel Huntington's 'clash of civilizations,' and Francis Fukuyama's 'end of history' (Barnett 2004). In the book and elsewhere, Barnett proposes a new map of the world to discuss the 'Non-Integrating Gap'— defined as those regions which 'remain fundamentally *disconnected*' from globalization's expanding web of connectivity (Barnett 2004, 121). Asserting globalization to be America's 'greatest gift to history' and the 'most perfectly flawed projection of the American Dream onto the global landscape', he then equates disconnectedness from global capitalism with dangerous criminals avoiding the rule of law. He declares disconnectedness from globalization as the 'ultimate enemy' (ibid. 50 and 124, respectively).

In contrast to the Gap, the Functioning Core is defined by Barnett as a region that accepts the 'connectivity' of globalization and welcomes the 'content flows associated with integrating one's national economy to the global economy' (125). He approvingly cites Thomas Friedman's distinction between the 'Lexus world' which embraces globalization and the 'olive tree world' that prefers to 'remain trapped in a simpler world' squabbling over meaningless gains (51). Huntington's 'fault-line wars' that result from a 'clash of civilization' makes an appearance in Barnett's work dressed up as the 'bloody' boundaries between the Core and the Gap. As Gillian Hart (2006) points out, these remarkably similar images of 'smoothed global space' not only downplay American dominance, but also enable the inequalities and asymmetries they try to obscure. To my knowledge, Barnett is not explicitly tied to the projects of world evangelization discussed in this chapter (but see Chapter 7, this volume). However, the undeniable resonance between the disconnectedness of the 'Non-Integrating Gap' and the unreachedness of the 10/40 Window constitutes a collective 'ensemble of representational practices that separate the world's components into bounded units, disaggregate their relational histories, turn difference into hierarchy and naturalize these representations' (Hart 2006, 22).

Reaching through English

Operation World defines 'unreached people' as follows:

> an ethnolinguistic people among whom there is no viable indigenous community of believing Christians with adequate numbers and resources to evangelize their own people without outside (cross-cultural) assistance. Other researchers have adopted the terms 'hidden people' or 'frontier people group'. (Johnstone et al. 2001, 759)

Unreached people are considered 'hidden' because it is believed that they have not been able to 'hear the gospel in their own language in a way that makes sense to them' (Moreau et al. 2004, 13). In other words, proselytization is seen as necessarily including the work of translation and cross-cultural communication. As such, the ethnolinguistics or ethnicity-language pairing stresses intelligibility and communicability. Long-term missionaries have been historically well-known for their language acquisition and Bible translation activities and the opportunities of language education are not lost on mission strategists:

> … Since the 1980s no professional skill has grown more quickly in use than that of teaching English… The demand for English language skills has swelled

tremendously over the past fifty years as English has evolved into a world language. North Americans, most of whom speak English as their first tongue, have a built-in advantage in teaching English to speakers of other languages. (Moreau et al. 2004, 197)

Nonetheless, the global terrain of English education remains a highly contentious. There was a long-standing belief, particularly among the British, in the supremacy of the English language and an innate relationship between English and Christianity, reason and rationality.[5] North American missionaries continue to occupy a privileged location in the global political economy of English and take advantage of the widespread demand for English language skills worldwide. Missionary instruction of English has been criticized as "a means to convert the unsuspecting English language learner" with "profound moral and political questions" (Pennycook and Makoni 2005, 139).

Reaching through Statistics

For years, Winter claimed that there are as many as 24,000 people groups in the world. In 2001, *Operation World* declared that there was now a 'reasonably complete listing of the world's peoples and languages,' and that 'for the first time in history,' they have come up with a 'reasonably clear picture of the remaining task for us to disciple the nations' (Johnstone et al. 2001, 15). The authors of *Operation World* revised the total estimate to 12,000 ethnolinguistic peoples in the world and another large-scale evangelical statistics project, *World Christian Encyclopedia* (2001) later arrived at a similar figure.

Out of these 12,000-plus people groups, over 9,000 are considered already reached by the Christian gospel. An estimated 3,000 to 3,600 peoples, however, are considered 'World A' people, defined as living in an area of the world that is less than half evangelized. This World A-B-C scale was invented by the *World Christian Encyclopedia* and is reproduced in various research and database projects including *Operation World*. In this scale, 'World B' is defined as over 50 percent evangelized but less than 60 percent Christian and 'World C' refers to areas that are over 95 percent evangelized, with over 60 percent of the population being Christian. In one particularly interesting illustration, the authors of *World Christian Encyclopedia* combine World A-B-C with a more commonly used tripartite imagination of the world—First, Second and Third World. If the 10/40 Window aligns missionary mandate with imperial and developmentalist targeting, the World A-B-C schema puts geopolitical dynamics and world evangelization as co-determinants of a country's location in the world.

---

    5 Charles Grant, Chairman of the British East India Company and Christian missionary in India, wrote in 1792: "The use and understanding of the English language would enable the Hindus to reason, and to obtain new and better views of their duty as rational and Christian creatures" (Clive 1973, 345 quoted in Pennycook 2003, 339).

Using a coding system also developed by the *World Christian Encyclopedia*, the Joshua Project—which directly grew out of the AD2000 and Beyond Movement and now operates as a part of Ralph Winter's U.S. Center for World Mission (Bush 1996a; Bush 1996c)—provides an elaborate racial classification system. With the claim that 'accurate, regularly updated ethnic people group information is critical' for the project of world evangelization (Bush 1996c, np), the system uses a three-part coding rubric with categories ranging from 'Australoid' and 'Mongoloid' to 'Negroid' and 'Arctic Mongoloid' (Joshua Project 2009).

Datasets thus generated are used by GIS specialists like Global Mapping International, who produce numerous maps pinning these ethnic/racial classifications onto the world. These statistical projects code the global space with a dizzying array of datasets, typologies, charts and graphs. Projects like *World Christian Encyclopedia* and *Operation World* are founded upon the faith in the need for quantification and enumeration of evangelization (Barrett et al. 2001; Johnstone et al. 2001) and argue that statistics provide 'solid factual basis for action' (Johnstone et al. 2001, xviii). They geo-code the world through demography, cartography and GIS, offering terms like 'missiometrics' and 'evangelistics' (Barrett and Johnson 2001). They laud the 10/40 Window as a 'brilliantly successful' focus, arguing that it has resulted in manifold increase in the deployment of missionaries to reach the unreached people groups (Johnstone et al. 2001, 6).

The enormity of these statistical projects has been likened to the impossible task of counting "every soul on earth" (Ostling and Matheson 1982, np). Despite the technical explanations and elaborate coding rubric, the underlying framework reflects the persistence of scientific racism and discredited anthropological classification systems. As a result, one theologian laments that the 10/40 Window concept and people group research are based on outdated anthropology, poor-quality geography and questionable theology, and comments that this combination has become "another colonial burden that has to be borne" (Rynkiewich 2007, 225).

Cultural Distance vs. Geographical Distance

The Unreached Peoples Program director of World Vision International likewise claims that a people group is the largest group within which the gospel can "flow along natural lines," and that unless the gospel comes from someone within one's own people group, it is considered foreign (Robb 1996, 16). Similar discourse is found in 'Perspectives on the World Christian Movement' (*'Perspectives'*), one of the most widely used mission curricula today. Developed by Ralph Winter in 1973, it is now a large-scale international training program operating out of Winter's U.S. Center for World Mission. Over 70,000 people in North America and over 30,000 elsewhere are said to have taken *Perspectives* and its training module has been translated and offered in numerous languages. Some mission agencies in South Korea even consider a certificate of completion in *Perspectives* a prerequisite for mission participation.[6]

At the core of *Perspectives* is an unsophisticated but elaborate typology of ethnolinguistic people groups. Claiming that the world can be best viewed in terms of people groups, the '16 Core Ideas' document from *Perspectives* states (2008, my emphasis):

> The gospel flourishes *amidst* a people group, and moves with more difficulty *between* people groups. [...] Seeing the world's population as individuals, or even as countries defined by geopolitical boundaries, can be helpful for some aspects of mission. But seeing humanity as a mosaic of people groups is strategically important when analyzing the entire task of world evangelization. Since every people group will require a unique approach and strategy, it becomes important to understand distinctive features of the peoples.

The idea that there is an in-group coherence that allows a 'natural flow' of Christianity is derived from Winter's widely-used typology of cross-cultural and frontier missions. He posited that E3 evangelism denotes mission activities 'among people just like us,' E2 evangelism engages those who are 'near cultural neighbors,' and that highest-priority E1 evangelism takes places 'among those who are culturally very different' (Winter 1981, 314). Privileging anthropology over geography, Winter differentiated geographical distance from so-called 'cultural distance,' categorizing for example Hispanic missions as E2 (neighbors), but Navajo missions as E3 (cross-cultural). While this formulation assumes a non-immigrant, white American subject as the sending missionary located at the E1 location.

While the rhetoric of cultural distance certainly reifies a static sense of cultural identity and a territorialized demarcation of cultural boundaries, the relations of distance depend on who is placed at the center. Using Winter's typology, the following illustration (see Figure 8.1) appears in *World Mission: An Analysis of the World Christian Movement* (1994).

The figure is designed to illustrate these relationships—and importantly, the relationalities—by showing the so-called cultural distance between the Highland Quechuas of Peru and others. As the Highland Quechuas are in focus, they occupy the center E-1 position. Because the Highland Quechuas are said to be farthest away from North Americans and Koreans because of significant cultural and linguistic barriers, those groups occupy the outermost E-3 distance. Meanwhile, the Mestizo and 'assimilated Quechuas' are considered to be cultural neighbors because of shared ancestry and customs and thus occupy the intermediary E-2 group. Arguing that in 'today's complex world, there are many 'Jerusalems', 'Judeas', 'Samarias' and 'uttermost parts'', the authors concede that one person's Samaria is another's Judea (Lewis et al. 1994, 7–10). They explain that the New Testament's apostle

---

6 Based on author's interviews and field research, 2005–2008.

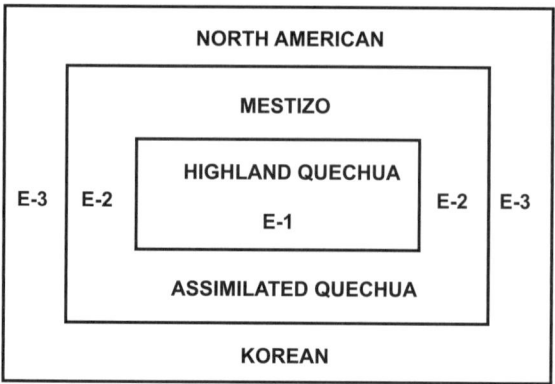

Figure 8.1 Cultural Distance Diagram

Note: Cultural Distances. Redrawn from *World Mission: An Analysis of the World Christian Movement* (1994) by Jonathan Lewis, Meg Crossman, Stephen Hoke.

Paul himself was ethnically a Jew but raised in a Gentile culture and that this background put him culturally closer to the Gentiles than the other apostles. This explanation of an intermediary cultural neighbor is offered in part to show why Paul is considered a particularly effective model of a cross-cultural missionary in the Bible. Similar rhetoric of cultural proximity and neighborliness is used today in numerous multi-ethnic congregations in North America, where immigrants and people of color are seen as uniquely well-suited and capable of reaching those who are culturally proximate to them. Through this frame of cultural distance as access and approachability, in other words, immigrants and people of color are seen as important resources for an otherwise declining North American Christianity.

To a certain extent, postcolonial criticisms of power and dominance and the struggles for self-determination in the post-War era of decolonization have heightened an attention paid to dynamics of cultural difference and importance of self-sufficient, local Christian leadership. Nonetheless, the rhetoric of cultural distance undermines the persistence of inequality or injustice. It reduces asymmetrical power relations to cultural miscommunication to be mitigated by diversity awareness or cultural sensitivity training, preferably informed by those deemed cultural insiders.

Reaching through Adoption

As discussed thus far, various methods are employed to reach the unreached peoples and foster connectivity. Maps and other visual tools are used to orient people to visualize and cultivate a sense of long-distance intimacy. Statistical projects render the target calculable and visible. Ideas around 'cultural distance' reinforce the idea that there are impermeable boundaries around discrete units of 'culture', that the distance between these discrete units are measurable and ultimately, that these distances and barriers are bridgeable through empirical knowledge.

In addition, publications such as *Adoption: A Practical Guide to Successfully Adopting an Unreached People Group* (1993) encourage people to consider 'adopting' a specific unreached people to encourage not only financial ties but also affective ties through the metaphor of family adoption. Reflecting the post-national geographies of world evangelization that conceive units of conversion in terms of ethno-linguistic groups, the so-called evangelism-as-adoption strategy likewise focuses its attention on people groups, not individuals or nation-states. The following explanation offered by the Adopt-A-People Campaign of the U.S. Center For World Mission reveals just what the people groups are being adopted into:

> What then does it mean to adopt a people? It means that a church, or a group within a church, works through their chosen and approved mission agency to provide informed, concerned, dedicated prayer and financial support for a specific unreached people group. It means maintaining that commitment until a church planting movement is started that can reach the rest of the people without needing outside cross-cultural help. We are not 'adopting' the people group into our congregation, denomination or mission agency. We are praying, giving, and serving to see a people adopted into God's family (Yoder 1999).

Conclusion

Understanding world evangelization not only as a complex of normative ideas and orienting imaginaries, but also as a set of concrete practices and proselytizing strategies designed to reach the so-called unreached people groups, this chapter has examined the racialized, pseudo-scientific geographies of difference and distance. From the unprecedented rise in popularity of maps and geographical literacy in the eighteenth century (Brückner 2006) and the imperial mapping that sought to govern the contact zones in the 'new world' (Pratt 1992), to the American popular geopolitics of the Cold War (Sharp 2000), geographical knowledge has long been complicit with imperial projects. This power-laden history of maps and mapping has been the subject of much critical scholarship (Pratt 1992; Pickles 2004; Brückner 2006). The logocentric ambitions of statistical methods and the Geographic Information Systems (GIS) and their shared faith in a fundamentally ordered (or orderable) world require continual and sustained scrutiny (Barnes 1994).

The "cartographic eye," David Matless writes, "is equated with the eye of power-as-domination" (Matless 1999, 193). To understand maps and mappings simply as representative practice for the goal of domination, however, would reduce the 10/40 Window to a cartographic embodiment of power, a mimesis or a simple reflection of geopolitical interests. Instead, I suggest that the 10/40 Window's application is more broad and varied, as part of a larger power-knowledge system that constitutes and renders the world legible for governance.

On one hand, the contemporary geography of evangelical Christian missions—spatial arrangements and geopolitical alignments that underpin missionary enterprises—illustrates clearly the militaristic penchant for envisioning the world, or particular regions, as a target (Chow 2006; Kaplan 2006). Missions are by definition deliberative and purposive. As demonstrated by the 10/40 Window, evangelicals are armed with a master narrative about the inexorable forces of Christianity and capitalist modernity and they continue to embark on a territorialized strategy for world evangelization. On the other hand, these strategies do not simply regurgitate imperial hostilities of the past or unquestioningly endorse American unilateralism. Instead, we see missionary subjectivities forged in the context of Third World liberation, global capitalism and hegemonic humanitarianism.

There has been much discussion of what William Connolly has called the 'bellicose ethos' of the American neoconservatives and the Christian Right (Brouwer et al. 1996; Diamond 1999; Buss and Herman 2003; Brown 2006; Runions 2007; Connolly 2008). While this analysis helps us understand the pugnacious ideologies and warmongering efforts coming from the particularly bellicose convergence between economic, religious and geopolitical forces in the United States, we still need more discussion of how the project of world evangelization is rationalized in terms of connectivity and generosity, producing what has been called the "new evangelical internationalism" (McAlister 2006, np). After all, even the diehard evangelicals today begrudgingly acknowledge the damaging legacy of colonialism and admit that the romance of missions has all but faded away amidst pleas for pluralism and tolerance. Instead, they vow to make missions less paternalistic and more equitable, more contextualized and culturally sensitive. The trouble is, these evangelization strategies together constitute a kind of an "anti-politics machine" (Ferguson 1994), depoliticizing poverty and inequality and demonstrating a startling ignorance of the historical and political realities of the people and places they purportedly intend to help. The scientific racism of unreached people discourse and the targeting of the 10/40 Window produce a cartographic gaze that privileges a distinctly hegemonic subjectivity, English-speaking and prosperous.

The project of world evangelization is fueled by an unapologetic defense and rationalization of territorial ambitions of political and religious expansion in all facets of contemporary life, i.e. "empire as a way of life" (Williams 1980, 5). The confluence of political and theological rationalities transforms the realities of territorial expansion, colonial conquest and military intervention into pious rhetoric about ending poverty and stopping religious persecution and ensuring political and religious freedom for all (Connolly 2005; Brown 2006). Whether we understand this convergence or articulation as an "evangelical-capitalist resonance machine" (Connolly 2005) or as "dreamwork" (Stuart Hall, cited in Brown 2006, 690), the crucial point here is that we understand the imbrication of political economic and religious rationalities not in terms of causality or manipulation (the contention that neoconservatives are simply disguising their intentions with religious rhetoric), or in terms of self-evident, congruent correspondence between ideology and practice (i.e. the idea that practice is nothing more than ideology in action). In other words, while there is a clear blending and harmonizing dynamic between evangelical missionary enterprises and projects of imperialism and colonialism, with respect to rhetoric, personnel and infrastructure, they certainly are not one and the same thing, nor do they cause or enable the other.

Proselytizing missions, as diverse as they are in intent and form, may seem regressive and repugnant to their critics, but they nonetheless exemplify a philosophy of praxis—believing by doing, synthesizing a unity of theory and practice (Forgacs 2000; Crehan 2002). As demonstrated by the rhetorical strategies of the 10/40 Window and unreached people groups, American evangelicals are armed with a powerful master narrative about the inexorable forces of Christianity and capitalist democracy and they continue to embark on territorialized strategies for world evangelization. Persuasive narratives can convince people that a certain course of history is not only possible, but also desirable and inevitable. Nonetheless, the power of persuasion lies not in the abstract realm of ideologies, narratives, or master blueprints, but in practice, as ideas become embedded in practical activity (Gramsci 1971, 330). The 10/40 Window itself may be a clever idea, but it has been persuasive precisely because it has been successfully embedded in the practical activity of evangelization efforts—through research, strategic planning and short-term and long-term missions. By obscuring their power-laden history, evangelization strategies like the 10/40 Window reveal an ensemble of representational practices through questionable science.

While geopolitical visions like the 10/40 Window certainly function as an ideological blueprint, it is not sufficient to critique it for being a 'normative universalist project' and for being too superficial to "foster an emotive connection to a particular place and its people" (Gerhardt 2008, 915). Such framework relies on a global-local duality that privileges local knowledge as inherently more authentic, more truthful and more intimate. It is true that the 10/40 Window is projected from a bird's eye point of view—or put more aptly, using the 'god trick', a conquering gaze from nowhere (Haraway 1991). But it also operates in concert with a host of other projects—like the *World Christian Encyclopedia* or *Operation World*—which are collectively committed to documenting and cataloguing every particularity and locality on earth, producing in the process the most detailed minutiae about every inch of the globe in order to grasp it, to reach it and to foster long-distance intimacies with it. All sorts of attachments and intimacies are created despite being long-distance, or even based on erroneous or patently false information; these connections create a mix of abstract and concrete encounters, affective ties both distant and proximate. Perhaps what matters the most is the nature of the connectivity, the vexed process of reaching itself.

Geopolitical visions are by definition deliberate and praxis-oriented. Rhetorical strategies like the 10/40 Window shape one's gaze, target and fix the object of mission, condition one's disposition towards the world and authorize concrete actions. The project of world evangelization must be understood as a simultaneously world-making and particularizing process, a far-reaching endeavor to absolve the powerful, absorb differences and incorporate the 'frontiers' of the world into a Christian totality.

Bibliography

Adopt-A-People Clearinghouse <http://www.adopt-a-people.org> (website), accessed on 21 January, 2009.

Ahmed, S. (2000), *Strange Encounters: Embodied Others in Postcoloniality* (London: Routledge).

—. (2006), *Queer Phenomenology: Orientations, Objects, Others* (Durham: Duke University Press).

Aikman, D. (1995), 'Global Consultation On World Evangelization (GCOWE'95) Backgrounder for The Media' <http://www.ad2000.org/gcowe95/aikman.html> (website), accessed February 2, 2005.

Banton, M. (2000), 'The Idiom of Race: a Critique of Presentism', in Back, L. and Solomos, J. (eds) *Theories of Race and Racism: A Reader* (London: Routledge).

Barnes, T. (1994), 'Probable Writing: Derrida, Deconstruction, and the Quantitative Revolution in Human Geography', *Environment and Planning A*, 26: 1021–40.

Barnett, C. and Land, D. (2007), 'Geographies of Generosity: Beyond the "Moral Turn"', *Geoforum*, 38: 1065–75.

Barnett, T. (2004), *The Pentagon's New Map: War and Peace in the Twenty-first Century* (New York: G.P. Putnam's Sons).

Barrett, D. and Johnson, T. (2001), *World Christian Trends, AD 30–AD 2200: Interpreting the Annual Christian Megacensus* (Pasadena: William Carey Library).

Barrett, D., Kurian, G.T. and Johnson, T. (2001), *World Christian Encyclopedia: A Comparative Survey of Churches and Religions in the Modern World—AD 30 to 2200* (London: Oxford University Press).

Bornstein, E. (2003), *The Spirit of Development: Protestant NGOs, Morality, and Economics in Zimbabwe* (New York: Routledge).

Bosch, D.J. (1991), *Transforming Mission: Paradigm Shifts in Theology of Mission* (Maryknoll: Orbis Books).

Brouwer, S., Gifford, P. and Rose, S.D. (1996), *Exporting the American Gospel: Global Christian Fundamentalism* (New York: Routledge).

Brown, W. (2006), 'American Nightmare: Neoliberalism, Neoconservatism, and De-Democratization', *Political Theory,* 34: 690–714.

Brückner, M. (2006), *The Geographic Revolution in Early America* (Chapel Hill: University of North Carolina Press).
Bush, L. (1995), 'GCOWE95 Welcome Speech', AD 2000 Movement. <http://www.ad2000.org/gcowe95/lbwelcom.html> (website), accessed 2 February, 2008.
—. (1996a), 'A Brief Historical Overview of the AD2000 & Beyond Movement and Joshua Project 2000'. Paper presented at the North East Asia AD2000/Joshua Project 2000 Consultation, Seoul, Korea. < http://www.ad2000.org/histover.htm>.
—. (1996b), 'Getting to the Core of the Core: the 10/40 Window', Global Mapping International. <www.gmi.org/products/1040_g.doc> (website), accessed April 1, 2008.
—. (1996c), 'Joshua Project 2000', in Johnstone, P., Hanna, J. and Smith, M. (eds) *Praying through the Window: The Unreached Peoples* (Seattle, YWAM Publishing).
—. (1996d), 'The Unfinished Task: It Can Be Done', AD 2000 Movement. <http://www.ad2000.org/tut0701.htm> (website), accessed 31 October, 2008.
Buss, D. and Herman, D. (2003), *Globalizing Family Values: The Christian Right in International Politics* (Minneapolis: University of Minnesota Press).
Camp, B. (ed.) (1993), *Adoption: A Practical Guide to Successfully Adopting an Unreached People Group* (Pasadena, William Carey Library).
Chow, R. (2006), *The Age of the World Target* (Durham: Duke University Press).
Comaroff, J. and Comaroff, J. (1991), *Of Revelation and Revolution* (Chicago: University of Chicago Press).
Connolly, W. (2005), 'The Evangelical-Capitalist Resonance Machine', *Political Theory*, 33: 869–86.
—. (2008), *Capitalism and Christianity, American Style* (Durham: Duke University Press).
Considine, J. (1925), *The Vatican Mission Exhibition: A Window on the World* (New York: Macmillan).
Cooper, F. (2005), *Colonialism in Question: Theory, Knowledge, History* (Berkeley: University of California Press).
Coote, R.T. (2000), '"AD 2000" and the "10/40 Window": A Preliminary Assessment', *International Bulletin of Missionary Research*, 24: 160–66.
'Core Ideas of Perspectives,' Perspectives on the World Christian Movement. <http://www.perspectives.org/site/c.eqLLI0OFKrF/b.3448385/k.7491/Core_Ideas_of_Perspectives.htm> (website), accessed 31 October, 2008.
Crehan, K. (2002), *Gramsci, Culture and Anthropology* (Berkeley: University of California Press).
Diamond, S. (1999), *Spiritual Warfare: The Politics of the Christian Right* (Boston: South End Press).
Editors (1978), 'Who Are the Evangelicals'. *Christianity Today*.

Ferguson, J. (1994), *The Anti-Politics Machine: 'Development,' Depoliticization, and Bureaucratic Power in Lesotho* (Minneapolis: University of Minnesota Press).

Forgacs, D. (ed.) (2000), *The Gramsci Reader: Selected Writings 1916–1935* (New York, New York University Press).

Gerhardt, H. (2008), 'Geopolitics, Ethics, and the Evangelicals' Commitment to Sudan', *Environment and Planning D: Society and Space,* 28: 911–28.

Goldberg, D. (2000), 'Racial Knowledge', in Back, L. and Solomos, J. (eds) *Theories of Race and Racism: A Reader* (London: Routledge).

Gramsci, A. (1971), *Selections from the Prison Notebooks* (New York: International Publishers).

Han, J. (2005), *Missionary Destinations and Diasporic Destiny: Spatiality of Korean/American Evangelism and the Cell Church.* Institute for the Study of Social Change. ISSC Fellows Working Papers. <http://repositories.cdlib.org/issc/fwp/ISSC_WP_01>.

—. (2007), Missionary Imaginations and Capitalist Deliverance: Korean/American missions in Uganda and Tanzania, Paper presented at the meeting of the Association of American Geographers, San Francisco.

—. (2009), *Contemporary Korean/American Evangelical Missions: Politics of Space, Gender, and Difference.* PhD thesis, University of California at Berkeley.

Haraway, D. (1991), *Simians, Cyborgs, and Women: The Reinvention of Nature* (New York: Routledge).

Harley, J. (1989), 'Deconstructing the Map', *Cartographica,* 26: 1–20.

Hart, G. (2006), 'Denaturalizing Dispossession: Critical Ethnography in the Age of Resurgent Imperialism', *Antipode,* 38.

Hiebert, D. (2000), 'Ethnicity', *The Dictionary of Human Geography* (Oxford: Blackwell).

Johnstone, P., Hanna, J. and Smith, M. (eds) (1996), *Praying through the Window III: The Unreached Peoples* (Seattle: YWAM Publishing).

Johnstone, P., Mandryk, J. and Johnstone, R. (2001), *Operation World: When We Pray, God Works* (Cumbria: Paternoster Lifestyle).

'Definitions and Terms Related to the Great Commission,' Joshua Project <http://www.joshuaproject.net/definitions.php> (website), accessed 31 October, 2008.

Kaiser, W. and Wood, D. (2001), *Seeing Through Maps: the Power of Images to Shape Our World View* (Amherst: ODT, Inc.).

Kaplan, C. (2006), 'Precision Targets: GPS and the Militarization of U.S. Consumer Identity', *American Quarterly,* 58: 693–713.

Keane, W. (2007), *Christian Moderns: Freedom and Fetish in the Mission Encounter* (Berkeley: University of California Press).

Kipling, R. (1899), 'The White Man's Burden', *Modern History SourceBook* <http://www.fordham.edu/halsall/mod/Kipling.html> (website), accessed 22 October, 2008.

Koop, D. (1995), 'Mobilizing for the Millennium: 4,000 Leaders Gather in South Korea'. *Christianity Today.*

Kritzinger, J. (2002), 'The Function of the Bible in Protestant Mission', in Wickeri, P. L. (ed.) *Scripture, Community, and Mission.* (Hong Kong: Christian Conference of Asia and The Council for World Mission).

Lausanne II: International Congress on World Evangelization <http://www.lausanne.org/nl/manila-1989/manila-1989-documents.html> (website), accessed 31 October 2007.

Lewis, J., Crossman, M. and Hoke, S. (1994), *World Mission: An Analysis of the World Christian Movement* (Pasadena: William Carey Library).

Marsden, G. (1987), *Reforming Fundamentalism: Fuller Seminary and the New Evangelicalism* (Grand Rapids: William B. Eerdmans Publishing).

—. (1991), *Understanding Fundamentalism and Evangelicalism* (Grand Rapids: Williams B. Eerdmans Publishing).

Matless, D. (1999), 'The Uses of Cartographic Literacy: Mapping, Survey and Citizenship in Twentieth-Century Britain', in Cosgrove, D. (ed.) *Mappings* (London: Reaktion Books).

McAlister, M. (2006), 'Evangelical Internationalism'. Under Fire <http://underfire.eyebeam.org/?q=node/509> (website), accessed 31 October, 2008.

McGavran, D. (1972a), 'Salvation Today,' in Winter, R. (ed.). *The Evangelical Response to Bangkok,* (Pasadena: William Carey Library).

—. (1972b), 'Will Uppsala Betray the Two Billion?' in Glasser et al. (eds) *The Conciliar-Evangelical Debate* (Waco: Word Books).

—. (1986), 'My Pilgrimage in Mission', *International Bulletin of Missionary Research* 10, 53–8.

Moreau, A., Corwin, G. and McGee, G. (2004), *Introducing World Missions: a Biblical, Historical, and Practical Survey* (Grand Rapids: Baker Academic).

Ostling, R. and Matheson, A. (1982), 'Counting Every Soul on Earth'. *Time Magazine* <http://www.time.com/time/magazine/article/0,9171,953480,00.html> (website), accessed 18 December 2008.

Ostrom, M. (2004), 'Clergy Members Protest Gay Vows: About 200 Rally in SF; Some Meet with Mayor'. *San Jose Mercury News* <http://www.mercurynews.com> (website) accessed 15 April 2004.

Pennycook, A. and Coutand-Marin, S. (2003), 'Teaching English as a Missionary Language', *Discourse: Studies in the Cultural Politics of Education* 24, 337–53.

Pennycook, A. and Makoni, S. (2005), 'The Modern Mission: The Language Effects of Christianity', *Journal of Language, Identity, and Education* 4, 137–55.

Pickles, J. (2004), *A History of Spaces: Cartographic Reason, Mapping and the Geo-coded World* (London: Routledge).

Piper, J. (2004), *Let the Nations be Glad!: The Supremacy of God in Missions* (Grand Rapids: Baker Academic).

Pratt, M. (1992), *Imperial Eyes: Travel Writing and Transculturation* (London: Routledge).
Robb, J. (1996), 'The Challenge of the Unreached Peoples', in Johnstone et al. (eds) *Praying through the Window III: The Unreached Peoples.* (Seattle: YWAM Publishing).
Roberts, S., Secor A. and Sparke, M. (2003), 'Neoliberal geopolitics,' *Antipode*, 35, 886–97.
Runions, E. (2007), 'Theologico-Political Resonance: Carl Schmitt between the Neocons and the Theonomists', *Differences: A Journal of Feminist Cultural Studies* 18, 43–80.
Rynkiewich, M. (2007), 'Corporate Metaphors and Strategic Thinking: "The 10/40 Window" in the American Evangelical Worldview', *Missiology: An International Review,* XXXV, 217–41.
Saayman, W. (1991), '"Who Owns the Schools Will Own Africa': Christian Mission, Education and Culture in Africa', *Journal for the Study of Religion* 4, 29–44.
Sharp, J. (2000), *Condensing the Cold War: Reader's Digest and American Identity* (Minneapolis: University of Minnesota Press).
Smith, C. (1998), *American Evangelicalism: Embattled and Thriving* (Chicago: University of Chicago Press).
Smith, E. (1972), 'Renewal in Mission', in Glasser, A. and McGavran, D. A. (eds) *The Conciliar-Evangelical Debate* (Waco, TX: Word Books).
Thomas, N. (1995), *Classic Texts in Mission and World Christianity: a Reader's Companion to David Bosch's Transforming Mission* (Maryknoll: Orbis Books).
Torassa, U. (2004), 'Thousands protest legalizing same-sex marriage. Asian Americans, Christians rally in Sunset District'. *San Francisco Chronicle* <http://sfgate.com/cgi-bin/article.cgi?file=/chronicle/archive/2004/04/26/BAG4R6B18L1.DTL> (website) accessed 26 April 2004.
'Transforming the 10/40 Windows nations through the power of prayer', Window International Network <http://www.win1040.com/the1040window.htm> (website) accessed 3 August 2008.
Van Biema, D., Booth-Thomas, C., Calabresi, M., Dickerson, J., Cloud, J., Winters, R. and Steptoe, S. (2005), '25 Most Influential Evangelicals in America'. *Time Magazine* <http://www.time.com/time/covers/1101050207/> (website) accessed 3 January 2007.
Wagner, C. (1989), 'Territorial Spirits: Identifying Demonic Spirits Over Geographical Domains', *Ministry Digest* 1:5, 42–5.
—. (1996), *Confronting the Powers: How the New Testament Church Experienced the Power of Strategic-level Spiritual Warfare* (Ventura: Regal Books).
Warner, T. (1991), *Spiritual Warfare: Victory Over the Powers of this Dark World* (Wheaton: Crossway Books).

Warren, R. (2002), *The Purpose-driven Life: What on Earth am I Here For?* (Grand Rapids: Zondervan).

Williams, W. (1980), *Empire as a Way of Life* (Oxford: Oxford University Press).

Winter, R. (1975), 'The Highest Priority: Cross-Cultural Evangelicalism', in Douglas, J. (ed.) *Let the Earth Hear His Voice.* (Minneapolis: World Wide Publications).

—. (1981), 'The Task Remaining: All Humanity in Mission Perspective', in Winter, R. and Hawthorne, S. (eds) *Perspectives On The World Christian Movement A Reader.* (Pasadena: William Carey Library).

—. (1989), 'Unreached Peoples: Recent Developments in the Concept', *Mission Frontiers* 18.

Wood, D. (1992), *The Power of Maps* (New York: The Guilford Press).

Yoder, S. (1999), 'A Strategy for Loving the Peoples of the World as Well as the Missionaries,' Adoption Guidance Program <http://www.ad2000.org/adoption/agpstrat.htm> (website) accessed 31 October, 2007.

Chapter 9

Between Armageddon and Hope: Dispensational Premillennialism and Evangelical Missions in the Middle East

Carolyn Gallaher

Protestant evangelicals[1] have been a core part of the 'New Right,' the term commonly used to describe the 'state-movement' alliance that developed in the late 1970s and early 1980s between the Republican Party and numerous right-of-center social movements (Diamond 1995). Left-wing movements such as feminism and civil rights have been better documented in (and out) of the academy, but the New Right has arguably had an equal, if not greater, influence in American politics (Brinkley 1998).

On the domestic front, Protestant evangelicals have used the political system to challenge some of liberalism's most sacred cows (Berlet and Lyons 1998; Diamond 1995). They have opposed the separation of church and state by championing prayer in public schools and the display of the Ten Commandments in government buildings. They have called the scientific method into question by advocating the inclusion of creationism/intelligent design into public school biology curricula And, they have confronted so called secular morals by calling for abstinence-based sex education (Dionne 2008; Gallaher 1997).

Evangelicals have also used the political system to influence American foreign policy, in the process crafting geopolitical views grounded in theology. There are multiple evangelical geopolitics, but the dominant[2] one, and the topic of this paper, is the evangelical geopolitical vision based in dispensational premillennialism.

---

1   In this chapter I use the term evangelical to refer to fundamentalist Protestants who adopt dispensational premillennial theology. It is possible to be an evangelical and not adopt this theology. However, contemporary uses of the term evangelical usually exclude these groups. As Frykholm (2004, 22) notes, "When the American media first embraced the term in the 1970s to describe a religious movement emerging from underground, it was broad enough to encompass both the 'born again' social activism of Jimmy Carter and the growing therapeutic movement of James Dobson. Gradually, the meaning has become increasingly narrow and identified with the religious and political right, leaving politically left-leaning but still theologically conservative Protestants without a name to call themselves".

2   According to Timothy Weber, dispensational premillennialism is a "minority position theologically" (Goodstein 2007). However, evangelical leaders on the New Right have tended to be advocates of dispensationalism, and as such have had greater sway

Dispensational geopolitics was developed and championed by founding members of the evangelical wing of the New Right, most notably Pat Robertson and the late Jerry Falwell. It continues to be promoted by contemporary leaders such as John Hagee.

Dispensationalism is based on three interconnected beliefs. The first belief is that human history can be divided into a series of epochs, or dispensations, in which God reveals part of his divine plan to humankind and tests them on it. The second and related belief concerns the 'end times,' a shorthand for the period leading up to the final dispensation (the millennium). Dispensationalists believe the end times will be marked by a series of turbulent events.[3] They will start with the Rapture,[4] when God calls the 'saved'[5] to heaven and will be followed by a seven year period of Tribulation in which billions of humans succumb to drought, famine, and disease. During the Tribulation dispensationalists believe an Antichrist will emerge calling for one world government as a solution to earth's problems (see Chapter 3, this volume). His rhetoric will sound helpful and peaceful, but his real goal will be to thwart Christ's return to earth. He will amass an army to take over the 'holy land' (modern day Israel and Palestine) and cast God's chosen people, the Jews, from it. Christ will return to earth and fight the Antichrist in the Battle of Armageddon. Christ's victory will mark the start of his millennial reign on earth. The third belief concerns Biblical prophecy about the end times. Dispensationalists believe the Bible maps out key events that must occur prior to the Rapture. The formation of a Jewish homeland in the Middle East in 1948, for example, is widely regarded as a sign the end times are near. So too is violence in the region (in Israel and beyond).

---

in crafting the geopolitical vision most often associated with the New Right (of which evangelicals are a part).

3 Other Protestant sects believe the millennium will emerge gradually, as a result of increasing conversions to Christianity.

4 Other Protestant theological traditions do not embrace the idea of the Rapture. It was first proposed by John Nelson Darby in the mid-1800s. Weber (2004) argues that Darby developed the idea of the Rapture in part to answer a puzzle in Protestant theology. In the Old Testament, Jews are described as God's chosen people. However, in the New Testament, which was written after the resurrection, the followers of Jesus (Christians) are also referred to as God's chosen people. This presented a conundrum for theologians who believed that God's promise to deliver a homeland to his chosen people had yet to be fulfilled. Darby's contemporaries generally believed God would fulfill that covenant during the millennium. However, they were uncertain how the covenant would be fulfilled with two sets of chosen people. Darby solved the puzzle through the idea of the Rapture where Christians would be taken to heaven and the covenant with the Jews fulfilled on earth during the millennium.

5 Evangelicals believe a person is saved if he/she accepts Jesus Christ as his/her personal savior. Most evangelicals can recite when and where they were saved. This view of salvation differs from other Protestant views of salvation. Calvinist ideas of salvation, for example, hold that God pre-ordains whether a person will go to heaven before birth.

The dominance of dispensational premillennialists within the New Right has traditionally translated into ardent support of Israel (Weber 2004).[6] However, the evangelical belief that a violence will mark the 'end times' has also led evangelicals to oppose peace efforts in the Middle East. Indeed, evangelical leaders often assert that working for peace in the Middle East during the end times is not only pointless, but works against prophecy (Weber 2004). Evangelicals supported Israeli military assaults against Hezbollah in its 2006 war, for example, even though its aggressive stance was widely viewed as counterproductive to both Israeli and American interests (Mearsheimer and Walt 2006). They also continue to be enthusiastic supporters of the Bush Administration's decision to go to war in Iraq, despite the instability it has brought to the region (Dyer 2003).

In many ways the evangelical geopolitical vision of the Middle East is at odds with the evangelical emphasis on spreading the 'good news', as the Bible and its message of personal salvation is often called. Dispensationalist theology tends to take a fatalist approach to politics in the Middle East. Its advocates believe that the end times are nigh and that a war in the Middle East is inevitable (for an example, see Hagee 2007). By contrast, the drive to evangelize (a tenet common to both dispensationalist and non-dispensationalist sects) is rooted in a more optimistic view of politics. Indeed, Christian mission work (especially by sects that follow dispensationalism) has been on the rise in Latin America (Garrard-Burnett and Stoll 1993; Gallaher 2007), Asia (Hunter and Chan 1996), Africa (Freston 2001), and the Middle East (Tamacke and Marten 2008) since the late 1970s and early 1980s. The increase in foreign missions is associated with the rise of the New Right in part because evangelicals believed they could make a difference not only in the state of potential converts' souls, but in the state of their societies (Diamond 1999). Indeed, mission work since the 1970s has been infused with a sense that wide scale conversions would set in place the conditions necessary for capitalist development and democratic political systems (Levine and Stoll 1997; see Chapter 7, this volume).

In the Middle East, the political imperatives of dispensationalism and evangelism come into especially stark contrast. Fatalism crashes into optimism. The idea that war in the Middle East is inevitable, that Christians should not stand in the way of prophecy, collides with the evangelical mission to convert the unsaved and by doing so improve the politics and economies of the societies in which they live. These competing directives create what I call here the *Middle East mission paradox*.

---

6 As I note above, dispensationalism is a "minority position theologically" (Goodstein 2007). However, the Republican Party's ties with evangelicals have tended to be with those who advocate dispensationalism. As such, when Republicans court evangelicals (especially in relation to Middle East Policy), it is this vision they most often reference. None of this is meant to imply, of course, that all Republicans are dispensationalist, or that the party's support is unconditional.

This chapter examines how evangelical sending agencies confront this apparent paradox and the degree to which their response supports (or challenges) the dispensational geopolitical vision. I begin with a brief overview of dispensational eschatology (for an extended discussion see the Introduction, this volume). I then discuss how dispensational eschatology has contributed to the creation of an evangelical geopolitical vision. In the third section, I outline a theoretical framework for situating the mission paradox. The framework I use, performativity, focuses on how individuals perform/act out subject positions such as gender, ethnicity, religion, and political affiliation, among others (see Bialasiewicz et al. 2007; Sturm 2006 for applications of 'theatricality' in political geography). Performativity provides a way to understand both how norms and expectations (often textual in nature) are translated into lived experience, and how social actors navigate the competing demands/expectations attached to the varied subject positions they occupy. In the fourth section I introduce three mission-centered organizations that work in, or sponsor, mission work in the Middle East—CrossGlobal Link, Moody Bible Institute, and Frontiers. I then explain how these mission-centered organizations deal with the contradictions that dispensational eschatology presents for their work and analyze how these strategies might affect the movement's geopolitical vision in the future. I conclude with some tentative assessments about the evolution of evangelical geopolitics in the Middle East and discuss the implications for the scholarship of evangelical geopolitics.

Dispensational Geopolitics

The evangelical geopolitical vision is based on three, interwoven tenets. The first is that Israel has a 'rightful' place among the world's nations. While Israel's right to exist has been approved by the UN and even some neighboring Arab states (e.g. Egypt and Jordan), evangelicals go a step further. Evangelicals active in the New Right often describe Israel as the fulfillment of God's covenant with the Jews for a homeland. In a speech at Furman University, for example, John Hagee argued:

> Israel has a Bible mandate to own and possess the land of Israel forever. That covenant cannot be amended by the United Nations. It cannot be revoked by Hamas or Hezbollah. It cannot be altered by the United States State Department. It cannot be amended by the President of the United States. It cannot be amended by President Carter. It belongs to the Jewish people now and forever (as cited in Hanson 2007, n.p.).

Other evangelicals are more cautious, describing the Israeli state as an important first step in the fulfillment of the covenant.[7] However, both groups view the creation of Israel as related to the covenant (whether as its total or partial fulfillment).

Evangelical leaders also tend to be unsympathetic to the Palestinian people who lost their homes and land when the state of Israel was created (Cimino 2005; Weber 2004). Indeed, most evangelical leaders view Palestinian demands (for land, statehood, and rights) as contrary to God's will and prophecy. They blame Arab countries in the region for the Palestinian problem. Evangelicals argue, for example, that the Haganah did not kick Palestinians out of their homes when the new state was formed; rather Palestinians fled because Israel's Arab neighbors threatened to kill them if they stayed (Dittmer 2007).

The prophetic place of Israel in dispensational eschatology means that many evangelicals believe that Israel, and Israeli interests, should be aggressively defended by the United States (Weber 2004). Some scholars refer to this as a form of Christian Zionism because like Zionists, evangelicals believe God has reserved Palestine for his chosen people (Mearsheimer and Walt 2006). Both also believe that territorial claims to the land by any other group are against God's will.[8]

The second and related tenet of the evangelical geopolitical vision is opposition to a peace agreement between Israel and Palestinians in the West Bank and Gaza (Weber 2004). Most evangelicals believe that the creation of the state of Israel is a marker of the end times. The Rapture will occur, most likely in this lifetime, and a seven year period of violence and strife will follow. It will culminate in the Battle of Armageddon. Working for, or supporting peace in Israel would entail working against God's divine plan.

This macro-level view of peace in the Middle East informs evangelicals' views on a variety of specific efforts to achieve peace. Evangelicals, for example, have largely opposed the Bush Administration's so called 'road map' to peace (even though they have tended to support other administration policies in the Middle East, like the war in Iraq) because the plan calls for the creation of a separate Palestinian state. For evangelicals, such a 'partition' would work against prophecy. As Weber (2004, 267) explains:

---

7   When Israel was first created, most dispensationalists did not view it as the complete fulfillment of God's covenant because the new state was smaller than the biblical Israel was thought to be.

8   The connection between Zionists and their evangelical allies is best understood as a marriage of convenience. Zionists welcome Christian support for Israel, but eschew Christian attempts to proselytize them. Likewise, while Christian Zionists are ardent supporters of Israel, their support has less to do with cultural affinity for Jews than their belief that God has instructed them to support the descendents of Abraham (Weber 2004). Indeed, many premillennial dispensationalists believe that Jews will go to hell for failing to accept Jesus Christ as their savior, even though they are chosen people (*Jerusalem Post* 2006).

They do not believe the road map will—or should—succeed. According to the prophetic texts, partitioning is not in Israel's future, even if the creation of a Palestinian state is the best chance for peace in the region. Peace is nowhere prophesied for the Middle East until Jesus comes and brings it himself. The worst thing the United States, the European Union, Russia, and the United Nations can do is force Israel to give up land for a peace that will never materialize this side of the second coming. Anyone who pushes for peace in such a manner is ignoring or defying God's plan for the end of the age.

Dispensationalists also oppose the division of Jerusalem because they worry that Palestinians will obtain control over the Temple Mount site. In Islam the site is revered as the place where the Prophet Mohammed ascended to heaven with the Angel Gabriel. Two of Islam's most important religious shrines—the al-Aqsa mosque and the Dome of the Rock—are located here. However, evangelicals believe that the site must remain in Jewish hands because they believe Jesus will rebuild his throne there when he returns to earth to begin his millennial reign. In a September 2006 sermon, for example, John Hagee explained evangelical opposition to dividing Jerusalem:

> Satan wants Jerusalem for his messiah, the Antichrist. God has promised it to King David, that his seed, Jesus Christ, shall rule over it forever and forever. People ask, why can't the Arabs and the Jews get together over this city? Listen to me: it is not about the two-state solution. It is not about money. It is not about land. It is about theology…Israel [has] an unconditional blood covenant with the God of Abraham, Isaac and Jacob, that the land shall be theirs forever and forever. God did not loan the land to Abraham, Isaac and Jacob. He gave it to them by blood covenant, and the Book of Genesis and that covenant still stands…I am telling you there will be no lasting peace in the Middle East until the Prince of Peace, Jesus Christ, returns to that sacred city and sets up his eternal kingdom.[9]

Dispensational opposition to peace between Israel and Palestine is mirrored in its approach to other wars in the Middle East (Dittmer 2007). Indeed, any war in the region, whether it involves Israel or not, is generally read as a potential prophetic marker of the end times. Moreover, because place names in the Bible do not always match contemporary ones, Dispensationalists have tended to take a geographically expansive view of the Holy Land to include areas outside of the contemporary borders of Israel, the West Bank, and Gaza. Dispensationalists were big supporters not only of Israel's war with Hezbollah in southern Lebanon in 2006, but also of the first and second Gulf Wars in Iraq (Mearsheimer and Walt 2006;

---

9 Excerpts of Hagee's sermon can be found in the archives of Tapped', a blog of the American Prospect: http://www.prospect.org/csnc/blogs/tapped_archive?month=02andyear=2008andbase_name=is_the_end_near_for_mccain.

Dittmer 2007). Charles Dyer, the Provost of Moody Bible Institute, for example, argues that Saddam Hussein's decision to rebuild the ancient city of Babylon in the 1970s is foretold in biblical prophecy and America's two wars in Iraq signal that the end times are approaching (Dyer 2003; see Sturm 2006). Many have also advocated a war with Iran. In his book *Jerusalem Countdown*, for example, John Hagee connects the battle of Armageddon to Iran. As Hagee sees it:

> There is a clear and present danger to America and Israel from a nuclear Iran. There will soon be a nuclear blast in the Middle East that will transform the road to Armageddon into a race track. America and Israel will either take down Iran, or Iran will become nuclear and take down America and Israel (Hagee 2007, 4–5.)

The third tenet of the evangelical geopolitical vision is suspicion of internationalism. In particular, evangelicals are suspicious of international organizations like the United Nations. Unlike other critics of the UN, however, who point to its lumbering bureaucracy and general ineffectiveness, evangelicals view the UN through a theological lens. Dispensational eschatology holds that the Antichrist will prepare himself for the Battle of Armageddon by consolidating political power on earth. To do so, he will form and lead a one world government. For Dispensationalists, the UN is the most likely vehicle for that to happen.

Dispensational eschatologists have been suspicious of the UN since it was first formed at the end of World War II (Diamond 1995). During the Cold War, dispensationalists often thought the Antichrist would emerge from the Soviet sphere of influence, and he would attempt to create a one world communist government using the UN (Dittmer 2007).[10] Since the attacks on the World Trade Center and the Pentagon on September 11, 2001, however, evangelicals have more often depicted the Antichrist as Muslim. John Hagee, for example, believes that the Antichrist will form a one world Islamic theocracy (Hagee 2007).

Dispensational opposition to the UN has manifested in a variety of ways. Some dispensationalists have lobbied for the United States to remove itself from the UN. Others have focused on opposing specific UN actions. In the run-up to the second Gulf War, for example, evangelicals argued the United States did not need to seek UN approval for its actions. Evangelicals are also critical of UN actions seen as benefiting Palestinians. Dittmer (2007) notes, for example, that evangelicals have long criticized UN Resolution 242, which stipulates that Israel must leave the occupied territories.

---

10 In the fictionalized *Left Behind* series, authors Tim LaHaye and Jerry Jenkins describe the Antichrist as emerging from Romania.

Performing Dispensationalism

Human behavior is full of paradoxes—a truism that most social scientists know firsthand. Whether they were digging through the archives or slogging through the proverbial field, most scholars can recount an instance when a subject or informant contradicted himself, took positions that appeared incongruent, or held different standards for different people/issues. Not surprisingly, social movements tend to manifest similar paradoxical qualities.

While some paradoxical behaviors are intentional (designed to mislead), most are better understood as the product of a complex social world that embeds and pulls people (and movements) in different directions. In this chapter I will use the concept of performativity to ground my understanding of what I label here as the mission paradox. Performativity is usually associated with Judith Butler's work on the constructed nature of gender (1990), and perhaps not surprisingly, feminists were the first to use the concept in geography (see Rose 1990). The concept describes the process by which social actors live out/perform the normative expectations that accumulate around the subject positions they occupy. Western men, for example, often perform masculinity by hiding their emotions, acting 'tough,' and solving conflict with violence rather than negotiation. Such expectations are not immutable. They vary across time and place. Recently, political geographers have adopted the concept to understand how states (and other political actors) interpret and perform the roles they craft for themselves in geopolitical narratives (Bialasiewicz et al. 2007; Gregory 2004; Sparke 2005; Sturm 2006).

Performativity provides two useful axioms for the analysis here. First, performativity provides a framework for understanding the materiality of discourse (Bialasiewicz et al. 2007). That is, while discourse is often understood as a rhetorical frame through which objects are defined and gain social meaning, discourse is also the process by which those meanings are fixed (however temporarily) in space. During the Bush Administration's first term, for example, the term 'Axis of Evil' was used to identify nations the administration considered particularly dangerous to U.S. interests. Using updated discourses of terrorism (which linked Islamic extremism and nuclear ambitions as the key threat confronting the Western world), Bush not only identified the geographic coordinates of 'evil' in three states—Iraq, Iran, and North Korea—it established the material constraints within which the U.S. could negotiate with them. Indeed, many observers noted that the administration's rhetorical framing of these countries as evil effectively took negotiation off the table as an option for dealing with them. Diplomacy with an 'evil' country was, for example, harder to legitimize than diplomacy with a non-evil one.

In the case of the mission paradox, the emphasis on the materiality of discourse draws our attention to how sending agencies carry end times discourses (about the certainty of war in the Middle East) into the mission field and how those discourses shape the actions missionaries take in the field. In particular, this axiom impels us to ask not only who, if anyone, is considered unreachable, but where, if anyplace, is off-limits to mission work. Understanding discourse materially also impels us to ask how dispensational beliefs about war in the Middle East prophetically affect the way missionaries address the potential violence converts face in the Middle East on a daily basis. Do missionaries acknowledge it and lend a sympathetic ear, or rather, do they share their view that the violence is necessary, even prophetic?

Performativity also provides a way to understand the paradoxes that stem from individuals holding different subjectivities. One of the key lessons to evolve from geographic (and other social science) forays into identity issues over the last twenty years is the idea that individuals often occupy multiple subject positions at once. These subject positions locate us *vis-à-vis* hegemonic categories of race, gender, sexuality, and so on. A white woman, for example, occupies a dominant racial category, but an 'othered' gender category. The multiplicity of subject positions that people can occupy, and the variety of combinations that can ensue (often a mix of hegemonic and 'other') as the previous example shows, create 'paradoxical space.' As Rose (1993, 159) notes with reference to the place of feminism within geography,

> This space is paradoxical because…it must imagine the position of being both prisoner and exile, both within and without. It must locate a place which is both crucial to, yet also denied by, the Same; and it must locate a place both defined by that Same and dreaming of something quite beyond its reach.

Political geographers have expanded our notions of positionality to examine how social movements employ political identities, which are often resisting some configuration of hegemonic interests and the spatial relationships that stem from them (see Bell and Binnie 2005, for example, on sexual politics). They have also examined how social movements contend with the often competing interests that arise from members who occupy multiple subject positions or deal with the competing demands of personal and political forms of identification (see Gallaher 2003 on how the U.S. militia movement deals with the competing interests of members' race and class backgrounds). In short, the focus on positionality within the literature on performativity allows us to locate a potential source for contradictions, giving them a logical character that might otherwise not be obvious.

In the case of dispensationalism, performativity impels us to examine not only how evangelicals position themselves in the American political realm but how individual evangelicals position themselves within evangelism. As I note above, dispensationalism is not the dominant eschatology within the broader tent that is evangelism, but its advocates have become the dominant players in crafting an evangelical geopolitical vision within the New Right. Thinking about evangelical geopolitics as a performance of subject positions compels us to ask how those who set mission 'policy' (broadly defined) confront contradictory positions within evangelism in a mission field some see as the geographic crucible of the end times. It also prompts us to ask how such individuals navigate the often competing demands of politicized and personal evangelical identities.

Mission Field or Mine Field?

Interview data from leaders at three organizations that teach and/or sponsor missionaries working among Muslim populations in the Middle East provides the starting point for further analysis. The first organization, the Moody Bible Institute, is a non-denominational organization devoted to a broad array of activities, including education, publishing, and broadcasting. Moody currently offers undergraduate and graduate degrees in Biblically-themed studies. Though diverse, the goal of Moody degree programs is evangelical. Its webpage informs visitors that the Institute "will teach you how to understand, apply, and communicate the truth of God's Word in today's world." Moody also has been one of dispensational eschatology's strongest institutional advocates. I interviewed Moody's provost, Charles Dyer, who has led mission trips to the Middle East, and has taught students interested in mission work in the Middle East. He is also the author of several books on the end times, including *The Rise of Babylon: Sign of the End Times* (1991).

The second organization, CrossGlobal Link, is an umbrella group of smaller sending agencies. Its goal is to provide a clearinghouse for information on mission work and provide a venue where missionaries can link up with one another. Several of its member groups focus exclusively on Muslim and/or Arab populations, including Arab World Ministries and Reach Across. As an organization, CrossGlobal Link emerged out of the so-called "faith mission movement," which was part of the fundamentalist resurgence of the early 20th century.[11] Most fundamentalist groups at the time were dispensationalist, and CrossGlobal Links was no exception. Today it remains firmly within the dispensationalist theological tradition. I interviewed the organization's executive director, Marv Newell.

The third group, Frontiers, is a non-denominational sending agency that works among Muslim peoples in (and out of) the Middle East. It sponsors short and long term mission work. The average Frontiers missionary is in the field for six years. The group does not have a theological position on dispensational premillennialism, but most of its members are associated with evangelical churches where dispensationalism is taught (though the degree to which it is emphasized obviously varies by church and pastor). I interviewed Jai Choi, a director at the Frontiers home office in Phoenix, Arizona.

Leaders, rather than missionaries, were interviewed in order to identify the macro-level policies that drive how missionaries associated with dispensational premillennialist theology approach and blend theology and politics in the mission field. Although the interview sample is small, the persons interviewed represent a much larger group of missionaries. CrossGlobal Link lists 87 sending agencies as members. Frontiers has 150 established mission communities in 8 distinct

---

11   When it was formed, CrossGlobal Link was called the Interdenominational Foreign Mission Association. The rise of fundamentalism in the U.S. was, in part, a reaction against the growing sway of liberalism within Protestant denominations.

regions, including Southeast, Central, and South Asia, the Balkans and Caucasus, the Middle East and Arabian peninsula, and North and Sub-Saharan Africa. Likewise, the Moody Bible Institute graduates thousands of young missionaries every year.[12]

Paradox or Not?

Initially, I assumed that missionaries would acknowledge what I label here as the Middle East mission paradox. Restated briefly, the paradox stems from the contradiction between premillennialism's apocalyptic view of the Middle East and evangelism's optimistic approach to social change. In particular, the emphasis on positive social change associated with overseas evangelism since the 1980s appears to contradict the dispensational notion that involvement in peace efforts in the Middle East works against prophecy (Weber 2004; see also Sturm 2008).

My interviewees noted that they see no paradox between their theology and their mission work in the Middle East. Dyer, the Provost at the Moody Bible Institute, explained, for example, "from my perspective and the majority of the scholars on it, we would not see a paradox. And the reason we wouldn't is what God has said." Dyer acknowledged, however, that "from someone looking from the outside it looks like there is. [But,] most evangelicals don't see it as a paradox." Marv Newell, a director at CrossGlobal Link, agreed, although he also acknowledged that at first glance, "it does seem paradoxical." Jai Choi of Frontiers concurred, noting, "there's not a paradox, but there is a lot of debate."

As evidence that a paradox did not exist, my informants all pointed to Acts 1:1–9, excerpted in a footnote below.[13] Summarized briefly here, the passage begins with Christ reappearing to his disciples after his resurrection to prove that

---

12   For more information on these groups see their respective homepages. Moody: http://www.moody.edu/. Frontiers: http://frontiers.org/. CrossGlobal Link: http://www.crossgloballink.org/.

13   Acts 1:1–9: In my former book, Theophilus, I wrote about all that Jesus began to do and to teach until the day he was taken up to heaven, after giving instructions through the Holy Spirit to the apostles he had chosen. After his suffering, he showed himself to these men and gave many convincing proofs that he was alive. He appeared to them over a period of forty days and spoke about the kingdom of God. On one occasion, while he was eating with them, he gave them this command: 'Do not leave Jerusalem, but wait for the gift my Father promised, which you have heard me speak about. For John baptized with water, but in a few days you will be baptized with the Holy Spirit.' So when they met together, they asked him, 'Lord, are you at this time going to restore the kingdom to Israel?' He said to them: 'It is not for you to know the times or dates the Father has set by his own authority. But you will receive power when the Holy Spirit comes on you; and you will be my witnesses in Jerusalem, and in all Judea and Samaria, and to the ends of the earth.' After he said this, he was taken up before their very eyes, and a cloud hid him from their sight. They were looking intently up into the sky as he was going, when suddenly two men dressed in white stood beside them. 'Men of Galilee', they said, 'why do you stand here

he is alive. One of his disciples asks if Jesus will now "restore the Kingdom of Israel" (most evangelicals believe this refers to Christ's millennial reign on earth). He responds that only his Father knows the time and the date, but the disciples will soon receive the Holy Spirit and be enjoined to spread his word. Jesus then ascends to Heaven as two angels inform the confused disciples that he will return to earth a second time.

The evangelical leaders I spoke with interpreted these verses in Acts to mean that since no one knows when Jesus will return, evangelism should continue apace until he arrives. This does not mean, of course, that these mission leaders have rejected the commonly held assumption in dispensationalist circles that humankind is currently living in the end times. All of them embraced this assumption in my interviews with them. However, they do not believe living in the end times changes the core mission to bring people to Christ, even in the Middle East. As Newell explained, "Too many things are happening that seem to be a culmination of prophecy. We're not date setters, but we are date watchers. But, we dare not say everything is set in place and it could happen late tomorrow. We just don't want to be presumptuous."[14] Likewise, Jai Choi of Frontiers noted that evangelism was a commandment. "It [being in the end times] doesn't affect the clear command that Jesus has given us [to evangelize]. Jesus has clearly said 'this [evangelism] you shall do.'"

Perhaps it is not surprising, given the daily performance of evangelism, that its directive to spread the good news trumps premillennial warnings about letting prophecy take its course. Evangelism is about expanding borders, about bringing people from other places into the literal and figurative boundaries of Christianity. Overseas mission trips materially reinforce evangelism in a way that theological interpretations (even when translated into foreign policy positions) can not.[15]

---

looking into the sky? This same Jesus, who has been taken from you into heaven, will come back in the same way you have seen him go into heaven.' (New International Version).

14 Other theologians also make this argument. Darrell Bock of the Dallas Theological Seminary argues, "The situation in the Middle East is exceedingly complex, even when it is viewed biblically. In fact, when Jesus made predictions about the end and told believers to keep watch for his return, he stressed that they should not obsess about trying to figure out when the exact time would be because that could not be known." Likewise, Wayne Brindle (2001) at Liberty University argues that Christians should embrace the idea of imminence—that there may be no "intervening signs or events" to signal the coming of the Rapture.

15 David Harriman, another director at Frontiers, made a similar point to me in an email conversation about my paper. At my request, interview informants were asked to read a rough draft of the paper to verify I had quoted them appropriately. Mr. Choi also asked Mr. Harriman to read the paper to ensure I had represented Frontiers correctly. In a personal correspondence (9/11/08) Harriman suggested that evangelicals are "more motivated by the clear commands of scripture (such as the command of Jesus Christ to 'go and make disciples') than they are by the less clear interpretations of theological viewpoints (such as dispensational eschatology)."

My informants also noted, in responding to my question of whether there was a paradox, that sending agencies have become more flexible in how they interpret dispensational premillennialism. Jai Choi explained, "Our theological stance is that we don't want to have a theological, non-negotiable [view] on premillennial dispensationalism."

Likewise, Marv Newell suggested that the sending agencies represented in his group cover a fairly wide spectrum of theological viewpoints. "Some [member groups] would be strongly, classically dispensational. Others are more moderate." When I asked him what 'moderate' meant in the context of dispensationalism he explained: "Classic dispensationalists believe in seven dispensations. Now, however, groups would say we don't need to be so rigid in dividing up the work of God in that way. Many groups today would say we're modified dispensationalists..." Newell suggested that many dispensationalists now believe there are only three dispensations—God's dealings with Israel in the Old Testament, His dealings with the church in the New Testament, and the millennial reign, which has yet to occur.[16] Newell concluded by noting, "This is a way many in the dispensationalist circles are leaning. Less emphasis on the dogma—that everything must be in order of the seven dispensations."

Newell also suggested that his organization contains groups who hold different views on the sequencing of the end times. Some groups, for example, hold the classic dispensationalist view that the Rapture will occur before the Tribulation while other groups believe it will occur midway through it. Still others believe it will occur after the Battle of Armageddon. These views are commonly referred to as pre-tribulation, mid-tribulation, and post-tribulation, respectively. Newell's organization provides a big tent under which all of them can work.

Newell also suggested that his organization shies away from making dogmatic interpretations of contemporary events as they may, or may not, relate to the end times. He even noted, "Zionism may be temporary." When I asked him to explain what he meant, he pointed to David Jeremiah, the former president of Christian Heritage College, as an example. Jeremiah has cautioned Christians not to vest too much significance in the restoration of Israel. Israel could fall and Christians need not worry. God will have another plan for ushering in the end times.

My informants also pointed to a variety of theories about the geography of end times violence. Indeed, though virtually all dispensationalists believe the Battle of Armageddon will occur within the borders of present day Israel, they do not agree about where prophetic events leading up to the battle will begin. Some dispensationalists talk only in general terms about war in the Middle East. Others assume that the violence will occur in and around Israel. Still others posit that Iraq is an important site for end times prophecy. As I note above, Charles Dyer argues that the ancient city of Babylon will be rebuilt in the end times as a throne for the

---

16   The Dallas Theological Seminary, a prominent source of dispensationalist thought for much of the last century, only identifies three dispensations in its doctrinal statement (see DTS 2008).

Antichrist. Dyer sees the decision by Saddam Hussein in the 1970s to build a new Babylon on the ruins of ancient Babylon in modern day Iraq as a clear indication that the end times are near (2003).

Such flexibility complicates the academic understanding of the influence of dispensationalism in evangelical geopolitics. Indeed, the variety of views expressed in my interviews suggest that there is often a significant gap between evangelism's public, often singular rhetoric (and academic representations thereof) and its more private, internal variation. In particular, this flexibility shows up in evangelical views about Middle Eastern geopolitics.

David Jeremiah's argument, for example, that end times prophecy can be fulfilled with or without a restored Israel represents a viewpoint rarely covered in academic and commentariat readings of dispensationalism. To be sure, Jeremiah's view is controversial for many dispensationalists, but it has gained enough acceptance within dispensationalist circles to represent a real fissure within the movement, and one worthy of representation in academic circles.

Moreover, some evangelical leaders are now willing to diverge from the ardent pro-Israeli positions of evangelical leaders like Pat Robertson, John Hagee, or the late Jerry Falwell. These leaders have tended to brook no complaint with Israel. In our interview, for example, Charles Dyer pledged his support for Israel, and its prophetic place in dispensational eschatology. However, he was also willing to criticize its actions.

> Israel, I believe, has been chosen by God when God made a covenant with Abraham. God and God alone made that covenant. In the Old Testament it was clear that Israel was God's chosen nation. But, God held them [Israel] to task when they mistreated others. God still holds them to a standard. Not everything Israel has done to the Palestinians is right.

When I asked Dyer for an example, he pointed to some (but not all) Israeli settlements.

> I believe Israel has a right to exist as a nation, to exist in the land. To have clear, free borders. But, some settlements, for example, are nothing but bedroom communities for Jerusalem, while others are in-your-face settlements designed to intimidate and drive away the local population.

Jai Choi, of Frontiers, did not explicitly criticize the Israel settlement policy, but he also suggested that people who worked for Frontiers did not have to automatically support them. As he explained,

> Settlements and peace agreements in the Middle East are not essential, or even theological questions. There are some things that scripture says about it, but it is not essential to theology. I would say that people who come into the mission field with that understanding, they would have those beliefs challenged.

While neither Dyer nor Choi explicitly rejects the standard evangelical line on settlements, their comments are quite radical when considered in context. Indeed, it is rare for evangelical leaders within the dispensationalist tradition to publicly speak out against Israeli actions because it is often equated with going against prophecy. Key evangelists like John Hagee and the late Jerry Falwell have suggested, for example, that ardent support of Israel is the only position a real Christian can take.

It is also useful to view these actions through the lens of performativity. While Charles Dyer's comments about Israeli settlements appear contradictory—ardent support for Israeli right to land on the one hand and sympathy for Palestinian anger over Israeli settlements on the other—performativity allows us to situate the contradiction at least in part to the different subject positions he holds. As a theologian and eschatologist, Dyer's job is to look for signs of gathering storms, wherever they may be and whomever they may affect. It is an abstract, even impersonal, job. As an educator he has to help young missionaries learn how to form human connections to Arabs and to demonstrate across all sorts of cultural barriers how a life spent following Jesus is worth it. Durable conversion requires missionaries to be sincerely sympathetic to the circumstances of their potential converts' lives. Being both a dispensationalist theologian and an evangelical is not impossible, but some contradiction is inevitable.

Before closing this section, however, it is worth noting the limits of dispensationalist flexibility. In some cases theological flexibility serves political ends. In particular, it can provide political cover for dispensationalists as they deal with outsiders. When I interviewed Charles Dyer, for example, I was surprised by how he situated his opposition to a divided Jerusalem. In our interview he argued: "It scares me when people in academic circles present simplistic views of how evangelicals view the Middle East. I think they've equated the eschatology [theory of the end times] and support for Israel. They've made a logical fallacy. They assume one is responsible for the other." When I asked Dyer to elaborate, he responded with an example. "I wouldn't be in favor of dividing Jerusalem or setting up a Palestinian state. It's not because of eschatology but because of Hamas. And, fear that Hamas would take over the West Bank."

I had expected Dyer to oppose the division of Jerusalem, but I was surprised he situated his opposition in political rather than eschatological terms. Indeed, the majority of his written work suggests eschatology drives his view of Jerusalem. In a February 10, 2008 blog posting, for example, Dyer (2008) outlines his opposition in strictly eschatological terms.

> I believe the Bible clearly establishes Israel's right to the city of Jerusalem. In Deuteronomy 12:11 Moses told Israel that God Himself would choose the place 'for His name to dwell.' David established Jerusalem as his royal city (2 Sam. 5:5), and Solomon built his Temple to the Lord there (2 Chron. 3:1). Jerusalem remained Judah's spiritual and political center from the time of David until the time of the Babylonian captivity. After the Babylonian captivity the

Jewish remnant returned to Jerusalem and rebuilt the Temple (Ezra 1:3; 5:1–2). It again became their chief city until it was destroyed by the Romans in A.D. 70. And Jerusalem has again become Israel's capital since the establishment of the modern State of Israel in 1948.

In the same posting Dyer acknowledges that Jerusalem will likely be divided. However, as the excerpt below indicates, he argues that it will be divided after the Rapture, when Christians will no longer reside on earth.

> But while the Bible clearly indicates the nation of Israel will someday possess Jerusalem, it also identifies an interim period of time when the Jerusalem will be divided…The picture painted in Revelation 11 is of a divided Jerusalem—one with a Jewish presence and Jewish center of worship but one that has also, at least in part, been given over to the ethnos, the Gentiles. As the chapter unfolds, it appears that the 42 months of Gentile control refers specifically to the first three-and-a-half years of the future seven-year period of trouble called the Tribulation.

What then, to make of such a contradiction? The concept of performativity is again useful here. In this case, performativity allows us to situate Dyer in relation to an audience. In comments to a self-identified member of the academy, he emphasized his opposition to a divided Jerusalem on 'secular' (i.e. political) grounds. In his blog, aimed at other dispensationalists, his emphasis remains eschatological. While some might see Dyer's varied justifications as intentionally misleading, my reading is that Dyer's varied justifications are indicative of the multiple subject positions he occupies, and the contradictions that stem from them. Indeed, Dyer would certainly not be alone among fellow Americans in mistrusting Hamas. And, though his religious views represent a decidedly minority view in the U.S., most Americans also generally support Israel over the Palestinians. As Rod Dreher (2002) notes, "there aren't enough dispensationalists in the United States to explain why so many American Christians feel a strong obligation to support Israel." Dreher argues that suicide bombings and American attitudes about Islamic terrorism since the attacks of September 11, 2001 likely drive many Americans' views of politics in the region.

The salient feature here is the elision of politics and eschatology and what it says about a politicized religious world view. Since the rise of the New Right in the 1980s, dispensationalists have had to navigate the dual and often contradictory directives that stem from their theology on the one hand and their politics on the other. Theologians can afford to explain world events in completely eschatological terms in ways politicized evangelicals, mindful of audience, cannot.[17] In the next

---

17 I situate Dyer's contradictions here in the conflicting roles he inhabits, but these sorts of contradictions can at times appear nakedly political. John Hagee, for example, has consistently portrayed his support for Israel in prophetic terms (2007). However, when he

section I examine how these divides affect mission work in the Middle East, and its potential impact on evangelical geopolitics.

Swimming Against the Current?

As I noted above, I assumed that missionaries would have to confront a theological paradox that pitted the call to evangelize against the fatalism of premillennial eschatology. However, missionary leaders saw no contradictions in their work. For them, Biblical passages suggesting that the date of the Rapture is unknown provide a basis for evangelism in the Middle East in spite of prophetic speculation about contemporary events there.

Although theology posed less of a burden for missionaries than I thought it would, politics, as the previous section suggests, posed a number of problems for them. Most missionaries come to the field with views that correspond to the evangelical geopolitical vision described in the second section of this chapter. However, when missionaries arrive in the mission field their world view does not always resonate with events on the ground. Some of the trouble is to be expected—fallout that comes with any cross-cultural encounter. Jai Choi from Frontiers notes:

> They have to deal with what they bring with being an American. There's a lot of perception [about what that means]. We tell people 'you're not necessarily going onto even ground.' There is baggage we bring in.

However, some problems missionaries confront stem from their reactions to politics on the ground. Who is right and who is wrong may seem crystal clear back home, but when missionaries arrive in the West Bank, Gaza, or in/near Palestinian refugee camps in Lebanon or Jordan, they confront checkpoints and travel restrictions and witness poverty and discrimination. Over time these experiences challenge ideas brought to the mission field. As Choi notes,

> Our organizations work incarnationally, hands on, working in the midst of the people. It changes your perspective from when you start to several years in on the ground…As they live and work amongst the people, a lot of them let go of their more firmer convictions.[18]

---

endorsed John McCain for president he declared, "our support for Israel has absolutely nothing to do with an end times, prophetic scenario" (Boyer 2008).

18 When I asked respondents to review my paper to ensure I had quoted them accurately, Mr. Choi requested [personal communication 9/05/08] that I clarify what he meant by 'let go.' "I want to make sure that you and your readers understand that our workers are not throwing out any biblical truth in the midst of living in very different contexts. What they are needing to do is challenge their prior held convictions and seeking to understand truth differently than in their home context norm … the truth that Jesus taught

Missionaries also have to deal with the baggage of a politicized evangelism back home. As Charles Dyer noted, evangelicals' ardent public support for Israel in domestic politics colors how potential converts view missionaries in the field.

> I know Moody graduates working in the Middle East and it comes across [to potential converts] that Christians are pro-Israel and anti-Arab. It looks like the God of Christianity is aligned with the God of Judaism against the God of Islam.

Dyer stressed that Moody's goals were not political, but they were often read that way in the Middle East.

In light of Dyer's comments, I asked him and my other informants what influence (if any) the apocalyptic rhetoric of oft-quoted evangelicals such as John Hagee, the late Jerry Falwell, and Pat Robertson had on mission work in the Middle East. That is, did the insistence of these preachers that the end times are nigh and that peace is impossible affect the ability of missionaries to do their work? All of the informants I spoke to suggested the rhetoric was unhelpful. Dyer argued:

> Those individuals you mentioned [Pat Robertson, Jerry Falwell, and John Hagee] are masters of the 20 second soundbite. I usually agree with 90 percent they say, but cringe at 10 percent. Falwell knew the nuances [of dispensationalism] but he also knew how the press worked and knew he needed a 20 second soundbite to get on TV. We're so polarized in this country because soundbites don't nuance.

Choi made a similar assessment:

> I can't say its neutral, it is probably more on the negative side. The reason is because, well, what they say may have elements of truth to it, but let's mention that in the context of the other things we've been commanded to preach, teach and live out.

Choi also noted that while the apocalyptic rhetoric of evangelical leaders like Falwell, Robertson, and Hagee may be the most dominant within the evangelical movement, they were not necessarily the most representative.

> They're loud, they also have a presence in the media. But, I would say they aren't reflective of the whole population. I would be hesitant to say they reflect how we would see our main responsibilities [as workers at Frontiers].

When I asked mission leaders how they dealt with the limitations that politics placed on their work, they offered theological and practical strategies.

---

is the guiding principle, but they have to rub away some of the cultural and personal-centric garments."

Theologically, my informants suggested a change in emphasis. Choi observed, for example, that:

> The nearness of the apocalypse is up for debate. But, there are other clear commands that are not debatable. For sure, Jesus does say the end will come. That's not debatable, but there are other responsibilities we must fulfill before we preach that, or at least in company with it.

Newell also noted that in his organization there is "less emphasis on the dogma." And, Dyer argued that what is important is:

> Knowing what are the majors and what are the minors. It's not getting them to dress western dress and adopt Western norms. It's about developing a relationship with God through Christ.

My informants also suggested practical solutions. These generally involved respecting the local culture and developing human connections to the people they were trying to convert. Dyer noted, for example,

> Thankfully in the Middle East people like Americans, even though they don't like America. If an American comes over and doesn't try to make them American, you can find ready acceptance.

Likewise, Choi argued,

> Its not what you know but who you are that will make a difference in the field Who you are in evaluating right and wrong, how you are with your family. It helps us dispel the presuppositions.

As these, and earlier comments suggest, there is a notable political gap between the eschatological views of evangelical leaders like Hagee, Robertson, and Falwell, on the one hand, and many evangelical missionaries on the other. What, then, does this means for the coherence of the evangelical geopolitical vision? At one level, the gap suggests the evangelical geopolitical vision is less coherent than scholars and commentators generally present it as being. It might, in fact, be more moderate than often assumed, and it may further suggest that the rigid discourse on Middle Eastern politics and the end times is losing its dominance within evangelical circles.

However, it is also worth noting that most discourses contain internal variance and contradiction. What is important is not whether there are cracks or fault lines within a given discourse, but how they are manifested. And, while my informants were willing to politely disagree with evangelical leaders who use apocalyptic rhetoric to lay out 'evangelical' positions on the Middle East, it remains to be seen whether their voices will coalesce into a cohesive, and competing discourse in its own right.

In July 2007 a counter-discourse seemed possible when a number of prominent evangelicals, some with attachments to historically dispensationalist organizations, sent a letter to President Bush voicing their support for a Palestinian state (Goodstein 2007). In the letter the signatories noted that Jews and Palestinians both have legitimate claims to the land. And, they acknowledged that both sides, rather than just the Palestinians, had committed violence in the region. While they maintained the Israeli right to security, they argued that a two state solution might bring peace to the region. It was, in many ways, a momentous occasion in evangelical circles. However, one letter does not a coherent opposition make. The relatively muted role of the group since may indicate that evangelical missionaries have chosen not to make too much an issue of their divergent views. They may believe that fighting such a battle would not be worth the cost—a public airing of internal differences that could undermine evangelical influence in political circles and incite ridicule of their leaders in the press. It is also possible that missionaries with divergent views find it difficult to completely part from the orthodoxy that has for so long informed evangelical geopolitics. Indeed, none of the groups examined here signed the letter. Adopting a flexible pose could be seen as a way to avoid making a firm stand for or against that orthodoxy.

In sum, such flexibility is certainly meaningful, but it is not necessarily transformative. Debate may represent the emergence of a counter-discourse, but it may also represent a form of cover for those not willing to select a side, or to fight for a view out of the mainstream of the movement.

Conclusion

The evangelical geopolitical vision is generally presented in newspapers, watchdog accounts, and academic work as a rigid view of the world. Evangelicals believe we are living in the end times. The formation of Israel marks the (partial) fulfillment of God's covenant with his chosen people, and prophesied events, including the Rapture and the Tribulation, are likely to occur in our lifetime. Christians must support Israel, oppose peace efforts in the Middle East, and protect the sanctity of Jerusalem so that Christ may return to earth to build a throne for his millennial kingdom.

This chapter has examined the degree to which the relative fatalism of dispensationalism and its concomitant geopolitical view clash with the optimism and change-agent imperative of evangelism. In particular, this chapter has examined how missionaries grapple with the competing directives of dispensationalism (that war in the Middle East is prophetic and Christians should not try to stop it) and evangelism (to convert individuals and in so doing improve their societies). I call these competing imperatives the Middle East mission paradox in this chapter.

The data collected for this paper, though partial, suggest that the mission field, and its relationship to evangelical geopolitics, is more nuanced than generally presented. To begin with, mission leaders do not see a paradox between their theology and their evangelism in the Middle East. They believe that the Bible is crystal clear about the need to evangelize while biblical references to the end times are more ambiguous and subject to interpretation. As such, missionaries argue that their evangelism does not work against prophecy even if the end times seem to be near. The leaders of the organizations interviewed for this paper also suggest that they prefer theological flexibility to theological dogma about the end times. Missionaries in the field are given wide leeway to interpret the connection between eschatology and political events as they see fit. The more important goal is helping potential converts become Christian. Mission leaders have also been willing to stray from evangelical orthodoxy regarding the morality of Israeli settlements, and they have been willing to reject, albeit quietly, the apocalyptic rhetoric of evangelical leaders like John Hagee, Pat Robertson, and the late Jerry Falwell. In fact, some believe that the heated rhetoric of leaders like these actually makes the conversion process more difficult for missionaries on the ground. It is difficult to know, however, whether these divergent views will coalesce into a coherent discourse that challenges the orthodoxy of the more apocalyptic version of evangelical geopolitics. Indeed, some mission leaders appear comfortable occupying paradoxical space, emphasizing their theology in dispensationalist circles while distancing themselves from it in others.

These findings have import for a number of debates, in and out of geography. Within geography, recent work by political geographers has been important for identifying and unpacking a uniquely evangelical geopolitical vision (Dittmer 2007; Sturm 2006). The data here build on that work while also presenting some of the internal divides within it. In particular, mission leaders within evangelical circles are less doctrinaire than evangelism's leaders suggest. This suggests that most commentators on the evangelical phenomenon (whether in the media or academia) are extrapolating the views of a few well known men (e.g. Hagee, Robertson, and Falwell) to the entire movement. This research indicates, however, that these men, though powerful, are not necessarily representative of the majority of those in the evangelical fold. More fruitful avenues of inquiry should focus on identifying internal variation and analyzing the power dynamics between different (and perhaps competing) viewpoints. In the case of the evangelical view on Israel, how and why did the views of Hagee, Falwell, and Robertson become dominant, even when they are not necessarily the majority view?

This work also complicates an ongoing debate in the field of international relations about the 'Israel Lobby' (Mearsheimer and Walt 2004). For those scholars who believe in the existence of the Lobby (a point not universally accepted, see Gergen 2006; Judis 2007), missionaries are often seen as the Lobby's most devoted foot soldiers. After September 11, 2001, many on the left were highly critical of Christian missionary efforts in Afghanistan and Iraq. Mission work was depicted as a softening agent—preparing the region for U.S. imperial actions to come. While missionaries may yet serve this role (wittingly or not), the evidence here suggests that mission leaders are less blindly supportive of Israel than the leaders typically quoted 'back home,' and decidedly less apocalyptic as well. That is, evangelical missionaries may actually serve to destabilize, however minimally, the Lobby's cohesiveness and its application on the ground. Some may actively work to support people or causes opposed by evangelical leaders back home. Others may simply refuse to publicly endorse the geopolitical views of evangelicals while overseas.

These findings also resonate with Mark Lacy's (2008) view that Mearsheimer and Walt's focus on the Israel Lobby is misplaced, and thus fetishized. My findings here indicate that the approach of mission groups presented in this chapter are too diffuse to act as a cohesive 'lobby.' Indeed, Lacy is correct that the wider intellectual movement of neoconservatism would be a better focus of analysis.

Finally, this work adds to literature on the history and evolution of the New Right. Some scholars have suggested that the wider New Right movement is coming undone. Dissatisfaction with corruption in the Republican-led Congress, frustration over the course of the Iraq War, and suspicion of the 2008 Republican presidential candidate Senator John McCain, are pulling evangelicals out of the state movement alliance they helped form in the late 1970s (Dionne 2008). Other scholars argue that the movement is changing, noting that so called second generation evangelicals are less conservative than their first generation counterparts (Wilcox 2007). Indeed, while the work here can not speak to the health of the wider state-movement alliance that is the New Right, it does suggest that mission leaders may well represent the more flexible, less dogmatic approach of the second generation evangelical. Hopefully, this research will inspire future work on mission leaders and missionaries. A larger 'n' study of mission leaders and missionaries would allow for more precise and detailed portrait of inter-evangelical diversity, and as a result more precise conclusions about the evolution of the New Right and evangelical geopolitics.

References

Bell, D. and Binnie, J. (2005), 'Editorial: Geographies of Sexual Citizenship', *Political Geography* 25:8, 869–73.

Berlet, C. and Lyons, M. (2000), *Right Wing Populism in America* (New York: Guilford Press).

Bialasiewicz, L., Campbell, D., Elden, S., Graham, S., Jeffrey, A., and Williams, A. (2007), 'Performing security: The imaginative geographies of current U.S. strategy', *Political Geography* 26(4), 405–22.

Bock, D. (2003), 'Some Christians See a Road Map to End Time', *Los Angeles Times*, 18 June. Commentary.

Boyer, P. (2008), 'Party Faithful: Can the Democrats get a Foothold on the Religious Vote?' *The New Yorker*. Available online at: http://www.newyorker.

com/reporting/2008/09/08/080908fa_fact_boyer?currentPage=1, accessed 29 September 2008.
Brindle, W. (2001), 'Biblical Evidence for the Imminence of the Rapture', *Bibliotheca Sacra* 158(630), 138–51.
Brinkley, A. (1998), *Liberalism and Its Discontents* (Cambridge: Harvard University Press).
Butler, J. (1990), *Gender Trouble: Feminism and the Subversion of Identity* (New York: Routledge).
Cimino, R. (2005), '"No God in Common:" American Evangelical Discourse on Islam after 9/11', *Review of Religious Research* 47:2, 162–74.
Cornerstone Church (2008), Homepage. http://www.sacornerstone.com/, accessed 22 September 2008.
Dallas Theological Seminary (2008), 'Doctrinal Statement', Available online at: http://www.dts.edu/about/doctrinalstatement/, accessed 15 August 2008.
Diamond, S. (1995), *Roads to Domination: Right Wing Movements and Political Power in the United States* (New York: Guilford Press).
—. (1999), *Spiritual Warfare: The Politics of the Christian Right* (Boston: South End Press).
Dionne, E.J. (2008), *Souled Out: Reclaiming Faith and Politics after the Religious Right* (Princeton: Princeton University Press).
Dittmer, J. (2007), 'Of Gog and Magog: The Geopolitical Visions of Jack Chick and Premillennial Dispensationalism', *ACME: An International E-Journal for Critical Geographies* 6:2, 278–303.
Dreher, R. (2002), 'Evangelicals and Jews Together', *National Review Online*, 5 April. http://www.nationalreview.com/dreher/dreher040502.asp, accessed 21 July 2008.
Dyer, C. (2003), *The Rise of Babylon: Sign of the End Times* (Chicago: Moody Press).
—. (2008), 'Dividing Jerusalem', http://web.mac.com/charleshdyer/Site/Blog/Entries/2008/2/10_Dividing_Jerusalem.html, accessed 7 October 2008.
Freston, P. (2001), *Evangelicals and Politics in Asia, Africa and Latin America* (Cambridge: Cambridge University Press).
Frykholm, A.J. (2004), *Rapture Culture: Left Behind in Evangelical America* (Oxford: Oxford University Press).
Gallaher, C. (1997), Identity Politics and the Religious Right: Hiding Hate in the Landscape. *Antipode* 29:3, 256–77.
—. (2003), *On the Fault Line: Race, Class, and the American Patriot Movement* (Lanham, MD: Rowman and Littlefield).
—. (2007), 'The Role of Protestant Missionaries in Mexico's Indigenous Awakening', *Bulletin of Latin American Research* 26:1, 88–111.
Garrard-Burnett, V. and Stoll, D. (eds) (1993), *Rethinking Protestantism in Latin America* (Philadelphia: Temple University Press).
Gergen, D. (2006), 'An Unfair Attack', *U.S. News and World Report*. 3 April. Opinions Page.

Gregory, D. (2004), *The Colonial Present*. (Oxford: Blackwell).

Goodstein, L. (2007), Coalition of Evangelicals Voices Support for Palestinian State, *The New York Times*. 29th July. Available online at: http://www.nytimes.com/2007/07/29/us/29evangelical.html, accessed August 7, 2008.

Hagee, J. (2007), *Jerusalem Countdown: A Prelude to War*. Updated and revised edition. (Lake May, FL: FrontLine).

Hanson, T. (2007), John Hagee Keynotes Night to Honor Israel at Furman. *The Times Examiner*. 5 March. Available online at: http://www.timesexaminer.com/content/view/106/1/, accessed 14 August 2008.

Hunter, A. and Chan, K-K. (1996), *Protestantism in Contemporary China* (Cambridge: Cambridge University Press).

Jerusalem Post Staff Writer. (2006), 'Hagee, Falwell Deny Endorsing "Dual covenant" Theology', *The Jerusalem Post*. 3 March. Page 1.

Judis, J. (2007), 'Split Personality', *National Review Online*, 8 February. http://www.carnegieendowment.org/publications/index.cfm?fa=printandid=19028, accessed 24 July 2008.

Lacy, M. (2008), 'A History of Violence: Mearsheimer and Walt's Writings from "An Unnecessary War" to the Israel Lobby Controversy', *Geopolitics* 13:1, 100–19.

Levine, D. and Stoll, D. (1997), 'Bridging the Gap between Empowerment and Power in Latin America', in Rudolf, S.H. and Piscatori, J. (eds) *Transnational Religion and Fading States*, (Westview: Bounder), 63–103.

Mearsheimer, J. and Walt, S. (2006), 'The Israel Lobby', *London Review of Books*, 23 March.

Rose, G. (1991), *Feminism and Geography: The Limits of Geographical Knowledge* (Minneapolis: University of Minnesota Press).

Sparke, M. (2005), *In the Space of Theory: Postfoundational Geographies of the Nation-state* (Minneapolis: University of Minnesota Press).

Sturm, T. (2006), 'Prophetic Eyes: The Theatricality of Mark Hitchcock's Premillennial Geopolitics', *Geopolitics* 11:2, 231–55.

—. (2008), 'The Christian Right, Eschatology, and Americanism: A Commentary on Gerhardt', *Environment and Planning D: Society and Space*, 26:5, 929–34.

Tamcke, M. and Marten, M. (2008), *Christian Witness between Continuity and New Beginnings: Modern Historical Missions in the Middle East* (Lit Verlag).

Weber, T. (1979), *Living in the Shadow of the Second Coming: American Premillennialism* (New York: Oxford University Press).

—. (2004), *On the Road to Armageddon: How Evangelicals Became Israel's Best Friend* (Grand Rapids, MI: Baker Academic).

Wilcox, K. (2008), *Social and Political Attitudes among Evangelical Christian Youth*. Unpublished senior honors thesis, American University.

Afterword:
The Geopolitics of End Time Belief in the Era of George W. Bush

Paul Boyer

What were George W. Bush's core beliefs? To the casual observer, Texas, Big Oil, and Major League Baseball might spring to mind. Though a self-proclaimed 'born again' Christian, Bush revealed little about the specifics of his faith. Throughout his years in public life, however, Bush did frequently assert that his political career, from the Texas governorship to the presidency, represented a direct fulfillment of a supernatural design. When still in Texas politics, he said: "I could not be governor if I did not believe in a divine plan that supersedes all human plans" (Bush's Messiah Complex 2003). Such pronouncements increased in earnestness following the September 11, 2001 attacks on the Pentagon and the World Trade Center, seven months into Bush's first term. Indeed, after 9/11 Bush repeatedly articulated an apocalyptic worldview that linked his personal destiny and that of the United States to God's cosmic plan for establishing righteousness in a wicked world. As he memorably vowed in his post-9/11 address to Congress and the nation, America would not only track down those responsible, but "rid the world of evil"—the ultimate goal of all who embrace the apocalyptic worldview (Bush 2001). This worldview, originating in mystical texts of the ancient Middle East, adapted by Jewish writers, and reaching its apogee in the Apocalypse of John, the concluding book of the New Testament, remained alive and well in the Bush White House.

As Bush's foreign-policy pronouncements took on an increasingly apocalyptic aura, he cast himself and the United States as central actors in a great world struggle between righteousness and wickedness. Although he stopped speaking of a 'crusade' when informed that Muslims associated the phrase with invading European armies marching under the banner of the cross, the underlying mindset remained. Embracing the absolute polarities of the apocalyptic genre, Bush insisted that in this struggle, no middle ground was tolerable: "You're either with us or against us in the fight against terror", he told the world in November 2001 (CNN 2001). His January 2002 State of the Union address called for a united front against the 'axis of evil'—Iraq, Iran, and North Korea (Bush 2002). Journalist Bob Woodward, interviewing Bush before the Iraq invasion, concluded that he was "casting his mission and that of the country in the grand vision of God's master plan" (Macintyre 2003).

A similar apocalyptic worldview pervaded the upper echelons of Bush's administration, including the neoconservative advisers who promoted the invasion of Iraq and set U.S. policy in other areas. In December 2003, when China proposed negotiating strategies for dealing with North Korea's nuclear-weapons program, Vice President Dick Cheney, widely viewed as the power-behind-the-throne in the Bush White House, snapped: "We don't negotiate with evil; we defeat it" (Kessler 2004). In a 2006 address to the American Legion, a veterans' organization, Defense Secretary Donald Rumsfeld accused Iraq War critics of "moral or intellectual confusion about what is right or wrong" (Boyer, Peter 2008, 31).

The baldest articulation of this religious apocalyptic undercurrent in the Bush Administration was that of Lt. General William G. (Jerry) Boykin, deputy undersecretary of defense for intelligence and a confidant of Donald Rumsfeld. In talks in evangelical churches in 2002–03, Boykin, in full-dress uniform, repeatedly voiced his basic message: Muslim terrorists hate America "because we're a Christian nation, because ... our roots are Judeo-Christian ..., and the enemy is a guy named Satan." During a tour of duty in Somalia, Boykin later reported, he taunted a captured Muslim warlord by dismissing Allah as a false god. "I knew my God was bigger than his", he boasted; "I knew that my God was a real God and his was an idol." The terrorists would "only be defeated if we come against them in the name of Jesus", Boykin told a church group in Oregon, adding: "[President Bush] is in the White House because God put him there." Amid protests from Muslim leaders and domestic critics, Bush distanced himself from Boykin's incendiary remarks. But Rumsfeld resisted calls to reassign him, and Boykin retained his Pentagon post (General Casts War in Religious Terms 2003; Thompson 2003).

Despite Bush's disavowals, many Americans clearly understood the 'war on terror' in religious terms. In CBS News Polls conducted between February 2002 and April 2006, the proportion of respondents holding an 'unfavorable' view of 'the religion called Islam' *rose* from 33 percent to 45 percent. Similarly, in ABC News/*Washington Post* surveys in 2002 and 2006, the number of respondents who agreed that 'mainstream Islam encourages violence against non-Muslims' increased from 14 percent to 33 percent (PollingReport.com).

Popular Prophetic Belief in Bush's America

Indeed, the apocalyptic configuration of the administration's post-9/11 foreign policy both reflected and strengthened the eschatology of millions of American evangelicals—a very substantial portion of the population. Polls in the 1980s found that some 40 percent of Americans viewed the Bible as "the actual word of God ..., to be taken literally, word for word", and more than 60 percent had "no doubt" that Jesus will return to Earth again (Boyer 1992, 2). In a 1996 opinion survey, 42 percent of the U.S. respondents agreed with the statement: 'The world will end in a battle in Armageddon between Jesus and the Antichrist.' Of self-identified 'conservative Protestants', 69 percent endorsed this statement (Canada/ U.S. Religion and Politics 1996, 80). In a 1998 *USA Today*/Gallup Poll, 39 percent of the respondents found it 'very likely' or 'somewhat likely' that 'the World will come to an end because of Judgment Day or another religious event in the next century' (*USA Today*/Gallup Poll).

As the Introduction to this volume notes, vast numbers embrace dispensational premillennialism, a system of prophetic interpretation formulated by the 19th century British churchman John Nelson Darby, a founder of the Plymouth Brethren, and widely disseminated in America by Darby himself and by influential U.S. preachers and theologians, including Cyrus Scofield, whose 1909 *Reference Bible*, published by Oxford University Press, gave a dispensationalist gloss to many passages. According to Darby, the present epoch, or 'dispensation', will soon end with the Rapture, when all true believers will join Christ in the air. Then will come the Tribulation, a seven-year interval when the Antichrist will rule the world, slaughtering all who reject his rule. As Antichrist's armies gather at Har Megiddo (Armageddon), an ancient battle site in Israel, to confront a vast army approaching from the east (Rev. 9:14–16), Christ and the raptured saints will return to vanquish the Antichrist, ushering in Christ's millennial reign (Boyer 1992, 86–100). Many contemporary dispensationalists further believe that the Bible also foretells a series of end time 'signs' signaling the imminence of the Rapture, and that with proper decoding these prophesies can be linked to current events, as well as to specific cities, nations, and world regions.

Much evidence underscores the pervasiveness of prophecy belief in contemporary America. Paperbacks linking the prophecies to current events sell millions of copies. *The* non-fiction bestseller of the entire decade of the 1970s, Hal Lindsey's *The Late Great Planet Earth*, a popularization of dispensationalism, cited biblical passages allegedly foretelling a coming nuclear war and Russia's invasion of Israel and subsequent destruction. In a chapter called "The Yellow Peril", Lindsey identified the eastern army destined for annihilation at Armageddon as Chinese communists. Adding geopolitical specificity to his work, Lindsey included two maps showing the precise routes by which Russian, Libyan, and "pan Arab" armies will attack Israel (Lindsey 1981, 70, 144, 148).

The books of many other prophecy popularizers flew off the shelves in the late 20th and early 21st centuries. Also spreading the prophetic word were internet websites; televangelists such as Jerry Falwell, Pat Robertson, Oral Roberts, and other luminaries of the electronic church; glossy magazines such as *Endtime; Moody Monthly, Midnight Call,* and *Prophecy in the News;* 'prophecy conferences' held in churches and resort hotels; and thousands of prophecy-preaching pastors in evangelical and Pentecostal churches, including suburban megachurches that spread across America in these years (Boyer 2008).

Left Behind (1995), a Rapture novel by Tim LaHaye and Jerry B. Jenkins, launched a phenomenally successful venture in prophecy popularization. Eventually extending to thirteen volumes published over the next twelve years, well into the Bush era, the series sold more than seventy million copies and produced a host of spin-off products, including a movie, DVDs, a video game, a juvenile version, calendars, greeting cards, t-shirts, and an internet chat room for fans.

Millions of Americans in the Bush years, in short, watched the TV news, read their newspapers, consulted internet news sources, and scanned world atlases from a prophetic perspective. This, in turn, had significant implications for American politics and international relations.

Eschatological Geopolitics in a Post-Cold War World

Ever attuned to shifting geopolitical realities, post-Cold War prophecy popularizers focused special attention on the Middle East, including Islam's prophetic role. This was, in fact, an ancient theme. The medieval Crusaders battled to free Jerusalem from Muslim rule to prepare for Christ's millennial kingdom. Later interpreters linked the Ottoman Empire to the Antichrist. This theme faded with the Ottoman collapse and the rise of the Soviet Union after World War I, but it surged back after the U.S.-led 1991 Persian Gulf War to expel Iraqi invaders from Kuwait. The Jews for Jesus organization, for example, published full-page newspaper ads proclaiming that the Iraqi strongman Saddam Hussein "represent[s] the spirit of Antichrist about which the Bible warns us" (Boyer 1992, 326–331; Boyer 2002, 326 (quoted passage).

Saddam's plans to rebuild ancient Babylon roused special interest, since Revelation portrays Babylon as the epitome of evil and prophesies its fiery destruction. One writer, calling Saddam's project "thrilling proof that Bible prophecies are infallible", went on: "If Babylon will be destroyed in the end times, who will destroy it? ... [The] United States is a major world power—how could it *not* play a major role in the last days?" (Dyer 1991, 14, 165, 166).

The Bush Administration's insistence on Iraq's alleged weapons of mass destruction and involvement in the 9/11 attacks gave this theme added urgency, and the 2003 Iraq invasion found prophecy believers well primed, viewing the action as part of God's end time plan. Hal Lindsey's website featured a cartoon of a warplane emblazoned with a U.S. flag and a Jewish Star of David and carrying a missile targeting 'Saddam.' The caption quoted Zechariah: "In that day I will ... destroy all nations that come against Israel" (Lindsey 2002). In a November 2002 sermon on the Trinity Broadcasting Network, the Texas televangelist John Hagee called the looming U.S.-led invasion "the beginning of the end", and a sign of Christ's second coming. As Hagee's sermon ended, Tom DeLay, the Republican majority leader of the House of Representatives, rose to proclaim: "[W]hat has been spoken here tonight is the truth from God" (The End is Nigh 2002).

The post-9/11 demonization of Islam to which General Boykin contributed also resonated with dispensationalist believers. Prophecy writer Michael Evans called Islam "a religion conceived in the pit of hell" (Evans 2003, 79). Franklin Graham, son of famed evangelist Billy Graham and himself a prominent evangelical leader, called Islam "a very evil and wicked religion" (Should Christian Missionaries

Heed the Call in Iraq? 2003). In *Jerusalem Countdown* (2006), John Hagee proclaimed: "This is a religious war that Islam cannot and must not win. ... The end of the world as we know it is rapidly approaching... Rejoice and be exceedingly glad—the best is yet to be" (Russell 2006). Such pronouncements from religious figures claiming the authority of Bible prophecy helped shape the increasingly negative public perceptions of Islam documented in the polling data, and provided a continuing core of support for President Bush as the Iraq venture bogged down in a morass of sectarian conflict.

Along with the attacks on Islam came fervent support for Israel's most hard-line and expansionist elements. Many U.S. prophecy popularizers cited biblical passages to buttress their backing of Israel's occupation of the West Bank, their opposition to a Palestinian state or shared governance of Jerusalem, and their enthusiasm for rebuilding a Jewish Temple on Jerusalem's Temple Mount, currently the site of two sacred Islamic shrines. God's promise to Abraham of all the land from the Euphrates to "the river of Egypt" (Genesis 15:18) was much quoted.

Like the anti-Islamic motif, this theme, too, had deep roots in dispensationalist thought. Indeed, the late nineteenth century Chicago real-estate tycoon William Blackstone, a leading dispensationalist, was among the earliest Zionists. Prophecy believers welcomed the 1917 Balfour Declaration and the establishment of the nation of Israel in 1948 as momentous end time signs (Boyer 1992).

The Bush Administration's short-lived 2003 peace initiative, calling for cutbacks in Israel's West Bank settlements, stirred fierce opposition among U.S. prophecy believers. The Christian Right leader Gary Bauer told AIPAC, the American-Israel Political Action Committee: "We believe God owns the land and he has deeded it to the Jewish people, a deed that cannot ... be amended – even by a president" (quoted in Schrag 2005, 6). Michael Evans agreed: "The only Road Map for peace is the Bible.... God gave [the Jews] that land and forbade them to sell it" (Evans 2003 94, 104). House majority leader Tom DeLay rushed to Israel to assure the Knesset of his unwavering support.

Millions of Americans who view the Bible as God's inerrant word embrace this view of God as the great real-estate agent in the skies, establishing national boundaries and apportioning land to favored groups. For them, the Genesis grant to Abraham and other biblical passages relating to Israel, Jerusalem, or Ishmael (Abraham's son by the maidservant Hagar, and mythic progenitor of the Arab peoples) trump all other considerations. Anyone who challenges these passages or views the Israeli-Palestinian dispute as open to negotiation and compromise is quite literally challenging God's supreme authority (Weber 2004).

In *Final Dawn over Jerusalem* (1998), John Hagee denounced any peace plan involving even partial Jewish withdrawal from the West Bank or shared governance of Jerusalem. Wrote Hagee: "The man or nation that lifts a voice or hand against Israel invites the wrath of God. ... There can be no compromise regarding ... Jerusalem, not now, not ever ... Israel is the only nation created by a sovereign act of God, and He has sworn ... to defend his Holy City" (Hagee 1998, 34, 131, 150).

The 2006 Washington rally of Hagee's political lobby, Christians United for Israel (CUFI), drew some 3,500 participants. Among the notables attending were Kenneth Mehlman, chair of the Republican National Committee and President Bush's 2004 campaign manager; Israel's ambassador to the United States; Congressman Tom DeLay; Republican senators Rick Santorum of Pennsylvania, a conservative Catholic, and Sam Brownback of Kansas, an evangelical Protestant. President Bush, a friend of Hagee's from his days as Texas governor, sent greetings (Posner 2006; For Evangelicals, Supporting Israel is God's Foreign Policy 2006). Declared Hagee: "What we have done is united all of this evangelical horsepower and said, 'We're not just going to Washington to stand on the grass and sing *Amazing Grace*. We're going into the halls of Congress to see the senators and … congressmen face-to-face and to speak to them about our concerns for Israel'" (Russell 2006).

At CUFI's 2007 Washington rally, Connecticut senator Joseph Lieberman, the 2000 Democratic vice-presidential nominee, compared Hagee favorably to Moses. Tom DeLay, asked by a reporter how his prophecy beliefs related to his support for Israel, responded: "[The Rapture] is what I live for…. And obviously we have to be connected to Israel … to enjoy the Second Coming of Christ" (Perlstein 2007).

Keenly sensitive to these powerful and eschatologically-rooted ideological currents in his evangelical base, as promulgated and amplified by media figures like Lindsey, Hagee, Falwell, Robertson, and others, Bush soon downplayed the 2003 peace initiative. Only near the end of his second term, and again in a low-key fashion, did Bush again propose renewed peace negotiations. In a 2008 address to the Knesset, Bush whole-heartedly supported Israel, insisted on the "moral clarity" of its claims, and denounced "appeasers" who advocated negotiations with Israel's hostile neighbors (Bush 2008).

Contemporary prophecy popularizers readily acknowledge that demands by Israeli hard-liners and their Christian Zionist supporters for Israel's exclusive control of Jerusalem and the West Bank, and even for a rebuilt Jewish temple on Temple Mount, could trigger horrendous conflict. Indeed, their scenario *assumes* just such an apocalyptic denouement to human history. They quote God's curse on Ishmael and God's blessing on Abraham's legitimate son Isaac (as recorded in the Hebrew scriptures) as proof of eternal conflict between Arabs and Jews. Like all apocalyptic struggles, they insist, this could end only in victory for one side, annihilation of the other (Boyer 1992, 193–199). In Hal Lindsey's 1996 prophecy novel *Blood Moon,* Israel responds to a threatened attack by launching a preemptive nuclear assault that destroys "every Arab and Muslim capital … along with the infrastructure of their nations" (Lindsey 1996, 312). Global genocide, from Morocco to Indonesia, finally fulfills God's end time plan.

Some Israeli leaders welcomed this support. When Prime Minister Benyamin Netanyahu, leader of the hard-line Likud party, visited the United States in 1998, he met first with Jerry Falwell, then went to Washington to see President Bill Clinton (Cohen 1998). On a later trip, Netanyahu told an audience of 3,000 prophecy believing Christian evangelicals: "We have no greater friends and allies than the people sitting in this room" (Stillman 2002)

But prophecy believers' support for Israel and its territorial aspirations had a darker side. Although Hagee and other Christian Zionists downplay the theme, dispensationalist doctrine holds that Antichrist will kill most Jews during the Great Tribulation, with only a tiny remnant who acknowledge Jesus as the Messiah escaping eternal damnation (Boyer 1992, 208–217). Paige Patterson, president of Southwestern Baptist Theological Seminary and the 1998–99 president of the Southern Baptist Convention (America's largest Protestant denomination), has asserted that during the Tribulation the Antichrist will persecute the Jews "almost to the point of extinction", unleashing "the worst wave of anti-semitism we've ever seen" (Forum: Eschatology nd).

Dispensationalists also teach that throughout history, God has harshly 'chastised' the Jews, his chosen but wayward people. Hal Lindsey in *The Late Great Planet Earth* described the Nazi Holocaust as "God's woodshed"—part of God's plan to punish the Jews for their unfaithfulness, and drive them back to their promised land (Lindsey 1981, 35). John Hagee was simply reiterating this familiar theme in the later 1990s when he described Hitler as a 'hunter' chosen by God to fulfill the divine plan for the Jews.[1] When the 2008 Republican presidential candidate John McCain sought Hagee's endorsement, he knew little of the Texan's eschatology. When Hagee's comments became a campaign liability, McCain dismissed theme as 'crazy', obviously unaware that Hagee was simply elaborating an end time belief system embraced by millions of American evangelicals.

As the dispensationalists' preoccupation with Israel and its boundaries makes clear, geography looms large in this belief system. The specific end-time role assigned to various nations and regions has shifted over time, reflecting evolving political realities, but the geopolitical component has remained strong. Europe, for example, has long figured prominently—and quite ominously—in the end time scenario. Citing prophecies in the Book of Daniel, interpreters have long taught that the Antichrist will arise in Europe, and organize a ten-nation alliance replicating the ancient Roman Empire. Believers thus followed with intense interest the post-World War II European-unification movement culminating in the European Union. Tim LaHaye's *The Europa Conspiracy* (2005) is a recent entry in this vast literature. The Bush Administration's disdain for European opinion during the build-up to the Iraq War, summed up in Donald Rumsfeld's contemptuous dismissal of 'old Europe', thus resonated with American prophecy believers' geopolitical worldview.

Russia, too, remains on prophecy believers' end-time map, though less prominently than during the Cold War. The Book of Ezekiel prophesies that a northern kingdom, 'Gog', will invade Israel and be utterly destroyed in consequence. During the Cold War, Hal Lindsey and other prophecy popularizers

---

1 Hagee also described the Roman Catholic Church as 'the Great Whore' of Revelation—another dispensationalist theme downplayed in recent years.

assumed that 'Gog' referred to Russia. With the Cold War's end and the shift of focus to the Middle East, this theme faded. It did not disappear, however, as many popularizers foresaw a combined Russian-Islamic attack on Israel in the last days (Boyer 1992, 152–180, 326).

The New World was unknown when the biblical prophecies were written, so finding America in prophecy has proven a challenge. Hal Lindsey thought it "quite possible" that an Ezekiel passage mentioning "those who dwell securely in the coastlands" refers to the United States (Lindsey 1981, 150). Others cite Isaiah 18, which speaks of a "land shadowing with wings" and "a nation meted out and trodden down, whose land the rivers have spoiled." Still others find America in the "great eagle" mentioned in Revelation 12 (Reagan nd). One ingenious interpreter suggested that the beast "with two horns like a lamb" in Rev. 13:11 refers to the buffalo that once roamed the West, and thus, allegorically, to the United States (America—Superpower of Prophecy 1996, 24–25). Another system of prophetic interpretation, British Israelism, assuming that one of the 'lost tribes' of Israel migrated to the British Isles, concludes that prophecies mentioning Israel really refer to Britain and its former colonies, including the United States.[2]

But if the quest for specific geographic allusions to the United States has proven frustrating, contemporary prophecy expositors (as Chapter 5 demonstrates) endlessly speculate on the nation's eschatological destiny. The rising tide of wickedness—pornography, gay marriage, legalized abortion, out-of-wedlock childbirth, Darwinian science textbooks, the banning of public-school prayer, and much else—many conclude, suggests that America no longer enjoys God's special favor, and that the United States will share the destruction prophesied for all evil-doers, with only a faithful remnant spared. At best, as Barkun suggests, contemporary America is viewed as the site of an apocalyptic struggle between good and evil, with the outcome far from certain. The New England Puritans' view of America as the New Zion, with a special prophetic destiny long survived, but today's prophecy believers see a tragically different reality.

Moving from a national to a global perspective, many prophecy believers view all supra-national movements, treaties, and organization as a prelude to the Antichrist's global rule. Pat Robertson's *New World Order* (1991) offered a deeply conspiratorial view of history, with the Masons, the Federal Reserve Board, the Rothschilds, the Rockefellers, the UN, the Beatles, the World Bank, and other organizations preparing humanity for Antichrist's dictatorship. The vast conspiracy, Robertson argued, even included the first U.S. Congress, which adopted the Great Seal of the United States with its motto from Virgil, 'Novo Ordo Seclorum', which he translated as 'New World Order.' The conspirators staged the entire Cold War,

---

2 A leading exponent of a version of British-Israelism encompassing the United States was Herbert W. Armstrong (1892–1986), a popular American radio preacher and founder of the Worldwide Church of God. In contemporary America, this belief is often espoused by racist, anti-Semitic, violence-prone, and anti-government movements (Boyer 1992, 191; Barkun 1994).

Robertson suggested, to divert attention from their nefarious aims. "[A] giant plan is unfolding", he wrote, "Everything perfectly on cue" (Robertson 1991, 35, 176)

This demonizing of the 'new world order' is deeply ironic, since the prophecy industry depends on the internet; the mass media; communications satellites; mass-market paperbacks; and multi-national outlets like Borders and WalMart to reach millions worldwide. The *Left Behind* series is available in thirty languages, from Afrikaans to Thai. After the success of this series, Tim LaHaye signed a $40 million-dollar contract for a new prophecy series with Bantam Dell, a division of the German media giant Bertelsmann (Donadio 2004). Hal Lindsey's first publisher, Zondervan, is now part of Rupert Murdoch's media empire. Mike Evans' publisher, WarnerFaith, is a division of AOL Time-Warner. In short, the end-time prophecy gurus are totally enmeshed in the same global system they denounce as a forerunner of the Antichrist's world empire.

All these themes converged in the *Left Behind* series, in which *all* institutions are suspect—the media; the government; most churches; and, of course, the UN. As the plot unfolds, the Antichrist, the charismatic Nicolai Carpathia, posing as a man of peace, becomes head of the UN, which he promptly moves to a rebuilt Babylon, so these twin embodiments of satanic power can be destroyed simultaneously, as Revelation prophesies. With this fictional series and its four-volume successor series, "Babylon Rising", Tim LaHaye, a veteran right-wing activist, gained a mass audience for his views and generated millions of dollars for his causes, including San Diego Christian College; the Creation Research Institute; the Council on Revival, dedicated to imposing 'biblical law' on the nation; Concerned Women for America, headed by his wife Beverly; the LaHaye School of Prophecy at Jerry Falwell's Liberty University; and the secretive Council for National Policy, which met with George W. Bush during the 2000 presidential campaign (Boston 2002).

As this overview of prophecy believers' geopolitical worldview suggests, the Bush Administration's nationalistic focus, ill-concealed contempt for the United Nations, suspicion of Europe; minimal engagement with the Israeli-Palestinian peace process; and manipulation of the domestic culture wars for political advantage closely matched the perspective of America's vast legions of prophecy believers. This is not to suggest that the administration's politics were explicitly formulated to fit a particular set of prophetic beliefs. But these beliefs permeated the political culture within which administration leaders operated, making it easier for them to pursue some policies, harder to pursue others. To the prophecy believers who comprised an integral part of Bush's political base, the homefront culture wars; the 'war on terror'; the Iraq conflict; Israel's West Bank expansion; and the administration's suspicion of the UN, 'old Europe', and international organizations and treaties all conformed closely to the end-time scenario they believe is rapidly unfolding.

Prophecy Belief and American Civic Culture

All this profoundly impacted America's civic culture. On one hand, prophecy belief, in principle at least, encourages passivity in the public sphere. *Individuals* can determine their destiny by accepting or rejecting Christ, but since history's overall trajectory lies in God's hands, beyond human control, civic engagement is pointless. Efforts to meliorate cultural differences, address global threats, or settle international conflicts by compromise are futile. Only Armageddon will bring resolution.

On the other hand, American prophecy believers also betray a characteristically activist impulse to fight for the right, and a patriotic desire to position the United States on God's side in the approaching final crisis. In a sense, they embrace both the *pre*millennial worldview, which assumes ever worsening conditions as the end approaches, and the more hopeful *post*millennial perspective, which holds that determined human effort can achieve a better, more righteous world in the present age. While insisting that all is in God's hands, politicized prophecy believers stoutly defend Israel's expansive claims, encourage opposition to the UN and international treaties, take up arms in America's domestic culture wars, and battle for policies that they believe may yet restore America to divine favor as the end approaches.

In all this activism, prophecy believers bring to bear the age-old apocalyptic worldview, in which all human history—indeed, all reality—dissolves into a cosmic struggle between opposing forces. This mindset inevitably leads to the kind of intransigent, absolutist, black-and-white thinking that George W. Bush so fully exemplified. Any talk of compromise, ambiguity, or 'shades of gray' simply obscures the underlying reality: the eternal conflict between the divine and the demonic. On one side are massed the agents of the Antichrist—Russia, Islam, the New World Order, the depraved and decadent persons undermining the foundations of American society, or all of the above. In opposition stand the forces of righteousness, embattled, but ultimately victorious, as promised in God's Word. While apocalyptic writers denounce the satanic forces seeking world domination, their own worldview is no less triumphalist, obsessively intent on ridding the world of evil. In the air-tight arena of apocalyptic belief, the opposing sides become, in reality, mirror images of each other.

Millions of Americans in the age of George W. Bush saw history's final crisis approaching—a cosmic denouement in which most of humanity will perish, but as a necessary cleansing, preparatory to the Millennium—and they formed political opinions and judgments about international affairs based on these beliefs. To be sure, most Americans are *not* dispensationalist prophecy believers, and Bush's policies roused vocal opposition as well as fervent support. But the mutually reinforcing symbiosis of apocalyptic belief and public policy in the Bush years was powerful indeed.

The Enduring Appeal of the Apocalyptic Worldview

Why does prophetic belief continue to attract so many? Certainly the apparent clarity of this worldview, in contrast to the ambiguities of everyday reality, helps account for its appeal. Resting on the authority of texts that many revere as divinely inspired, prophetic belief gives history meaning. History is not just one thing after another; it is teleological, moving toward a final consummation.

Further, Rapture doctrine offers believers the promise of escape from the terrors of Antichrist's reign, and once the world has been purged of evil at Armageddon, there looms a glorious future of righteousness, justice, and peace—humanity's age-old utopian dream. The word itself dates only from Thomas More's *Utopia* of 1516, but the longing it embodies arose in antiquity. And in the prophetic worldview, this longing, too, has a firm geographic rootedness. Thomas More set his ideal society on an island off South America. The author of Revelation pictured it as a city—but a city visible only to the eye of faith:

> And I John saw the holy city, new Jerusalem, coming down from God out of heaven, prepared as a bride adorned for her husband. And I heard a great voice out of heaven saying, Behold, the tabernacle of God is with men, and he will dwell with them, and they shall be his people.... And God shall wipe away all tears from their eyes; and there shall be no more death, neither sorrow, nor crying, neither shall there be any more pain, for the former things are passed away. (Rev. 21: 2–4)

Over the ages, this vision has drawn men and women to a prophetic view of the future, and it continues to do so. Unquestionably the Falwells, LaHayes, Hagees, and Boykins of our age have much to answer for. Their apocalyptic worldview exacerbates cultural conflicts, undermines efforts to address global problems, and snarls attempts to resolve regional disputes such as those that bedevil the Middle East. Yet, at the grassroots level, and the level of the individual psyche, this worldview binds together communities of believers and invests human history—and thus individual lives—with meaning and purpose.

Answering to such deep human needs, the prophetic beliefs that helped shape early-twenty-first century American politics and foreign policy exert a potent appeal. To fail to understand both the pervasiveness and the enduring strength of these belief is to fail to understand contemporary American culture and the geopolitical worldview of millions of American citizens.

A Glance Forward

What of the future? Will geopolitics remain as central to prophetic belief as it has been in recent decades? This, in turn, raises still broader questions: Will interest in Bible prophecy continue to loom large in American culture? Will premillennial dispensationalism persist as the most popular system of prophetic interpretation? And beyond these puzzlers lies the key public-policy question: Will these beliefs remain as potent politically as they have been in the recent past? While historians are notoriously chary about making predictions, I'll venture a few speculations, based on extrapolating recent trends into at least the short-term future.

As for the continued strength of prophetic belief, this depends on the future of the evangelical movement, which grew so dramatically after 1970. Given the vast expansion of evangelicalism's institutional base in recent decades, from home-schooling families to sprawling suburban megachurches, the movement seems likely to remain robust for the foreseeable future. U.S. history is littered with failed predictions that secularism, or at least liberal religion, would inevitably become dominant. Of course, secularism or religious indifference pervade sectors of American society, particularly in academia and the upper-socio-economic levels, but fundamentalism and evangelicalism—the soil in which prophecy belief flourishes—remain powerful. Millions of children and youth live wholly within a subculture in which evangelicalism is not only a doctrinal matter, but an all-encompassing social milieu.

As for prophetic belief itself, it, too, seems unlikely to fade away anytime soon. Given the centuries-long tenacity of these beliefs; the psychological comfort they provide; their rootedness in texts millions revere as divinely inspired; the mass-media megaphones that popularize them; and their social embeddedness in families, interest-group networks, and congregations across the nation (as well as in Latin America and sub-Saharan Africa where evangelical missionaries have penetrated), these beliefs seem well positioned to survive and flourish.

Further, given many believers' literal readings of the prophecies, the practice of applying them to contemporary geopolitical realities is likely to remain strong as well. The desire of pastors, televangelists, paperback popularizers (and their publishers), and even the managers of Internet websites to keep their prophetic message up-to-date reinforces this tendency to find eschatological significance in the current configuration of world politics.

The specifics of the interpretive scenario will doubtless continue to evolve, however, reflecting shifting patterns of global power and crisis spots. Since tensions in the Middle East, and Israel's conflict with the Palestinians and its Arab neighbors, show little sign of easing, this geopolitical region will surely remain a central preoccupation of prophecy believers. The ever-rising level of wickedness—another key end time sign—will surely continue to preoccupy prophecy popularizers as well.

China's emergence as a superpower suggests the likelihood of even more attention to those cryptic passages in Revelation describing an army of "two hundred thousand thousand" led by 'the kings of the East crossing the Euphrates River to do battle at Armageddon (Rev. 9:16 and 16:12). As the 'ProphecyHotLine' website noted in 2007: "The meteoric rise of China from a backward communist nation to an economic and military juggernaut appears to fit in with the biblical alignment of nations in the end times" (Event 10: China Flexes Its Awesome Power

2007). The prospect of 200 million soldiers tramping across China, Pakistan, and Afghanistan, crossing the Euphrates in Iraq (according to Revelation, the river will miraculously dry up), and then slogging on through Iran, Turkey, and Jordan to meet their doom at Armageddon boggles the skeptical mind. But true prophecy believers take seriously even the most improbable scenario if it can plausibly be grounded in some passage of scripture.

With the European Union increasingly integrated economically and politically, and emerging as a major trading bloc, the ancient theme of the Antichrist using a revived Roman Empire as his launching pad for world rule seems likely to receive increasing play as well. Since a ten-toed statue in the Book of Daniel allegorically represents successive world empires, prophecy expositors long held that the Roman Empire's end-time reincarnation would have ten member nations. With EU membership currently at twenty-seven countries, and more poised for admission, this notion has been quietly dropped.

As the prospect of catastrophic climate change related to global warming looms ever larger, along with other environmental concerns, one may also expect considerable attention to the cosmic disruptions foretold in Revelation, including a darkened sun and moon, falling stars, giant hailstones, monsters emerging from the earth, horrible sores on people's bodies, and the waters of the earth becoming "as the blood of dead men" (Boyer 1992, 331–37). Since Al Gore won the 2007 Nobel Peace Prize for his environmental warnings, the Internet has been abuzz with speculation that he could be the Antichrist, including calculations by which his name adds up to 666, the number of the Beast (Rev. 13:18) (Fortuna 2007). In short, given the popularizers' resourcefulness in adapting the prophecies to the twists and turns of current events, their end-time scenario will doubtless remain ominously timely and up-to-date in the years ahead.

But will prophetic belief, and the evangelical revival that sustains it, retain the political power that it wielded in recent decades? On this, the portents were mixed as George W. Bush's term ended, with some observers predicting a waning of the Religious Right's political clout. Younger evangelicals, some claimed, were rejecting aging leaders like Dobson, LaHaye, Robertson, and Falwell (until his death in 2007), and instead turning to figures like Jim Wallis of the liberal-leaning *Sojourners* movement and Rick Warren, pastor of California's 22,000–member Saddleback Church, who stresses social-justice themes and had the audacity to invite Barack Obama to speak. These new evangelicals, it was said, were downplaying the Religious Right's divisive cultural agenda, and instead embracing such issues as the environment, poverty, and peacemaking. If this liberalizing trend does, indeed, transform American evangelicalism, the politicization of prophecy that loomed large in the era of Lindsey, LaHaye, and others who used the dispensationalist scenario to buttress their cultural and geopolitical agenda, will likely diminish.

But all such reflections must remain speculative. As John C. Green of the Pew Forum on Religion & Public Life noted in 2008, the future of religion in American politics depends on many unknowns, including the intensity of the culture wars,

Republican and Democratic electoral strategies, trends within the evangelical camp, and the state of the economy (Green, 2008). Unlike prophecy believers, most historians do not profess to have a key to the future. If one does not view history as fixed and predetermined, then the future appears as an unfolding tabula rasa, shaped for better or worse by human actions and decisions, individual and collective.

References

'America—Superpower of Prophecy' (1996), Magazine-format publication, taken from the writings of Ellen G. White. Place of publication not specified.
Barkun, M. (1994), *Religion and the Racist Right: The Origins of the Christian Identity Movement* (Chapel Hill, NC: University of North Carolina Press).
Boston, R. (2002), '"Left Behind", Americans United for Separation of Church and State', *Church & State,* February 2002. www.au.org/site/News2?/page=N ewsArticle&id=5601&abbr=cs-, accessed 23 January 2006.
Boyer, P. (1992), *When Time Shall Be No More: Prophecy Belief in Modern American Culture* (Cambridge, MA: Harvard University Press).
—. (2002), 'The Middle East in Modern American Popular Prophetic Belief', in Amanat, A. and Bernhardsson, M. (eds), *Imagining the End: Visions of Apocalypse from the Ancient Middle East to Modern America* (London: I.B. Tauris).
Boyer, Peter J. (2008), 'One Angry Man', *New Yorker,* 23 June 2008.
Boyer, P. (2008), 'The Evangelical Resurgence in 1970s American Protestantism,' in Schulman, B. J. and Zelizer, J. (eds), *Rightward Bound: Making America Conservative in the 1970s* (Cambridge, MA: Harvard University Press).
Bush, G. (2001), 'President's Remarks at National Day of Prayer and Remembrance', www.whitehouse.gov/news/releases/2001/09/20010914–2.html, accessed 2 July, 2008.
—. (2002), 'President Delivers State of the Union Address', www.whitehouse.gov/news/releases/2002/01/20020129–11.html, accessed 2 July 2008.
—. (2008), 'President Bush Addresses Members of the Knesset', www.whitehouse.gov/news/releases/2008/05/20080515–1.html, accessed 3 July 2008.
'Bush's Messiah Complex' (2003), *The Progressive*, February 2003. Online at www.progressive.org/~progress/?q=node/1344, accessed 2 July 2008.
CNN (2001), 'You Are Either With Us or Against Us', 6 November 2001. http://archives.cnn.com/2001/U.S./11/06/gen.attack.on.terror/, accessed 18 June 2008.
'Canada/U.S. Religion and Politics', Angus Reid Group Cross-Border Survey, print-out of results dated 11 October 1996, provided to the author by Professor Mark Noll.
Cohen, D. (1998), 'Liberty University to Send 3,000 Students on a Study Tour of Israel', *Chronicle of Higher Education,* 25 September, 1998, A51.

Donadio, R., (2004), 'Apocalypse, Nu?' *The New York Observer,* 26 September 2004. www.observer.com/node/49809, accessed 21 June 2008.

Dyer, C.H. (1991), *The Rise of Babylon: Sign of the End Times* (Wheaton, IL: Tyndale House Publishers).

Evans, M. (2003), *Beyond Iraq: The Next Move* (Lakeland, FL: White Stone Books).

'Event 10: China Flexes Its Awesome Power' (2007), Prophecy Hotline, 1 June 2007. www.prophecyhotline.com/?cat=18, accessed 30 June 2008.

'For Evangelicals, Supporting Israel is God's Foreign Policy', *New York Times,* 24 November 2006.

Fortuna, Q. (2007), 'Know Your Antichrist Candidates: Al Gore', Armageddon Cocktail Hour, 15 October 2007. http://armageddoncocktailhour.wordpress.com/category/a-gore/, accessed 30 June 2008.

'Forum: Eschatology' (undated). A discussion sponsored by the journal *SBC Life.* www.baptiststart.com/print/eschatology_panel.html, accessed 2 July 2008.

'General Casts War in Religious Terms' (2003), *Los Angeles Times,* 16 October 2003, online at www.commondreams.org/headlines03/1016–01.htm, accessed 18 June 2008.

Green, J.C. (2008), 'The Faith-Based Vote in the United States: A Look Toward the Future', in DeMuth, C. and Levin, Y. (eds), *Religion and the American Future* (Washington, D.C.: AEI Press).

Hagee, J. (1998), *Final Dawn Over Jerusalem* (Nashville, TN: Thomas Nelson Publishers).

Kessler, G. (2004), 'Impact From the Shadows: Cheney Wields Power with Few Fingerprints', *Washington Post,* <www.washingtonpost.com/wp-dyn/articles/A7036–2004Oct4.html>, accessed 2 July, 2008.

Lindsey, H. (1981[orig. 1970]), *The Late Great Planet Earth* (New York: Bantam Books).

—. (1996), *Blood Moon* (Palos Verdes, CA: Western Front Publishing).

—. (2002), 'Hal Lindsey Oracle' website, 20 Nov. 2002. www.hallindseyoracle.com/.

Macintyre, B. (2003), 'Bush Fights the Good Fight, with a Righteous Quotation', [London] *Times,* 8 March 2003. www.timesonline.co.uk/tol/comment/article1117203.ece, accessed 2 July 2008.

Perlstein, R. (2007), 'Rapture Ready', 26 July 2007. (video of 2007 CUFI conference). http://commonsense.ourfuture.org/rapture_ready, accessed 2 July 2008.

PollingReport.com (2008), CBS News Polls, February 2002 and 6–9 April, 2006 (N=899 adults nationwide, margin of error ±3 percent); ABC News/*Washington Post* polls, January 2002 and 2–5 March, 2006. www.pollingreport.com/religion.htm. Accessed 18 June, 2008.

Posner, S. (2006), 'Lobbying for Armageddon', *AlterNet,* 3 August 2006, www.alternet.org/story/39748/, accessed 6 Aug., 2006.

Reagan, D. (undated), 'America the Beautiful? The United States in Bible Prophecy', Rapture Ready website. <www.raptureready.com/featured/reagan/dr10.html>, accessed 20 June 2008.

Robertson, P (1991), *The New World Order* (Dallas: Word Publishing).'Should Christian Missionaries Heed the Call in Iraq?' *New York Times,* 6 April 2003, 14.

Russell, A. (2006), 'U.S. Evangelist Leads the Millions Seeking a Battle with Islam', [London] *Daily Telegraph,* 5 August, 2006. www.telegraph.co.uk/news/1525660/U.S.-evangelist-leads-the-millions-seeking-a- battle-with-Islam.html, accessed 2 July 2008.

Schrag, C. (2005), 'American Jews and Evangelical Christians: Anatomy of a Changing Relationship', *Jewish Political Studies Review,* 17:1–2 (Spring 2005). www.jcpa.org.il/cjc/cjc-shcrag-s05.htm, accessed 2 July, 2008.

Stillman, D. (2002), 'Onward Christian Soldiers', *The Nation,* 3 June 2002, 26–29 (Netanyahu quote, 27). www.thenation.com/doc/20020603/stillman, accessed 2 July 2008.

'The End is Nigh' (2002), *The Texas Observer,* 6 December, 2002, online at www.texasobserver.org/article.php?aid=1192, accessed 2 July 2008.

Thompson, M. (2003), 'The Boykin Affair', *Time,* 27 October 2003. Online at cnn.com/2003/ALLPOLITICS/10/27/timep.boykin.tm/, accessed 18 June 2008.

USA Today/Gallup Poll, '21st Century Millennium Poll', 30 September.-1 October, 1998, N=1000 national adults, margin of error ± 3 percent. Supplied to the author by George H. Gallup, Jr.

Weber, T. (2004), *On the Road to Armageddon: How Evangelicals Became Israel's Best Friend* (Grand Rapids, MI: Baker Academic).

Index

abortion xii, 33, 51, 61, 169, 240
Abraham 11, 15, 67, 86, 134, 145, 148, 213–4, 222, 237–8
absolutism 158, 160–1, 171, 173, 176, 242
abstinence 169, 172, 209
activism xiv, 8, 64, 209, 241–2
Afghanistan 145, 148, 229, 245
Africa
    American Evangelical Tradition 86–7, 113
    Barack Obama as African American, 73, 89–91
    as Cush 139, 142, 144
    as Ham 86–87
    missions in, 157–158, 162–6, 169–70, 174–7
    as Tribe of Dan 88
AIDS 36, 58, 136, 169, 172, 183
AIPAC 5, 237
Al-Qaeda 110, 148
America
    11 September 76
    capitalism xiii, 188
    culture wars 242
    dispensational premillennialism and, 7, 9, 18, 142, 235, 242
    elections 5, 73
    evangelicals xi, xiii, 3–4, 6–8, 10–11, 15–16, 17–19, 86, 108–09, 124, 136, 157, 166, 172, 177, 184, 187, 190, 201, 234, 245
    exceptionalism 17, 80–1, 99, 101, 105–06, 110–111, 124
    foreign policy 2, 16, 79, 100, 102–03, 112–13, 134–36, 150, 200, 210, 236, 243
    frontier 12
    fundamentalism 50–4, 119, 166, 187
    geopolitics 3, 85–6, 100, 114, 145, 150, 199, 239
    identity 16, 80, 85, 87, 93, 99
    imperialism xv, 17, 194
    Israel relations 211, 215, 224, 237
    as Judeo-Christian 15, 136, 234
    lobbies 5
    manifest destiny 18
    masculinity 104–106
    megachurches in 235
    Middle East relations 215, 227, 234
    millennial role 129–132, 240
    missions 169, 171, 185, 190, 195
    Mormonism 49–50, 59–60, 64, 66–8
    nationalism 18
    Native 81
    neoconservative 100, 102, 134, 200
    New England 29, 79
    'New Right' 210
    postmillennialism 10
    revolution 9, 84
    race and, 87–93, 197, 198,
    security 105, 114, 124
    south 13, 27, 32–34, 38–9, 42, 44, 87
    U.S.S.R. and 14, 141, 144, 148
Americanism 16, 18, 50, 53, 58–9, 63, 74, 80–1, 85, 99, 101, 148
amillennialism 10
Amish 37–8
Anglicanism 13, 84
animism 5, 170
anthropology 8, 190, 192, 196–7, 199
antichrist 243, 224–6, 239–43, 245
    America and 59, 60, 81
    Barack Obama as 16, 62–63, 74–75, 77–79, 88–92, 111
    as deceiver 59, 82, 122
    European (Roman Empire) 239
    Left Behind 56, 146, 215, 241; see also Carpathia, Nicolae
    Mohammed as 139, 234
    Mormon rendition 50, 58–60, 65

'New World Order' 122–127, 129, 131, 240, 242
  Saddam Hussein 236
  son of perdition 73, 83
  third temple and 9, 214
  world government/economy and 8–9, 56, 59–60, 81–93, 119–120, 146–147, 210, 215, 235, 239, 241
anti-Semitism 83, 85, 121–4, 126, 192, 239–40
apocalyptic xv, 6–7, 93, 139–43, 145–7, 175, 188, 238
  America and 42–3
  apocalypticism 7, 89
  Book of Revelation 83, 125
  fiction 17
  geopolitics 135, 143, 146, 150
  Islam and 85, 219
  Mormon 53–4
  non-fiction books 14, 17–8
  numerological calculations 82
  rhetoric and 1, 4–5, 78–81, 133, 135–7, 149–150, 226–27, 229–30, 233–34, 240, 242–3
  secular 6
  timeline 129, 134
Arabs 14, 18, 54, 58, 60, 63, 67, 74, 85–6, 89, 104, 124–5, 129, 139–43, 146, 149, 212–4, 218–9, 223, 226, 235, 237–8, 244
Arafat, Yassir 85, 147
Armageddon 8–9, 14, 119–20, 122, 125, 133, 135, 143–4, 147, 149, 209–11, 213, 215, 217, 219, 221, 223, 225, 227, 229, 234–5, 242–5
Asia 7, 125, 139, 145, 148, 167, 184, 186, 190, 211, 219
atheism 90, 140
Augustine xiv, 10, 82, 150

Babylon 9, 80, 82, 139, 145–8, 215, 218, 221, 222–3, 236, 241
Babylonian Captivity 82, 223
Balfour Declaration 141, 237
Baptists 13, 27–8, 65, 185, 239
Bauer, Gary 126, 237
Beast
  of Revelation 9, 129, 240

  mark of 59, 77, 83–4, 126–7
  number of 83, 89, 245
  system of governance 120, 126–7, 129
Beck, Glenn 63, 91
Bede, Venerable 82–3
Beth-Togarmah 9, 139–40, 144
biology 192, 209
blogs 74, 104, 138, 214, 223–4
bombings 59, 110, 224
borders 9–10, 13, 32–3, 42, 77, 92–3, 101–2, 105, 126, 128, 134–5, 137, 142, 144, 148, 150, 194, 197–8, 214, 220–2, 237–9, 241
Boykin, Lt. General William G. 234, 236
Britain 58, 81, 84, 110, 112, 124, 141, 185, 195, 235, 240
Bush, George W. xii, 1–6, 15, 19, 54, 79–80, 99–100, 107, 111, 133, 144, 148–9, 163, 170, 211, 213, 216, 228, 233–5, 241–2, 245
Bush, George H.W. 15, 121

Canada 7, 14, 145, 235
capitalism xiii, 5, 18, 39, 85, 102, 136, 159, 164, 185, 194, 200–1, 211
Carpathia, Nicolae 146, 241
cartography 15, 30, 140, 183–4, 186, 196, 199–200
Catholicism xii, 11, 18, 50–2, 65, 83–4, 91, 106, 125, 164, 188–9, 238–9; see also Vatican
Cheney, Richard 148, 234
China 59, 105, 124, 129, 142, 145, 234, 244–5
Choi, Jai 218–23, 225–7
Christianity 5, 7–14, 59, 81–4, 88, 90, 93, 123–4, 126, 129, 133, 140, 148, 161, 165–7, 177, 188–90, 192–5, 197–8, 224, 226, 228, 230, 238
  in Africa 157, 165
  in American South 29–31, 35–6, 41–4
  Bush, George W. and, 3, 99
  capitalism and xiii, 200–1
  conflict with Islam 99–100, 112–3, 135–6, 157–8, 169, 174–5
  conservative 171–2
  conversion to 56, 107, 169, 185–6, 210, 220

dispensationalist 76
Kant, Immanuel 9
Lee, Robert E. and 30
mainstream Protestantism xi
medieval 135
millennial 1
as Mormonism 49
nationalism 29, 31, 40
nature 37–8
premillennialism 12
Promised Land 100
prosperity 186
slavery 113
supernaturalism xiv
Trinitarianism 32

CIPAC 5
citationality 17, 133–5, 138, 140, 146–50
colonialism 28, 79, 84, 174, 200–1, 240
communism 5, 54, 99, 124, 135–6, 138, 141, 143–4, 146, 148, 150, 187, 215, 235, 244
computers 37–8, 59, 122, 125
Confederacy 13, 16, 27, 29–32, 36, 41–2, 44, 88
conservativism
  Catholic 238
  eschatology 49–51, 55, 166
  foreign policy 16–7, 169–170, 177
  morality 30–2, 166
  Mormons 57
  Muslims 174, 176
  neoconservative 63, 99–100, 102–3, 106, 111, 113–14, 171, 200–1, 134
  Pentecostals 10
  Protestants 121–2, 124, 131, 145
  religious beliefs 27, 145, 157–9, 161, 167, 169–70, 172, 189
  theology 209, 235
  women and, 172, 174–7, 189, 209, 230, 235, 238
constitution 10, 50, 64, 80, 130
conversion 8–9, 13–4, 51, 56, 59, 61, 66–8, 107, 138, 166–70, 184–5, 187–9, 193, 195, 199, 210–1, 217, 223, 226–9
corporations 39, 61, 64–5, 76, 124, 131, 159

corruption 11, 33, 42–3, 81, 84, 111, 230
covenants 8–12, 18, 29, 53, 67, 76, 82, 84, 130, 140, 190, 210, 212–4, 222, 228
CrossGlobal Link 212, 218–9
crucifixion 30, 36, 140
crusade 4, 7, 80, 140, 150, 157, 161, 176, 233, 236
CUFI 6, 238
cults 51, 64
curricula 196, 209
Cush 9, 139

damnation 171, 186, 239
Dan, Tribe of 83, 87–9
Daniel, Book of 14, 53, 77, 83, 107, 139, 239, 245
Darby, John Nelson 12, 14, 52, 82, 122, 128–9, 140, 210, 235
Darwinism 85, 187, 240
David, King 128, 214
decadence 33–4, 42, 81, 99, 103, 110, 111, 242
democracy xi, 1, 15, 58, 73, 77, 102, 130, 136, 160, 171–4, 176–7, 201, 211, 238
Democrats 91, 230
demography 162, 196, 199
demons 10, 122, 167, 188, 242
demonization xiii, 40, 85, 241, 236
denomination 13–4, 16, 50, 193, 199, 218, 239
Deseret Books 57, 67
desire xv, 33, 60, 67–8, 126–7, 163, 242, 244
détente 9, 50
diplomacy 19, 85, 101, 106, 112, 157, 165, 170, 172–3, 216
disciples 192, 195, 219, 220
disease 186, 210
dispensationalism 6–9, 11–4, 16–8, 52–3, 55, 68, 75–7, 79, 82, 85, 92, 119–20, 122–5, 127–30, 133–5, 137–46, 150, 166–7, 175–6, 187, 209–24, 226, 228–9, 235–7, 239, 242, 244–5
divorce 37, 61
Dobson, James 14, 149, 209, 245

doctrine xi, xiii–xiv, 11, 14, 50, 55, 64, 81, 107, 149, 171, 185, 192, 239, 243
dogma 18, 157, 221, 227, 229–30
Dyer, Charles 134, 141–2, 145–7, 211, 215, 218–9, 221–4, 226–7, 236
dystopia 2, 136

economics 120, 149
    American hegemony of 15, 86–7, 102, 111, 162–3, 165, 187, 244
    anti-Semitism 83, 102
    change 133, 147
    development 176
    egalitarian 50–1, 172
    European Economic Community 142–3, 147, 149, 245
    geoeconomics 141, 147
    international 110–1, 143, 162, 200
    liberalism 187–9, 200, 244–5
    microeconomics 136
    political-economy 159–60, 200
    Protestant 162–5, 172
    socio-economic crisis 28, 244
ecumenism 56, 120, 127, 188–9
education 42, 145, 194–5, 209, 218, 223
Egypt 130, 134, 139, 142, 212, 237
elections 49
    evangelical influence on 5, 16, 42, 89
    of Obama 19, 73, 81, 86, 93, 111, 113
emotions 17, 29, 36, 66, 216
empire 2
    American 112, 130
    British 81, 84, 112
    decline of 111–2
    'empire of disorder' 162
    evangelical 200
    evil 133, 145
    Hardt and Negri 159
    media 241
    Ottoman 140, 236
    Roman 9, 83–4, 89,112, 120, 122, 136, 142, 239, 245
    Soviet 143–4
    Tacitus and 149
    world 241
emplotment 75–9, 86, 89, 91–2

England 28–9, 33, 84, 137, 140–1, 159, 192, 194–5, 200, 240
entrepreneurialism 38, 164
environment 2, 8, 38–9, 79, 90, 175, 245
Ephraim, Tribe of 59, 64, 67
Episcopalianism 1, 49, 185
epistemology 51, 62, 66, 102, 113, 137
epistles 82–3, 85, 122, 129
Equator 167, 183, 186
eschatology 6–13, 15–6, 49–50, 52–8, 63–5, 67–8, 75, 79, 81, 83, 89, 91, 119, 128–30, 148, 161, 166, 175, 184, 212–3, 215, 217–8, 220, 222–5, 227, 229, 234, 236, 238–40, 244
essentialization 11, 32, 40, 87
ethics 18, 44, 51, 164, 175, 184
Ethiopia 9, 56, 139, 144
ethnicity 15, 17, 58, 60, 66, 85–6, 140, 164, 167, 171, 173–4, 177, 187, 191–2, 194, 196, 198, 212
ethnography 2, 16–7, 164, 184
ethnolinguistics 191–2, 194–5, 197
Euphrates River 134, 145, 237, 244–5
Europe xii, 2, 51, 56, 58–9, 105, 113, 124–5, 139–41, 145, 161, 171, 186, 239, 241
Europeans 6, 10, 56, 58, 63, 83, 87, 89–90, 107, 111, 120, 122–3, 142–3, 174, 214, 233, 239, 245
evangelicalism xi– xiii, xv, 1–8, 10, 12, 14–9, 27–32, 44, 49–57, 61–3, 65, 67–8, 74–5, 77–8, 82, 85–7, 89–93, 106, 108, 112, 124–5, 133–5, 138, 141, 147, 149, 157–9, 161, 163–77, 183–9, 191, 195, 200–1, 209–5, 217–8, 220, 222–3, 225–30, 234–6, 238, 244–5
evangelicals
    in Africa 165–170, 175, 177
    African American 86
    American 1–2, 4–8, 10, 13–9, 28, 31, 43, 49–53, 55, 62–6, 68, 74, 76, 79–80, 86–7, 91, 93, 119, 126–7, 133–4, 136, 138, 141, 149–50, 157–9, 161, 163–71, 173–7, 183–5, 188–91, 200–1, 209–15, 217,

219–20, 223–4, 226, 228, 230, 234, 238–9, 245
British 141
definition of 7–15, 18
domestic policy 19, 31, 43, 49, 52, 133, 209
as fascists 1
foreign policy 4, 49, 68, 133, 158, 170–1, 185, 211, 230
Islam and 17, 176
liberal forms xiii, 15, 51, 62, 86, 189, 245
'New Right' and 209, 210, 212
pluralism and xiii, 183
popular culture 6
race and 87, 136
evangelistics 192, 196
evangelization 17, 126, 166–7, 170, 183–5, 187–97, 199–201, 211, 220, 225, 229
evil 1, 4, 7, 10–1, 13–4, 17, 58–9, 61–3, 65, 80, 82, 84, 86–9, 99, 101, 104, 108, 110, 123, 126, 131, 133–45, 148–50, 157, 174, 176, 216, 233–4, 236, 240, 242–3
evolution xii, xiv, 30, 135, 187, 212, 230; see also Darwinism
exceptionalism, American 17, 80–1, 99, 101, 103, 105–7, 109, 111, 113, 119, 121, 123–5, 127, 129, 131
existentialism xv, 77
Ezekiel, Book of 5, 9, 14, 16, 53, 56, 77, 81, 85, 101, 106–8, 110, 125, 134, 139, 141, 144, 147, 239–40
Ezra, Book of 224

faith xi–xii, xv, 5, 37, 43, 62, 66, 78, 84–5, 90, 108, 129, 138, 143, 157–8, 160, 163–7, 169, 171–2, 175–6, 185–6, 188, 196, 199, 218, 233, 243
Falwell, Jerry 14, 51, 85, 126, 210, 222–3, 226–7, 229, 235, 238, 241, 243, 245
famine 8, 60, 189, 210
fatalism 10, 14–5, 17–8, 42–3, 133, 211, 225, 228
feminism 18–9, 61, 80, 86, 209, 216–7

fiction 1, 5–6, 17, 55–7, 100, 104, 106–7, 109, 111, 114, 138, 146–7, 215, 235, 241
finance 7, 58, 105, 123–5, 129, 163, 193, 199
Forrest, Nathan Bedford 34, 41
Forsyth, Frederick 101, 110
freedom 59–60, 80, 92, 113, 169–70, 200
Friedman, Thomas 193–4
frontiers 3, 12, 28, 50, 101, 106, 157, 184, 190, 194, 197, 212, 218–20, 222, 225–6
fundamentalism 7–8, 10, 13–4, 19, 34, 39, 44, 50–1, 85, 119, 122, 127, 144, 158, 160, 166, 173, 184, 187–8, 209, 218, 244

Gaza 4, 128, 160, 213–4, 225
Genesis, Book of xiv, 83, 137, 214, 237
genocide 85, 177, 192, 238
gentiles 198, 224
geoeconomics 141, 147
geopolitics
    alternative 113
    apocalyptic 135, 146, 150
    classical 2
    critical 1–3, 5–7, 100, 115
    evangelical 2, 4–6, 92–3, 209, 212, 217, 222, 225, 228–30
    formal 3–4
    neoliberal 193
    practical 4–6, 100, 137
    popular 6–7, 16, 55, 100, 199
    religion and 2–3, 29, 100
    tabloid 16, 101, 104, 138
Germany 1, 139, 141, 145, 241
Gesenius, Wilhelm 138, 140, 142, 144
Gilpin, William 130–1
GIS 167, 196, 199
globalization 3, 77, 127, 159, 162, 164, 193–4
Gog 9, 14, 81, 139–140, 142, 148, 239–40
Gomer 9, 139–41, 144–5
Gorbachev, Mikhail 121, 143
governance 77, 129, 158, 170–1, 173, 200, 237
governmentality 17, 135, 158–63, 169–71, 173, 176

grace xiv–xv, 11, 113, 166, 238
Graham, Billy 8, 14, 15,
Graham, Franklin 170, 236
grassroots 52, 54–5, 57, 65, 243
Great Commission 157, 166–7, 169, 175, 192

Hagar 86, 237
Hagee, John 5, 17, 63, 91, 123–6, 210–2, 214–5, 222–4, 226–7, 229, 236–9, 243
Ham 86–7, 142
Hamas 160, 212, 223–4
heaven xii, xiv, 7–10, 12, 14, 79, 87, 119, 175, 210, 214, 219–20, 243
Hebrew 15, 238
hegemony 3, 10, 15, 51, 78–9, 81, 85, 87, 92, 114, 135–6, 158–9, 172–4, 177, 200, 217
heritage 11, 30–1, 33–5, 38, 41–2, 44, 90, 192, 221
Herodotus 138, 141–2, 150
Hezbollah 211–2, 214
Hinduism 186, 195
Hitchcock, Mark 11, 133–4, 138, 141–7
Hitler, Adolf 10, 84, 125, 239
Holocaust, the 10, 239
homosexuality xii, 52, 60–1, 65, 240
humanitarianism 29, 158, 160, 163, 169–70, 184, 200
Huntington, Samuel 4, 136, 145, 176, 193–4
Hussein, Saddam 84–5, 89–90, 145–6, 215, 222, 236

ideology 7, 31, 37, 44, 49, 51, 62, 121, 135–6, 144, 158, 160, 174, 177, 200–1, 238
Illinois 13, 74, 106
Illuminati, the 123–6
imams 59, 65
immigration 32–3, 84–5, 136,176, 183–4, 197–8
immorality 30, 33–4, 43
imperialism 60, 100, 111–4, 167, 191, 195, 199, 200–1, 229
    American 17

messianic xv
neo-imperialism 54
tabloid 112
India 59, 142, 145, 189, 195
individualism 13, 28, 30
Indonesia 74, 238
internationalism 123, 200, 215
internet 6, 16, 32, 74–5, 77, 79, 89, 91–3, 235–6, 241, 244–5
intersubjectivity xi, xiii, 16
intertextuality 16, 79, 82, 100–1, 106, 144
Iran 4, 17, 56, 108, 139, 142–3, 145–9, 215–6, 233, 245; see also Persia
Iraq 103, 113–4, 125, 135, 139, 143, 145–9, 165, 211, 213–6, 221–2, 229–30, 233–4, 236–7, 239, 241, 245
Isaac 141, 148, 214, 238
Isaiah 9, 53, 240
Ishmael 86, 137, 141, 237, 238
Islam
    American opinion of 234
    clash with West 160
    crusade against 4
    as evil 58–9, 85, 143–5, 236–7
    fundamentalist 158
    as Magog 140
    movements xi
    Nation of, 78, 87
    Obama and, 74, 89–91
    radical militarism 147–8, 173
    as threat xii, 2–3, 17, 99, 105, 107, 112, 136–8, 150, 168–9, 173–7, 186, 214–6, 224, 226, 240, 242
Islamism 3–4, 86, 160, 171–4, 176
Israel
    America as 7, 44, 54, 79, 100, 113, 130
    Arab opposition to 85, 88, 120, 129, 142
    as centre of the world 150
    Israelite 140
    lobby 4–6, 124, 134, 211, 228–30, 237–8
    peace/conflict with Palestine 106, 109, 113, 128–30, 142, 138
    relation with evangelicals 1, 18, 107, 124–26, 134, 148–50, 184, 210–5, 239

relations with Mormons 52–4, 56, 62–7
as sign 5–9, 11–5, 53–4, 59–60, 63, 77, 81, 83, 108, 120, 125, 138, 143–4, 148, 176, 210–5, 219–26, 235–42, 244
shift to right 19

Jacob 148, 214
Japan 105, 142
Japheth 87, 142
Jefferson, Thomas 7, 38
Jenkins, Jerry 9, 16, 55–6, 91–2, 146, 215, 235
Jeremiah, David 221–2
Jerusalem 4, 7, 10, 13, 52–3, 63–4, 67, 79–80, 82–5, 111–4, 128, 141, 149, 184, 197, 213–5, 219, 222–4, 228, 236–8, 243
Jesus xiii, 15, 170, 186, 223, 225
antichrist as mirror of 88
atonement 66
blood sacrifice 85, 193
divinity of 82–3, 85
incarnation of 81–2
Israelites and 140
'Jesus-trick' 136
Jewish rejection of 11
Jews for Jesus 236
LDS 49
as Messiah 239
of Nazareth 76, 112–3
Nero as mirror of 83
resurrection 210
return of 32, 83, 148, 161, 175, 184, 214, 220, 227, 234
Roman occupation and 114
as Savior 157, 166, 185, 189, 213
Son of God xii, 35
'What would Jesus do?' 41
Jews 9–14, 49, 53, 56, 58–9, 66–8, 76, 80, 82–3, 85, 89, 121, 123–5, 128–9, 140–1, 148, 176, 184, 192, 198, 210, 212–4, 224, 228, 233, 236–9
jihad 105–7, 157
Jordan 139, 189, 212, 225, 245
Joshua Project, the 167, 196
Judah 59, 223
Judaism xii, 226

Judea 148, 197, 219

Kaduna Peace Declaration 174–5
Kazakhstan 58, 144
Keller, Catherine xiv–xv
Kennedy, John F. 85, 88
Kenya 169, 176
Kissinger, Henry 84–5
Knesset 237–8
Korea
North 216, 233–4
South 17, 144, 184, 191, 197
Kremlin 107, 148; see also Russia, Soviet Union

LaHaye, Tim 9, 14, 16, 55–6, 62–3, 91–2, 134, 138, 142–3, 145–6, 149, 215, 235, 239, 241, 243, 245
language 1, 64, 67, 73, 127, 133, 145, 177, 183–4, 191–2, 194–6, 241
Latin America 211, 244
Lausanne, International Conference on World Evangelization in 188, 190, 191
LDS 49–52, 54–68; see also Mormonism
Lebanon 139, 214, 225
Lee, Robert E. 27, 30, 41
liberalism 10, 14–5, 17, 50–1, 85, 99, 102–3, 106, 158, 160, 162, 165, 171–4, 176, 188, 209, 218, 244–5
liberationism 186–200
liberty 4, 111, 130, 220, 241
Libya 56, 139, 144–5, 235
Likud 5, 238
Lindsey, Hal 9–10, 17, 109, 134, 137–8, 141–7, 149, 235–6, 238–41, 245
linguistics 135, 140–1, 190, 197, 199
literalism 10, 51, 62, 68, 147
Lucifer 58, 60, 63, 65, 123; see also Satan
Lund, Gerald 58, 66–7
lynching 34, 86

mabus 79
Magog 9, 14, 81, 139–41, 144, 147–9
mahdi 59, 65
marriage 52, 65, 67, 190–1, 213, 240
Martin, Andy 73–4
martyrdom 83, 146

masculinity 41, 106, 110, 216
Mason-Dixon line 13, 43
Masonry 125, 240
Matthew, Book of 7, 9, 192
McCain, John 74, 91, 125, 214, 225, 230, 239
McGavran, Donald 189–90
media xi–xii, 6, 10, 15, 56, 74, 78, 100, 102–4, 131, 136, 138, 175, 209, 226, 229, 238, 241, 244
megachurches 235, 244
Megiddo 9, 235
Melville, Herman 130–1
Meshech 9, 139–40, 144, 148
messiah xv, 13, 74, 125, 130, 148, 214, 233, 239
metanarrative 74–7
metaphysics 171, 174
Methodism 7, 13, 27–8, 65, 185
methodology 55, 57
militarism 3, 60, 148, 185, 190, 200
millenarianism 56, 119–7, 129–31
millennialism 1, 7–10, 28–9, 53–4, 56–7, 59, 61, 64, 68, 79, 91, 93, 112, 119–20, 122, 124, 126–31, 138, 145, 167, 210, 214, 220–1, 228, 235–6
millennium 8–12, 14, 52–4, 79, 84, 130, 139, 210, 242
Miller, William 119, 127–8
missiometrics 196
missionaries 13–4, 17–9, 51, 63, 108, 158, 164–6, 169, 174–7, 183–7, 189–90, 193–8, 200–1, 216–9, 223, 225–30, 236, 244
Missouri 50, 53, 61, 63, 65–6
modernity 8–9, 12, 38–9, 44, 80, 84, 93, 103, 111, 184, 187, 192, 200
Mohammed xii, 85–7, 186, 139, 214
money 39, 73, 124, 163, 166, 170, 214
Moody Bible Institute 4, 12, 145, 212, 215, 218–9, 226
Moody, Dwight L. 12, 187
Moore, Roy 10
morality 17, 30–1, 33, 35, 106, 110, 114, 229
Mormonism 16, 49–55, 57–68; see also LDS

Morocco 139, 238
Moscow 107–8, 139–40, 148; see also Russia, Soviet Union
Moses 10, 60, 223, 238
Muslims 7, 14, 39, 74, 85–6, 90–1, 104, 136–7, 140, 142–4, 150, 158, 161, 174, 176, 184, 186, 233–4; see also Islam
mythology 11, 93, 136

narrative 16–17, 19, 73–93, 100–3, 107, 110, 113–4, 147, 200–1, 216
nationalism 7, 18–9, 27, 29–31, 33, 35, 37, 39–41, 43–4, 100, 126, 150, 157, 171, 241
Nazism 192, 239
neoconservativism 99–100, 102–3, 106, 111, 113–4, 171, 200–1, 230, 234
Newell, Marv 218–21, 227
New York 19, 44, 74, 119, 124, 142, 143, 145, 146, 230
NGOs 4, 19, 166, 168, 170
Nigeria 165, 174, 176
Noah 86–7
nontheism xi, xiii–xiv
normativity 55, 57, 134, 158, 160, 170–2, 176–7, 185, 199, 201, 216
Nostradamus 79, 93, 147
nuclear weapons and war 4, 58–60, 107–8, 120, 147, 149, 215–6, 234–5, 238
numerology 82, 90, 93

Obama, Barack 1, 15–6, 63, 73–9, 81, 83, 85–93, 111, 113, 245
Occidentalism 39, 40, 76
occult 79, 120
ontology 76, 113, 135, 176, 192
Orientalism 16, 32–6, 38–40, 43–4, 76, 137, 145, 150, 183, 185, 198
Ottoman Empire 140, 236
Oxford University Press 12, 235

pagans 37, 130, 174, 185
Pakistan 59–60, 145, 148, 245
Palestine 8, 15, 62, 66–7, 106, 109, 113, 141–2, 210, 213–5, 222–5, 228, 237, 241, 244
paramilitary 121, 131

paranoia 12, 74, 99
patriarchy 28, 61, 65, 169
patriotism 102, 111, 125, 242; see also nationalism
peace 8, 53, 56, 59, 63, 73, 85, 106, 129, 137, 149, 173–6, 211, 213–4, 219, 222, 226, 228, 237–8, 241, 243, 245
peacemaking 56, 113, 245
Pentagon, the 107, 125, 190, 193, 215, 233–4
Pentecost 10, 138
Pentecostalism 10, 13, 149, 165, 185, 188, 235
performance 1, 11, 19, 86, 217, 220
performativity 18, 212, 216–7, 223–4
Persia 9, 139, 141, 148–9, 236; see also Iran
philanthropy 130, 189
Philippines, the 61, 190
philosophy xiii, 30, 39, 54, 109, 138, 201
pluralism xi–xiii, 134, 158, 172–3, 175, 177, 200
Plymouth Brethren 12, 235
polygamy 50, 191
Pope, the 56, 84–5, 122, 161
populism 91–2, 102, 129, 136, 230
pornography 51, 61, 240
postmillennialism 9, 10, 14, 42–3, 53, 79–81, 84–5, 123, 175, 242
postmodernism 15, 77
poverty 1, 31–2, 163, 186, 188, 200, 225, 245
praxis 185, 201
prayer 37, 183, 184, 193, 199, 209, 240
predestination 107, 133, 138
prejudice 34, 36, 86
premillennialism 6–12, 14, 16, 42, 52–3, 80–1, 84–5, 89, 92, 119–20, 122–3, 130, 133, 143, 175, 187, 209, 211, 213, 218–21, 225, 235, 242–3
Presbyterianism 13, 49, 185
progressivism xii, 10, 51, 136
prophecy 1, 5, 7, 11–4, 17–9, 50–1, 53, 55–7, 59–60, 62–8, 73–4, 76–81, 83, 87, 89, 92–3, 99–100, 107–9, 120–5, 127–9, 133–8, 141–8, 150, 186, 210–1, 213–5, 217, 219–25, 228–9, 234–45
proselytization 11, 51, 59, 67, 157, 170, 175–6, 184–5, 190, 192, 194, 199, 201, 213
Protestantism xi, 8, 11–3, 27, 32, 44, 50–1, 91–2, 121–2, 124, 129, 164–6, 185, 188, 209–10, 218, 235, 238–9
Protocols of the Elders of Zion 83
providence xv, 18, 84, 161
puritans 7, 17, 44, 80, 99, 129, 240
Putin, Vladimir 84, 148

racism 16, 31–2, 34–6, 38–40, 64, 73, 75, 86–8, 90, 136, 145, 185, 191–2, 196, 199–200, 217, 240
radio 64, 108, 176, 240
Rapture, the 8–9, 11–14, 19, 56, 58, 62–3, 81, 92, 99, 108, 119–20, 122, 128, 175, 210, 213, 220–1, 224–5, 228, 235, 238, 243
rationality 73, 75, 76, 159, 160, 162, 195, 200
reachedness 186–8, 193, 195
Reagan, Ronald 1, 5, 102, 104, 109, 133, 149, 240
realism 112, 135, 145, 150
redemption 13, 30, 53, 59, 67, 80, 85, 113, 129, 131
Reid, Harry 49, 54
relativism 102, 171, 174
representations 11, 59, 129
    academic 222
    elite 133
    fictional 100–2, 104, 111, 115
    geographic 137–8, 176
    geopolitical 3
    missionary 194, 201
    of Muslim World 136, 140–1
    Orientalist 33
    of the South 34
    'War on Terror' 4
Republicans 5, 49, 73, 103, 123, 125, 133, 149, 157, 209, 211, 230, 236, 238–9
resistance 41–2, 64, 84, 112, 173, 177

resonance xiii–xiv, 4–6, 11, 18–9, 55, 64, 107, 134–5, 148, 161, 176, 194, 200, 225, 230, 236, 239
restoration 10, 42, 50, 185, 221
resurrection xii–xiii, 8–10, 63, 188, 210, 219
reterritorialization 134, 142, 145, 148
Revelation, Book of
   America in 240
   Author of 146, 243
   Ecology and 60
   Evangelical interpretation of xv, 9, 77, 83, 91, 107, 119, 224, 236, 241
   Evil armies of 14, 145–6, 245
   Great whore of 239–1
   Hagee, John and 125
   Millennial movements' interpretation 127–9
   as obscure 134, 244
   Thomas Jefferson and 7
Revelation, divine 7, 50, 53, 65
revival
   Council on 241
   evangelical 28, 158, 166, 169, 172–3, 175, 245
   fatalism 14
   non-Western 157
   postmillennialism 10
   Whitefield, George 7
revolution 5, 9, 13, 42, 54, 80, 84, 123, 139, 140, 143
rights
   Catholic League for Civil and Religious 125
   civil 27, 31, 189, 209
   human 18, 102
   individual 177
   last-days 67
   Palestinian 213
Robertson, Pat 14, 17, 80, 123–6, 147, 210, 222, 226–7, 229, 235, 238, 240–1, 245
Romania 56, 146, 215
Rome 9, 56, 58, 81, 83–4, 89, 112–4, 120, 122, 125, 136, 139, 142, 145, 149, 189, 224, 239, 245
Romney, Mitt 49, 52, 54
Rosenberg, Joel 16, 99, 101, 104, 106–11, 134, 138, 141–2, 146–9
Rosh 9, 13, 139–41, 144
Rumsfeld, Donald 234, 239
Russia
   Arab coalition with 124–5, 129, 136, 138–44, 146–50, 242
   conflict with Israel 120, 214, 235
   dissolution of U.S.S.R. 128, 134, 239
   Ezekiel and 5
   Joel Rosenberg 107–8
   in *Left Behind* 56
   Mormon interpretation of 58
   revolution 13
   as Rosh, Gog, Magog, 13, 140–2, 144, 239–40
   Tom Clancy and 104–5
Sabbath xv, 82
sacrifice xiv, 10, 30, 60, 86, 87, 110, 111
Saddleback Church 15, 245
Sahel 169, 174
Salem Witch Trials 84–5
salvation xii–xiii, 43, 51, 67, 80, 86, 108, 131, 166, 171, 175, 187–90, 210–1
Samaria 170, 197, 219
Satan 8–10, 60–1, 73–4, 83–4, 87–8, 91, 122–7, 131, 145, 175, 186, 188, 214, 234, 241–2; see also Lucifer
Satanism 120, 125
Saudi Arabia 58, 148
savior 52, 157, 166, 184–5, 189, 210, 213
scale 2, 18, 76–7, 93, 108, 110, 169, 172, 177, 188, 195–6, 211
Schmitt, Carl 17, 158–62, 172–3, 177
science 51, 196, 209
   of evangelism 185, 192, 199–201
   evolutionary 13, 240
   fiction 56, 172
   natural xiv–xv, 138, 187
   political 2
   social xi, 3, 11, 75, 190–2, 216–7
Scofield, Cyrus 12–3, 122, 235
scripture xiii– xv, 7–8, 12, 36, 50–1, 53, 56, 60, 63, 68, 82–3, 89, 119, 122, 129–30, 134, 146, 148–50, 185, 220, 222, 238, 245

secularism xi–xiii, xv, 6, 9, 16, 28, 42, 80, 86–7, 110, 121, 126, 131, 143, 170–4, 177, 189, 209, 224, 244
security 29, 88, 93, 99–105, 107, 111, 114, 121, 126, 228, 230
sexuality 31, 61, 67, 80, 86, 172, 190–1, 209, 217, 230
sharia 173, 175
sin 28, 35, 62
Solomon 9, 134, 223
soul 14, 17, 19, 61, 78, 130, 158, 186, 196, 211
South Carolina 32, 41
southern U.S. 13–16, 27–44
southernization 27, 31
sovereignty 2, 9–10, 32, 122–3, 126–7, 159–61, 237
Soviet Union 13, 54, 58, 80, 85–6, 99, 102, 108, 121, 128, 136, 138–40, 142–5, 149, 215, 236
statistics 135, 188, 195–6, 198–9
Stewart, Chris 56, 58, 60, 63, 65
subjectivities 3, 75, 93, 136, 159, 164, 200, 217
Sudan 4, 139, 144, 170, 174–5
suicide 107, 110, 224
superpower 102, 240, 244
supersessionism 53, 140
Syria 139, 146

Tanzania 164, 169, 176
technology
  American superiority in 104–5
  flow of 162
  governmental 79, 92, 101, 123, 166, 169–70, 184–5
  information 127, 129
  mark of the beast and 77, 127
  of warfare 80, 107–10
technothriller 100, 104, 107, 114
teleology 76, 171, 243
televangelists 123, 235–6, 244
temple
  of God 73
  Mount 214
  second 82
  Solomon's 9, 134, 223

third 7, 13, 53, 83, 120, 128, 148, 224, 237–8
temporality 75, 81
territory 57, 125, 140, 144–5, 161, 185
terrorism 66, 85, 103, 105–7, 110, 120, 144, 147, 175, 216, 224, 234
Texas 124, 233, 236, 238
theatricality 62, 212
theocracy 50, 84, 161, 215
theology 4–5, 13, 50–2, 60–1, 64, 78, 80, 84–5, 88, 93, 100, 106–7, 109, 143, 145, 185, 192, 196, 200, 214–5, 239
  America in 113
  covenant 29, 53
  creedal xiii, xv,
  critique of 1
  dominant Christian form xiv, 29
  eschatology and 8, 15
  foreign policy and 209–11
  formal 140–1
  intersubjectivity and 16
  liberal forms xiv
  millennial forms 8, 112–4, 123
  missions and 218–27, 229
  Mormon 52
  prosperity 188
  replacement 11, 140
  theology of becoming xv
Thessalonians, Epistles to the 8, 73, 83
Tobolsk 139–40
toponyms 137, 139–41, 145, 150
transcendence xi, xv, 30, 41, 73, 136, 159
treaties
  of Westphalia 2, 126
  Antichrist peace-pact 9, 56, 85, 240–2
  rejection of in American South 32
  of Tilsit 140
tribulation 8–9, 11–4, 53–4, 56, 68, 92, 109, 119, 122, 124, 128–9, 143, 147, 210, 221, 224, 228, 235, 239
Tubal 9, 139–40, 144
Turkey 108, 139–40, 144–5, 245
Tyndale House 6, 106
tyranny 79, 125, 130

unilateralism xv, 165, 200

universalism 17–8, 64, 158, 160, 162, 171, 201
unreachedness 167, 183–5, 187–91, 193–201
unsaved 56, 187, 211
Utah 57–9, 63–7
utopia 2, 80, 171, 243

values
   American 51, 59–60, 103
   Christian 37, 157
   family 16, 51, 164–5
   Judeo-Christian 136
   liberal 103, 123
   spiritual 39
   Southern 33, 36, 42
Vatican, the 186, 189; see also Catholicism
Venerable Bede see Bede, Venerable
Vietnam 66, 189
vigilantism 17, 106
violence 14, 17, 172, 177
   American 99–101, 103–6
   bodily 3–4
   eschatological 60
   exceptional 159–60
   fictional 108–9
   geopolitical 114
   imperial 60, 113
   insecurity and 111–4
   Islamist 3, 174, 234
   Israeli/Palestinian 113, 210–1, 228
   Left Behind 56, 62–3
   masculinity and 106, 216
   Mormon 54
   religious sanction of 100, 213, 217, 221, 240
   Southern 32
visions
   apocalyptic xv, 125, 130
   geopolitical 3, 18, 29, 32, 42, 62, 74, 103, 167, 172, 184, 201–2, 209–13, 215, 217, 225, 227–9
   global 2, 158, 170–2
   of Joseph Smith 49–50
   neoconservative 106, 111–2
   prophetic 76, 54, 64, 123, 160–1, 233, 243
   utopia 171

visualities 15, 17, 101, 168, 183, 198

Wahhabism 74, 148
Wal-mart 88, 241
Walvoord, John 134, 138, 141, 143–44
Wang, Thomas 190–1
war xii
   in Afghanistan 148
   against poverty 163
   American Civil 4, 7, 29–30, 34, 36, 38, 42, 79, 160, 170
   American war machine 150
   Carl Schmitt and 173
   Cold 3, 13–4, 17, 54, 85, 99, 102, 104, 120–1, 134–6, 138–45, 148, 150, 188, 199, 215, 236, 239–40
   Crimean 140
   culture 19, 102, 241–2, 245
   on drugs 105
   Ezekiel 9, 85
   fault-line 194
   on Gaza 4
   George Orwell and 137
   Gulf War 121, 128, 214–5, 236
   Hezbollah (2006) 211, 214
   holy 84
   of ideas 30
   imperial 112
   inevitability of 111–2
   Iran–Iraq 149
   in Iraq 145, 213, 215, 230, 234, 239, 241
   with Islam 74, 137, 237
   Israel/Palestine 58, 128–9
   nuclear 120, 234
   preemptive War with Iran 149, 215
   prophetic 8, 29, 53, 79, 81, 124, 136, 138, 142–3, 148, 174, 211, 214, 216–7, 221, 228
   and rumors of wars 7
   Six-Day 128, 133, 141
   spiritual 188
   on Terror 3–4, 99–102, 104, 108, 110, 112, 136, 160, 165, 234
   Thirty Years 161
   virtuous 102, 111
   world 7, 13, 51, 54, 80, 107, 122, 125, 141, 145, 149, 187–8, 198, 215, 236, 239

warriors 54, 64, 105, 110
Washington 4–5, 19, 60, 91, 93, 103, 106, 234, 238
Westphalia, Treaties of 2, 126
Wilson, Woodrow 79–80
Winter, Ralph 190–1, 195–7
witchcraft 6, 84–5, 93
witnessing 8, 78, 113, 148, 163, 172, 174, 219, 225
wrath 13, 148, 237

Yankees 30, 31, 40
Young, Brigham 54, 65, 67

Zion, city of 50, 52–3
Zionism 237
   Christian 1, 4–5, 93, 113, 125, 213, 221, 238–9
   Jewish 141, 138